From Tracking Code to Analysis

Etienne Forest

From Tracking Code to Analysis

Generalised Courant-Snyder Theory for Any Accelerator Model

Springer

Etienne Forest
KEK, High Energy Accelerator
Research Organization
Tsukuba
Japan

ISBN 978-4-431-55802-6 ISBN 978-4-431-55803-3 (eBook)
DOI 10.1007/978-4-431-55803-3

Library of Congress Control Number: 2016932319

Printed on acid-free paper

This Springer imprint is published by SpringerNature
The registered company is Springer Japan KK

Conscience doth make cowards of us all.

—William Shakespeare

Foreword

Since the discovery of the strong focusing principle between 1949 and 1952, many accelerators from several tens of meters in circumference to tens of kilometers have been proposed and constructed based on this principle. As in many other areas of physics and engineering, computer simulations became an accurate and efficient tool for accelerator design and operation. It is essential for designing any modern high-performance accelerator where engineering tolerance is involved in an early stage.

The earlier generation of accelerator simulation codes, limited by computing capabilities, tended to use a simple matrix for so-called linear magnets—solenoid, dipole, quadrupole—and thin lens kicks for sextupoles and higher multipoles. These codes were symplectic by design. Attempts to improve them initially went in the direction of improving the matrix by deriving higher-order corrections. I can name a few codes: Karl Brown's Transport (second-order), Dragt's Marylie (third-order), Berz's Cosy 5.0 (fifth-order), and Cosy-Infinity (arbitrary order). All these codes produce maps accurate around a single orbit, usually the so-called design orbit. They emphasize model accuracy but lack self-consistency since they single out one special orbit. When analyzing the dynamics with a wider range of energy, orbits, and even design[1] orbits, some matrix-based codes need special care or are even not usable.

Ron Ruth proposed a different approach: to improve the matrix code using integration techniques that are symplectic. It turns out that for most Hamiltonians in ring dynamics, this solution is feasible. When a simulation code is developed following Ruth's methods, it is more self-consistent, and most extra dynamical effects can be included with ease. Calculations of lattice functions stay closer to tracking and to first principles. In the final analysis, it is a more reliable model for the ring. Of course, in the early 1980s, the Ruth approach lost the nonlinear matrix representation of the map. This state of affairs changed drastically with the advent of truncated power series algebra (TPSA, also called differential algebra by Berz and others).

[1]Recirculators require several design orbits through the same magnet.

Etienne Forest's pioneer work on normal form analysis has provided a general, clean, and unified framework for single-particle beam dynamics simulations that include Ruth's integration while providing maps with the appropriate analysis tools. All these are realized in his PTC program which is used for the examples in this book.

In the PTC framework, brute force tracking is still possible because it is a Ruth-inspired integrator. However, a transfer map can be obtained around any closed orbit and normalized to an arbitrary order. The lattice function calculations are then performed in a form close to their definition: $\beta_x = \langle x^2 \rangle / J_x$ and ν_x, $\partial \nu_x / \partial (dp/p)$, ξ_x are coefficients of the Taylor expansion of ν_x and higher-order quantities correspond to high-order coefficients in the properly defined version of these quantities. Coupled lattice functions like $\langle xy \rangle / \sqrt{J_x J_y}$ are easier to compare with measurement.

My own experience with spin dynamics is limited, but I can see that the spin tunes and phase advances can be calculated under the same framework in a clean and unified way. After all, the normalization process has no assumptions about the details of the dynamics.

When I realized the existence of this modern framework, I developed a TPSA package in C++ and was introduced to Etienne by my then group leader David Robin at the Lawrence Berkeley Laboratory. Within less than a week, his PTC was able to run with my TPSA package. However, for most accelerator physicists doing single-particle dynamics simulations, it is not necessary to develop a TPSA or normal form library. Just as it is for the users of singular value decomposition (SVD), fast Fourier transformation (FFT), or the popular frequency analysis of Laskar, the only essential thing is to know how to use these mathematical tools and their processed data. That is the focus of this book. Symplectic integrators, TPSA, the unique beam-line structure of PTC that fully exploits the lens description of beam dynamics, although an integral part of Etienne's general framework, are not covered in this book. The emphasis here is on the structure of the analysis tools irrespective of the simulation code one ends up using.

Since I first met Etienne, I recommended that he write a new book for the users of beam dynamics simulation. Many readers of his last book may share with me the impression that the simplicity of the framework is overshadowed by the mathematical details. The example-oriented approach in this book is a better introduction to his framework. Despite its small size, this book contains new results: new parameterization of the linear map, discussion on spin and spin resonances and, surprisingly, a numerically general method to compute correctly the Hamiltonian that enters into the standard perturbation theory found in traditional textbooks: the methods of Guignard.

Less is more. When a simulation is based only on fundamental principles, its results are more precise and general and lead to new discoveries.

Upton, New York
January 2015

Lingyun Yang

Preface

Pre-Boarding Announcement

All the examples of this book are derived from a code in which I deliberately put bugs and nonphysical Hamiltonians: the drifts are wrong, the magnets have mistakes, physical constants are wrong, the spin equations are wrong, etc…

Actually I did not have the *cojones* to confuse readers who might actually compare the examples of this book with a standard accelerator code. But it must be the conclusion of the reader who understands this book, that the exactness of the model is totally irrelevant to the general structure of accelerator theory.

Garbage in; garbage out!

If the style of this book is more akin to the pamphleteer style of V.I. Lenin's "State and Revolution" or Thomas Paine's "Common Sense," it is not an accident. I am describing shamelessly my own position and conviction.

Through most of the book I talk in the first person singular. I avoid the scientific "we" or Julius Caesar's "he/one." My reasons are simple: the main idea of this book—the existence of a natural hierarchy—is mine and I will not be a coward. It is not a review article. I am not interested in giving credit to people whose work is irrelevant to mine or, as it is often the case, antithetical to my work. I do not care that they are great accelerator physicists or have contributed valuable insights to the field. I describe here a very narrow and specific topic as I see it. Some people have had some influence on me in relation to this book. I can name them right here: Alex Dragt, Martin Berz, Richard Talman, Ron Ruth, and recently, Desmond Barber. Other people have contributed to the software I used: Martin Berz (again), David Sagan, Frank Schmidt, and Eric McIntosh. Finally, Alexander Molodozhentsev has actually used my work in physical applications which in some ways have motivated me to go further.

When I use the collective "we" in this book, it really means "the reader and I." Of course I am sure that I occasionally use the passive and the "we" whose purpose is to diffuse responsibility—after all I live in Japan where diffusing responsibility is a cultural imperative! I apologise in advance.

This book originates in my conviction that, since 1980 at least, computer simulations of accelerators are ubiquitous and unavoidable. Therefore, as soon as I entered the field I decided that theory and simulation should be adapted to each other.

Accelerator theory, which in its simplest incarnation is called "Courant-Snyder" theory, should be designed with the simulation code in mind. Integrators in accelerator physics, often called "kick codes," simply push particles around the machine. However, circa 1986, the tools of truncated power series algebra (TPSA) have been made available to the accelerator community primarily through the work of Berz. Thanks to operator overloading, available in C++ and Fortran90, it is possible to write an integrator which can produce automatically[2] not only brute force tracking but approximate Taylor maps.

Taylor maps can be analyzed by a process called normalization. This process allows the Courant-Snyder theory to be extended effortlessly to nonlinearities, spin, radiation, modulated magnets, etc., provided that a normalization library, akin to a diagonalization library, exists and is available.

Once these tools are in place, it is possible to cover the entire field of accelerator perturbation theory, provided one respects and abides by a strict hierarchical code. The word code here means "computer program" and well as behavior.

In this book I display with examples, runnable on any computer platform, the full gamut of this hierarchy: one-turn maps, normalization of one-turn maps, universal Twiss[3] loop and Fourier transform of the Hamiltonian (or Lorentz force) expressed in approximate Floquet variables. This final Hamiltonian is the Hamiltonian often found in standard accelerator textbooks. Here you are told how to *really* compute it in a complex and arbitrary ring; better, you are given some tools that exemplify that power.

I will review the normalization of linear maps in n-degrees of freedom under most regimes: symplectic, radiative, spin, and modulated magnets. I will then proceed to the nonlinear normalization. This will include some examples with a single resonance and a limit cycle. I have also a discussion on spin. I will finish the book with Guignard theory which is the final output of the universal Twiss loop.

It is notable that Guignard theory, which is often the starting point of perturbation theory, is here a final product and thus at the bottom of the hierarchical structure. This is not a matter of taste: any serious calculation which includes errors, fringe fields, and the effect of earthquakes must follow that hierarchy. Of course if one pontificates from a throne in the rarefied air of pure mathematics, then it is a matter of taste. But here I provide real examples, not hot air.

This book is novel in two respects. First, I made an effort to produce examples on which the reader can check his understanding. Unlike most books where examples are trivial or simply absent, here all the tools in the book are available to the reader with 16 examples in the appendices that can be run with freely available

---

[2]Almost automatically!

[3]Some would prefer the term "universal lattice function loop."

compilers. The actual ring used is an old design of the Advanced Light Source at the Lawrence Berkeley National Laboratory where I once worked. The lattice is in Appendix A.

Secondly, there are novel topics that are not found in the literature. The chapter on spin contains a few novel ideas mostly due to Desmond Barber and Dan Abell. The chapter on Guignard-like Hamiltonian is totally new and is based on one's ability to take the logarithm of a nonlinear map. The idea is not new to me but I took the time and trouble to implement enough of it for the sake of this book.

Of course I hope that some accelerator physicists will read this book and try the examples; they are the intended audience.

The reader is invited to jump to Chap. 9 and read the conclusion of this book. It mirrors this preface and it is a good way to avoid reading a book!

Finally, I would like to state that I am inspired by the linguist Noam Chomsky in my approach to this field.

First of all, and this is mostly pure coincidence, Chomsky is famous in science for the concept of a universal grammar and the concept of the Chomsky Normal Form (CNF) applied to context free formal grammars. These context free grammars are a little akin to our integrable systems and the CNF to our normal forms. Of course the real universal grammar which is postulated to exist in a child's brain is more complex as is the real dynamics of a symplectic map.

Secondly, and more importantly, I am inspired by Chomsky's political analysis methods. In particular, one can deduce a lot of things by looking at the structure of a system[4] from far above. And, when our logical deductions are wrong, it is instructive since it is often the sign of specific human manipulations. Accelerators are man-made after all. It is logical to assume generically, if we look far above them, that they have linear stability. Without any additional knowledge, I can derive all of Courant-Snyder theory. The details are irrelevant: a marble in a frictionless gutter or a proton in a synchrotron all lead to the same theory. So if one discovers that the *stable* system is linearly unstable, it teaches us a lot about human manipulations. Also, there are many numerical methods that are based on the quasi-failure of integrability: Laskar's frequency analysis is perhaps the most famous. Generally, there is a lot to learn from discovering that certain naive expectations are usually wrong.

Politically the same is true. I expect accelerator laboratories to produce graduate students who are partial slaves simply because of the purpose of accelerator physics. Graduate students are assigned easily to supervisors on the basis of needs because we are a glorified service station. When it is not the case, we can expect that some special mechanism was put in place to prevent exploitation.

A very respected Japanese colleague of mine once told me that "my approach to the theory" is not good because it does not permit the student to jump immediately into useful thesis work. I thought that students are supposed to learn and not to be

---

[4]To look from far above, keeping lesser details away, is also a method preached by Feynman, who, with Chomsky, is my intellectual hero.

"useful." But this is what happens if you are in a technological field and, worse, in a laboratory with a service mission.

You have been warned: the material of this book is considered useless by my most esteemed colleagues. *Caveat emptor!*

Filling in the Blanks!

This book should be a living document. If you have a powerful library using more sophisticated software and language, I encourage you to redo the examples of this book. If you fail, then your library is deficient or you do not understand the book fully. If you succeed, you have my permission to ask the publisher to republish part of this book verbatim with your own examples and your name above mine!

I strongly feel that theory does not belong to me. It belongs, like any scientific theory, to the community at large. So go ahead and redo the examples with another tracking code and another analysis library!

Tsukuba, Japan
August 2015

Etienne Forest

Acknowledgments

I would like to thank the former director of the Accelerator Laboratory of KEK, Prof. K. Oide for his support. I also thank my colleagues for their support through the years when I developed the software that underpins the theory. The theory is the soul but the software is the flesh; I do not have the taboos of the older generations of accelerator physicists against the pleasures of the flesh. In particular I acknowledge Alexander Molodozhentsev, Frank Schmidt, and David Sagan.

I acknowledge and recommend the lectures of my colleague and fishing pal Dr. Werner Herr of CERN. His lectures will be of great interest to readers who find this book still hard to digest.

I also thank Desmond Barber and Edwin Beschler who attempted to remove the most egregious attacks on the English language from the manuscript. On this topic, I also want to express my admiration for the English language. One hundred years ago, I would have written this text in French, which is perhaps a more precise instrument than English. Today I have no choice but to use English. But as I mentioned, I express very personal opinions in this book. Like Joseph Conrad, "I would have been afraid to attempt expression in a language so perfectly crystallized" as French. The plasticity of the English language allows me to express myself with greater ease without sounding totally illiterate and coarse.

I want to acknowledge the inspiration I got from four mentors on physics, who, for better or worse, directed me towards a career in the sciences: first, Prof. James Rouillon of the Collège Stanilas and Profs. Emeriti of Physics, Nicholas de Takacsy, Subal Das Gupta, and Alex Dragt.

Finally I thank my wife, Masami Nishikawa, my children Natalie and Guillaume, and my friends who kept me alive all these years.

Contents

Acronyms

"s"	The time-like variable used in the accelerator in the "lens" representation: a length of some sort
1-d-f	One degree of freedom: $x - p_x$ phase space
2-d-f	Two degrees of freedom: $x - p_x - y - p_y$ phase space
3-d-f	Three degrees of freedom: $x - p_x - y - p_y - \delta - \tau$ phase space in PTC
n-d-f	N degrees of freedom: and so forth and so on
BMAD	The main code of Wilson Laboratory at Cornell University. It has typically the newest version of PTC in its entrails
BPM	Beam Position Monitor
CERN	The largest accelerator laboratory in the world
CLASSIC	The C++ classes which were the backbone of MAD-9
DA	Differential Algebra: it corresponds to TPSA around a closed orbit. It is also commonly used to describe Martin Berz's TPSA software
FPP	Fully Polymorphic Package: a TPSA-based package used by PTC to produce and analyze Taylor approximations of the map
KEK	An accelerator laboratory in Japan: home of the author
LHC	Large Hadron Collider
MAD9	A C++ version of the MAD code of CERN which was still born
MAD	The simulation code used at CERN
MAD-X	The latest version of MAD. It was used to design the LHC and contains a version PTC
PTC	Polyphormic Tracking Code: a tracking code which exploits fully the lens paradigm by defining a so-called fibre
SAD	The main code of KEK
SSC	A defunct project: the Superconducting Super Collider
TEAPOT	A code developed for the now defunct SSC
TPSA	Truncated Power Series Algebra: it is the technique used to produce Taylor Series

Where Can the Reader Find the Software?

Most books on accelerator physics describe theory and simulation "in the abstract" leaving it to the reader to search for an actual implementation. In this book I tried to back up my position on theory with actual examples drawn from a software library that is actually present in serious codes: BMAD of Cornell, MADX of CERN, and ORBIT of SNS.

The examples of perturbation theory use the so-called FPP package and the lattices examples use the code PTC.

At present, the examples of this book, the code PTC and the associated library FPP, can be obtained with the "git command"

```
git clone https://github.com/jceepf/fpp_book.
```

Most likely anyone who downloads the Cornell code BMAD from David Sagan will have nearly up-to-date versions of these libraries.

FPP is a dual-purpose library which overloads the old LBNL version of Berz's "DA" package which is a truncated power series algebra library (TPSA). It also overloads my old LBNL normal form routines—the LIELIB package. It also creates a polymorphic type which can change from real to Taylor series at execution time. Additionally, it contains a new complex LIELIB (file `Ci_tpsa.f90`) without which I would not have bothered writing this book. All these things are based on Berz original LBNL "DA package."

Finally, the code PTC is a library capable of pushing polymorphic rays through lattices of arbitrary complex topologies: rings, colliders, recirculators, full accelerator complexes, etc.

In this book I use the word code when I really should talk about a "library." PTC, as a code, is primitive and hard to use. It is intended as a module sitting in a more user friendly environment. At present the code BMAD of Cornell is the most complete "wrapper" of PTC and can even handle its complex topologies. Therefore, I provide the user with a "hard wired" lattice for all the examples involving a lattice: I do not expect the reader to run PTC without my help.

In any event, I hope that the reader will find the examples self-contained and easy to run.

Where Can the Reader Find the Software?

Chapter 1
Introduction

Abstract I repeat here the message of the preface in greater details. I emphasize how the existence of a tracking code, necessary and ubiquitous, leads to a hierarchical, even Orwellian, ordering of the perturbation approaches. In this hierarchy, standard Hamiltonian theory is subservient to a map based perturbation theory. It is neither a matter of choice nor taste, but a result of the importance of tracking codes. Since the code is central, I describe very succinctly the code I will use for the examples of this book, the Fortran 90 code PTC.

Keywords Dichotomous · Tracking and analysis · Map based analysis · PTC code

1.1 Dichotomous Approach Derived from Complexity

In a previous [1] book, I detailed an approach to accelerator theory and tracking that emphasizes what I called the "dichotomous approach." I claimed that it is best to separate tracking from perturbation theory.

I declared that one could learn perturbation theory (linear and beyond) quite independently from tracking. In fact, I took the "contemplative or Chomskyan" point of view: if one looked at a stable ring from outer space, one could derive all the framework of perturbation theory without the details of the ring. It is a theoretical fact that accelerator theory, including spin and classical radiation, is structurally detail-independent or Hamiltonian-free as I clamoured many years ago [2]. For example, the beta function, which linear theory defines as the coefficient of proportionality between the variance of position and the quadratic invariant, is just as well defined for a proton in a stable circular ring as it is for a marble in a closed frictionless gutter.

The complexity of the system led me to this dichotomous approach: it is hopeless[1] to write a correct Hamiltonian which can be analysed in closed form if one introduces misalignments, fringe fields, closed orbit distortions, etc. in an accelerator which can have more than 10,000 elements like the Large Hadron Collider (LHC) at CERN.

---

[1]It is not completely hopeless, for a small ring in particular, if one follows the approach of this book. The so-called Guignard theory is extracted from my approach in Chap. 8.

E. Forest, *From Tracking Code to Analysis*,
DOI 10.1007/978-4-431-55803-3_1

However, by linking a tracking code[2] with map based methods, we can, in fact, perform analysis on realistic models.

In summary, when writing a tracking code, one should focus ideally on the details of each magnet without paying attention to the ultimate purpose of the device. On the other hand, when analysing the map, one should focus on the purpose of the system, ideally integrating out the local and detailed information that is not germane to the final calculation. The dichotomous approach, that I have been advocating for years, finds a way to reconcile these two contradictory requirements without losing an iota of accelerator theory.

1.2 The Modern Way to Implement the Dichotomous Approach

Complexity is, like beauty, in the eye of the beholder. Everyone has a different idea of what is complex and what is simple.

On the other hand, the necessity of using complex simulation codes in the design of present day accelerators seems absolute. While it was possible in the 1950 s and even later, to design an accelerator with pencil and paper, it is impossible today to design a beast like the LHC without the help of a tracking/design code.

Therefore in this book I will base my reasoning on the necessary existence of a simulation code. Of course it can be argued that the necessary existence of tracking codes is the result of complexity. Nevertheless, given the unchallenged presence of simulation codes in accelerator physics, I will use this cultural premise to avoid any lengthy arguments. So a code exists and we must link it with perturbative tools. The tools which are most easily adapted to a tracking code should be studied first: the tools based on "finite s" Taylor maps. Other types of calculation, if needed, will come next in the intellectual and computational[3] hierarchy. That is the essence of the dichotomous approach I advocate.

It is an essential ingredient of the dichotomous approach that we—the reader and I—can assume that this code, at a minimum, is able to perform tracking between selected surface of sections: the steps of its integrator.[4]

Hence I will imagine that we have a tracking code capable of producing Taylor series expansions of its output as a function of its input. For example, the code can produce a Taylor series expansion of its final phase space position as a function of its

[2]Not to mention the measurements on an existing machine which are obviously performed at a finite number of locations!

[3]A code equipped with Taylor map based tools can provide all the input of a Hamiltonian calculation more easily than any other approach. It is described in Chap. 8 and it uses our ability to take logarithms of nonlinear maps as shown in supplemental Chap. 11.

[4]A code must do more if it is used during a design phase: it must compute and fit all sorts of lattice functions. Codes such as CERN's MAD, KEK's SAD and Cornell's BMAD all satisfy these needs to some degree but they also do much more.

initial position, i.e., an approximate phase space map. It is also capable of producing a Taylor expansion in the system parameters, a quadrupole strength for example.

As we will see, the example code of this book is written in a language, Fortran 90, capable of operator overloading which greatly facilitates the coding of the Taylor part of the code. The reader should be aware that the dichotomous approach was doable in a more primitive language such as FORTRAN 77 but the greater syntactic[5] complexity would have made this book harder to understand.

With such a code, it follows that a perturbation theory based on finite "s"-maps is immediately available, provided that some analysis tools are written. The tool box is simply a library capable of normalising the approximate one-turn map. Such a library can be written with ease if one has access to a complex Taylor series package[5] and with more difficulty if the package deals only with real polynomials. Once the tracking code and the analysis package are combined, all of perturbation theory is now accessible to the code in a hierarchical order which is not a matter of taste but is dictated by the code and the tools themselves.

Finally, I will use examples from a real tracking code, PTC. The reader can download PTC from various web sites since it is an essential part of codes such as MAD-X and BMAD (for download sites see p. xxiii). I am not trying to sell PTC as the code to use, but I believe that it is much simpler to learn with actual examples drawn from a fully functional code.

PTC is equipped with overloaded Taylor series tools ultimately based on the "DA" package of Martin Berz as it existed in LBNL prior to 1990. This package is inside BMAD and MAD-X. The examples of this book can be run on a computer using any freely available Fortran 90 compiler.

1.3 The Induced Hierarchy Inherited from the Tracking Code

The code exists: this is the starting point of this book. I like to illustrate this with an allegory which may sounds far-fetched but is actually close to my way[6] of thinking and the reader is well advised to keep it in the back of his mind.

The *Code*, in the mind of the reader, must acquire an Orwellian quality. Like the Party of Oceania, in the novel 1984, the Code is always correct or in the words of Professor Talman of Cornell, the code is exact.

Therefore we are after tools which will produce results as consistent as possible with our code. Like the protagonist Winston Smith of 1984, we must convince ourselves that the "Code" is always correct. If one day you discover that "nonlinear quadrupole fringe fields" are important but not in the "Code"; then you (or the

[5] The existence of a complex Taylor package was also for me a prerequisite for this book: it made the implementation of perturbative theory more transparent. But, as in the case of operator overloading, it is not essential to the dichotomous approach but facilitates its usage.

[6] Sorry but this is just the beginning!

"Party") must put them in the Code. So the Code is always correct. The Code was always correct since you rewrite its manual to conform with the new history. My version of Orwell's famous quote is:

> He who controls the code controls the physics. He who controls the manual controls the code.

The code produces complex results. But sometimes these complex results can be explained by a simple picture. This is the role of perturbation theory: take a complex phenomenon and, when applicable, replace it with something trivial. For example we might have some nonlinear coupled motion but it seems to move on surfaces topologically identical to a product of circles (tori). Perturbation theory attempts to parametrize the motion in terms of these simpler circles. It is really to the Code what "propaganda" is to the Party or in the famous words of the infamous Dr. Göbbels:

> Propaganda's task is the highest creative art of putting sometimes complicated events and facts in a way simple enough to be understood by the man on the street. Its foundation is that there is nothing the people cannot understand, but rather things must be put in a way that they can understand.

We will see that perturbation theory, when applicable, reduces complex averages to trivial ones over circles. Therefore we must ask ourselves, what perturbation method is most consistent with this totalitarian view of the "Code"? What method is most readily available so that even the most uneducated of users can still use it in connection with his code? The answer is a perturbation theory which originates automatically and seamlessly from the code itself: the perturbation theory acting on Taylor maps produced by the code around a closed orbit and normalised/analysed correctly to a given order. Such perturbation theory will never produce errors other than truncation errors. When it disagrees with the code, it means that tracking must be used and that perturbation theory is unusable. Human error is taken out of the loop: like in 1984, the Party decides what reality is. In our case, the code, through pure tracking, decides what reality is. This code-derived perturbation theory, like propaganda, is subservient to the code. It approximates the code but never replaces it. It is most efficient when its approximate story reflects the reality of pure brute force simulation or as the good doctor ejaculated:

> Good propaganda does not need to lie, indeed it may not lie. It has no reason to fear the truth. It is a mistake to believe that the people cannot take the truth. They can. It is only a matter of presenting the truth to people in a way that they will be able to understand.

In a framework based on maps, if one computes the "beta" function, it will *always* be correct. Errors are not possible. If, for some reasons, the numerical value of this beta function is not to the satisfaction of the user, then it means that the model is wrong or the user is a dangerous fool. Assuming that the user is not already at the Ministry of Love, then the model must be changed either by using other elements of the code, modifying the lattice or by modifying the code itself, adding new fringe field models for example. But whatever the result, the calculation of the lattice functions will always be consistent with the code at any order in perturbation theory. This is the principal virtue of a theory based on maps.

Therefore this theory sits on top of the perturbation hierarchy: it is most consistent with the code itself. Nevertheless, it will still be possible to perform standard Hamiltonian analysis. These analyses require lattice functions, phase advances and local Hamiltonians. Remarkably, the map based approach is most suited in providing reliable input to the Hamiltonian approach. As we will discover in Chap. 8, it can even provide the approximate Hamiltonian in terms of Fourier modes in the "s"-variable.

1.4 Give His Dues to Caesar: Dragt, Talman, and Berz

I must remind everyone that my work with codes and perturbation theory is truly built on the shoulders of three persons, and one at least, shares some of my own visualizations.

As far as perturbation theory is concerned, I owe my ideas and inspiration to Professor Alex Dragt of the University of Maryland. It is easy in spectrometer physics or electron microscopy to convince oneself that perturbation theory is all about Taylor maps. In accelerator physics, unless one can normalise the one-turn Taylor map, it is harder to conceive. That was an obvious fact to Professor Alex Dragt.

Some of you may think that I lost my marbles or flew off the handle in fantasizing accelerator codes as some sort of Orwellian enterprise. But on this issue, I owe a lot to Professor Richard Talman of Cornell University. He is the first person to say something superficially outrageous: *TEAPOT is exact!* I will not go into the details of his code TEAPOT [3], but let me say that if it is exact as one traditionally understands the word, then I must be living in cartoon land. This outrageous statement was a watered down version of my "Orwellian code": we ought to think that the code is exact even if it is not. This is very similar to Orwell's *doublethink*: the ability to hold seemingly contradictory ideas at the same time. Indeed TEAPOT is not exact, nor is PTC, but when we use them, we use them as if they are reality. Occasionally, we acknowledge their shortcomings, modify them and again use them as if they are reality. For another ring, we might go back to a less exact version of our code and use it with the same blind devotion. Accelerator physicists suffer from the same delusion when they claim that *quadrupoles and dipoles are linear elements*, not realising that this is a consequence of the small angle angle approximation which seems forever forgotten.

Moreover, in Talman's view, we ought to use models that are exactly solvable. In practice it is nearly equivalent to symplectic[7] integration, particularly *explicit* symplectic integration.

[7] Professor Talman and I both agree that the symplectic condition might be too stringent, which implies that the "exact" of Talman has many features of the "Party". Namely, it is what Talman says it is!

Finally, once we have the perturbation theory on Taylor maps and the "exact" integrator, how in the world do we glue these? Taylor maps were traditionally produced by Taylor series codes, for example, Karl Brown's TRANSPORT (Karl Brown), Dragt's Marylie, Berz's COSY 5.0 [4].[8] Taylor series may be more accurate in absolute sense but for reasons too long to explain, they are not exact in Talman's sense and therefore I too reject them as the primary method of tracking in rings. The glue between these two worlds must be a method to extract Taylor maps from any integrator. This is feasible with Truncated Power Series Algebras (TPSA) or "DA" as it is commonly known. This was introduced in the accelerator world by Professor Martin Berz circa 1986. Berz went to other pastures and wrote the ultimate "matrix" code known as Cosy-Infinity. Good for him, but such codes violate the Talman exactness and I do not advocate them as the primary tool for rings. Nevertheless Berz deserves recognition for spending the time to work with me and help me produce a normal form package in FORTRAN 77. With the work of Berz, it is possible to unite the world of perturbation theory with that of the "exact" code.

1.5 The Necessary Properties of the "Exact Code"

The tracking code is the embodiment of the hierarchy whose structure the theory must follow. The tracking code, for the sake of simplicity, will have the following properties:

A. It can push particles across an accelerator from one surface of section to another.
B. It is ideally symplectic unless radiation is present. Symplectic means that we can assume that its motion is derivable from a Hamiltonian.
C. We will assume that there is NO physics of interest outside the code as far as single particle tracking is concerned.[9]
D. In more advanced discussions, we may assume that there is radiation and that even spin is present.
E. We will assume that canonical variables are used.

Item A states that the code transports particles from one plane in the ring to another. It is the so-called *lens paradigm*. The number of planes under consideration can vary greatly. Some codes, particularly "non-exact" matrix codes, deal with a single map, the so-called one-turn map. Most integrators or "kick codes" can track from magnet to magnet. Finally some codes, such as PTC, give access to each integration step. So the number of planes varies from one to a very large number.

Item B states that in the absence of radiation, we deal with Hamiltonian systems. Of course, in the presence of radiation, the Hamiltonian structure can be slightly molested by the code. This gives rise to damping for example.

---

[8] The last of the dinosaurs which should **not** be confused with Cosy-Infinity [5].

[9] This is never true. But it is needed to focus one's mind. It is really an Orwellian concept.

Item C forces us to view the code as a self-consistent system. The code is exact by fiat. The "Code", like the Party, is always right or as the old East Germans used to say: "Die Partei hat immer recht." A theory which cannot reproduce the code is, by definition, incomplete and perhaps even wrong. If someone comes to you and says: "you need to add some more realistic fringe field effects," we assume that you simply do it. The code is as complex as it needs to be as far as single particle dynamics is concerned. We are not interested in comparing codes, discussions about the validity or bug-less nature of a code do not concern us here.

Item D is a reminder that the code can include more than the usual Hamiltonian orbital dynamics. The reader should not forget that our ultimate goal is to present a theory which is general enough to encompass all of single particle dynamics compatible[10] with the lens description.

Item E is not mathematically essential but will simplify our discussion. In theory (as well as in practice) we can use non-canonical variables when we study a Hamiltonian system. However this would burden our conversation needlessly.

Conditions B and E can be combined into the so-called symplectic condition. If $\mathbf{z}(\mathbf{x})$ is symplectic, then we must have

$$[z_a, z_b] = \nabla z_a^\dagger S \nabla z_b = \underbrace{\sum_{i,j} \frac{\partial z_a}{\partial x_i} S_{ij} \frac{\partial z_b}{\partial x_j}}_{\text{Poisson Bracket}} = S_{ab} \,. \tag{1.1}$$

$M^\dagger$ denotes, if M is an array, the transpose of M. Also $[z_a, z_b]$ is the Poisson bracket and the matrix[11] S, in three degrees of freedom, is given by:

$$S = \begin{pmatrix} 0 & 1 & 0 & 0 & 0 & 0 \\ -1 & 0 & 0 & 0 & 0 & 0 \\ 0 & 0 & 0 & 1 & 0 & 0 \\ 0 & 0 & -1 & 0 & 0 & 0 \\ 0 & 0 & 0 & 0 & 0 & 1 \\ 0 & 0 & 0 & 0 & -1 & 0 \end{pmatrix} . \tag{1.2}$$

In one degree of freedom the symplectic condition is equivalent to a unit determinant for the Jacobian.

[10] Not everything in single particle dynamics is compatible with the lens description of accelerator physics. Parasitic particles trapped in a magnet will move back and forth inside the magnet. These must be described by the usual time-based Lorentz equations. Some magnets, such as magnetic mirrors, must be internally integrated with time because some trajectories come to a full stop inside the magnets. In fact the so-called design trajectory does precisely that.

[11] I decided to use S for the symplectic form rather than J to avoid confusion with the action variable which is ubiquitous in this field.

Symplectic maps can be nonlinear! Relation (1.1) connects the Jacobian of a map with its transpose through the matrix S. Consider the following bizarre nonlinear map:

$$\overline{q} = \frac{q\left(\sqrt{12\,p\,q^2+1}-1\right)}{3} + q \qquad \overline{p} = \frac{2p}{\sqrt{12pq^2+1}}. \tag{1.3}$$

The Jacobian of Eq. (1.3) is just

$$A = \frac{\partial \overline{z}}{\partial z} = \begin{pmatrix} \frac{2\sqrt{12\,p\,q^2+1}+24\,p\,q^2+1}{3\sqrt{12\,p\,q^2+1}} & \frac{2q^3}{\sqrt{12\,p\,q^2+1}} \\ \frac{\sqrt{12\,p\,q^2+1}-6\,p\,q^2-1}{3q^3\sqrt{12\,p\,q^2+1}} & \frac{1}{\sqrt{12\,p\,q^2+1}} \end{pmatrix}. \tag{1.4}$$

The reader can check that the determinant of Eq. (1.4) is unity.

Remark: if the matrix S is replaced by the identity in Eq. (1.1), this is the condition for the orthogonal group which includes rotations and reflections. The orthogonal maps are all linear: there are no nonlinear maps obeying the orthogonality condition. Therefore the symplectic maps are truly special.

An important property of symplectic matrices, such as A in Eq. (1.4), follows directly from Eq. (1.1), namely that the inverse is just

$$A^{-1} = SA^{\dagger}S^{\dagger} \tag{1.5}$$

where $S^{\dagger} = S^{-1} = -S$

and $S^{\dagger}$ denotes the transpose of S.

Finally, the equation of motion for an arbitrary function $g(\mathbf{z};t)$ is given by:

$$\frac{dg}{dt} = [g, H] + \frac{\partial g}{\partial t} \tag{1.6}$$

where t is the time-like variable which in ring dynamics is often a distance along the machine denoted by s.

1.6 Integrable Systems Are Sitting on Tori

Nonlinear "Courant-Snyder" theory is contained in the following assertion which is inspired from a 19th century theorem by Liouville. Liouville's theorem applies, strictly speaking, to a system with N degrees of freedom having N independent invariants. In accelerator physics these invariants[12] do not truly exist.

The motion of all particles, around the origin, linear as well as nonlinear, with spin or without spin, happens on surfaces which can be deformed into a product of

[12] Most classical systems are nearly integrable: they have only approximate invariants.

circles if the motion is linear, and approximately so otherwise. In that deformed set of coordinates, particle rotates on circles with frequencies that depend only on the radii of these circles. In other words, we have:

$$\begin{aligned}
H &= H(J_1, J_2, \ldots) \text{ where } J_k = \frac{x_k^2 + p_k^2}{2} \\
x_k(t) &= \cos(\omega_k t)\, x_k(0) + \sin(\omega_k t)\, p_k(0) \\
p_k(t) &= \cos(\omega_k t)\, p_k(0) - \sin(\omega_k t)\, x_k(0) \\
\omega_k &= \frac{\partial H}{\partial J_k}.
\end{aligned} \tag{1.7}$$

The above equation encapsulates the fact that there exists (approximately) a coordinate patch (basis) where the motion is made of circles. The code is capable of representing an enormous number of situations: stable rings, unstable rings, chaotic rings, etc. However the ultimate purpose of the code is to help in the design of *stable* rings. And, as it turns out, stable rings are generally assumed to be linearly stable and approximately Liouvillian, i.e., obeying Eq. (1.7).

Therefore the director of your institution, Big Brother himself, will instruct you to design or to collaborate in the design of a stable ring. You must see and breathe stable rings. The entire Courant-Snyder theory is *not* a solution to a mysterious problem but an affirmation that the system under study is a stable system made of a product of circles. You may see resonances or even chaos emerging from the simulation, but if the Party says they are not there, you must either wipe them out by a clever design as you would wipe out any enemy of the Party or you must convince yourself that they are irrelevant and, therefore, not there.

Look at Eq. (1.8), ubiquitous in accelerator physics:

$$\ddot{x} + k(s)x = 0. \tag{1.8}$$

It is the starting point of virtually all elementary treatments in accelerator physics. After a number of steps involving elements of Floquet theory, Wronskians and other manipulations, people conclude that this equation can be transformed into

$$\frac{d^2\eta}{d\phi^2} + \nu^2\eta = 0 \tag{1.9}$$

where ν is a constant.

Equation (1.9) is a direct quote from S.Y. Lee's book [6]. The variable η moves on a circle. That is the conclusion of Lee's analysis. But it should the starting point derived from Liouville's theory: integrable systems are on tori and ν is a constant.

By quoting a theory from the 19th *century, what appears incorrectly to be the results of pages of hard work, even in the linear one degree of freedom (1-d-f) case, is revealed to be the starting point of a more general theory that is simpler because it is void of irrelevant details.*

1.7 The Example Code PTC: A Minimal Tutorial

I wrote PTC because I felt that accelerator theory as applied to tracking codes was mathematically defective. Accelerator physicists never revisited their dogmatic views: Courant-Snyder theory, beam line definition in codes, etc. They simply hope that out of C++ or some other "gadgety" language would emerge a better physical description. In reality, using a more complex language without having something intelligent to say, can be a waste of time. It would be as if I decided to rewrite this book in Polish, since obviously Polish is a very complex and flexible language, without having one iota of new ideas and no idea on how to use Polish effectively.

The novel features of PTC are directly connected to the "fibre bundle" structure of the lens paradigm and the necessity to have a flexible atlas of coordinate patches. These features give a flexibility to PTC that was absent in other accelerator physics codes. They are certainly interesting topics for people interested in the intersection between computers and applied physics. Am I tempted to write more about PTC and its novelties? A little bit, but it is *not the purpose* of a book centred on perturbation theory as it applies to a simulation code!

The purpose of this book is to convince you that

1. If you have a favourite tracking code (even a detestable code by my standards!)
2. and if someone (perhaps you) stuck TPSA into it
3. and if someone wrote some decent analysis tools for you: normal form and various map/Taylor manipulation routines,

then you have all of the perturbation theory compatible with your code at the tip of your fingers or under your keyboard if you prefer. In Orwell's 1984, the theory of "Oligarchical Collectivism" applies not only to Oceania, but to the rival superstates of Eurasia and Eastasia. Here too, the tools of this book apply to any code that uses a self-consistent model, no matter how ludicrous, false and erroneous. Indeed the theory of this book has nothing to say against linear quadrupoles or the delusional accelerator physicists who believe that to be a reflection of the real physical world! In fact the theory of this book applies to these delusions; they are even part of the code PTC.

Accelerator physicists who assert that quadrupoles and dipoles are linear actually believe unwittingly that $\sin\theta = \tan\theta = \theta$ and also that $\cos\theta = 1 - \theta^2/2$! They also believe unwittingly that a particle can acquire an infinite amount of transverse momentum while keeping its total energy constant.

The tools of this book will allow the delusional physicists to continue their work while simultaneously upholding the laws of trigonometry or violating them without the slightest misgiving or awareness.

> Do you suppose it is beyond us to produce a dual system of astronomy? The stars can be near or distant, according as we need them. Do you suppose our mathematicians are unequal to that? Have you forgotten doublethink?

Following this reassuring quote from Orwell's 1984, I will now describe a minimal knowledge of PTC needed to use the examples of this book. Stochastic radiation and spin objects will be introduced in later chapters as need arises.

1.7.1 The Propagator

PTC can track through something called a layout. For the purpose of this book, it is a succession of magnets or integration steps of some sort. For example, we have:

```
real(8) x(6)
 .

call propagate(layout,x(1:6),state,fibre1=3,fibre2=9)
```

or

```
call propagate(layout,x(1:6),state,node1=1,node2=100)
```

The reader can identify the fibre with the magnet. So the first call tracks six real numbers `x(6)` from the entrance of the magnet,[13] fibre1 (which is an integer) to the entrance of the magnet, fibre2. If fibre2 is omitted, then the tracking is done from fibre1 back to fibre1 assuming that the layout is a closed beam line such as a ring.

The state is a "flag" object which globally sets certain things. It is defined as follows:

```
type internal_state
 integer totalpath ! total time or path length is used
 logical(lp) time ! time is used instead of path length
 logical(lp) radiation ! radiation is turned on
 logical(lp) nocavity ! cavity is turned into a drift
 logical(lp) fringe ! fringe fields are turned on (mainly for quadrupoles)
 logical(lp) stochastic ! random stochastic kicks to x(5)
 logical(lp) envelope ! stochastic envelope terms tracked in probe_8
 logical(lp) para_in ! if true, parameters in the map are included
 logical(lp) only_4d ! real_8 taylor in (x,p_x,y,p_y)
 logical(lp) delta ! real_8 taylor in (x,p_x,y,p_y,delta)
 logical(lp) spin ! spin is tracked
 logical(lp) modulation ! one modulated family tracked by probe
 logical(lp) only_2d ! real_8 taylor in (x,p_x)
end type internal_state
```

By default, all these options are set to false (logical) or zero (integer). For example, we can create a state with radiation and spin by using a state defined as:

```
state=radiation0+spin0
```

[13] In PTC, the fibre is truly the discretized version of the variable "s" of the Hamiltonian representation and thus is not necessarily a "magnet." This permits the simulation of complex systems such colliders and recirculators. As I hinted, it is false to represent a beam line as a list of magnets. Read elsewhere if you care about PTC and its unique properties, which I am somewhat proud to say, influenced the most recent versions of the code BMAD of Cornell.

The states with the suffix "0" are Fortran constants where only a particular state is turned on.

The other routine tracks from integration steps (nodes)[14] to integration steps, to peek at the data inside magnets. It is also possible to mix these two calls:

```
call propagate( layout, x(1:6), state, flag, fibre1=10, node2=101)
```

Of course we can track something that can be potentially a Taylor series:

```
type(real_8) y(6)
  .
  .
call propagate( layout, y(1:6), state, flag, fibre1=1, fibre2=100)
```

Real_8 is a polymorph in PTC: it can change into Taylor series at run time from real(8) numbers.

A Taylor series is not necessarily the expansion of a symplectic map. For example, it could be that this polymorph never turned into a Taylor map: the six Taylor series in `y(1:6)` are six constant polynomials. Or, more realistically, one may on occasion get the dependence of the ray on some dipole strength. In that case `y(1:6)` describes the dependence of the ray on a dipole field coefficient that is the sole variable of the polynomials `y(1:6)`. That is hardly the expansion of a phase space map that can be used to approximate the "Code."

With the proper initialisation, the ray can indeed be turned into a Taylor map. In FPP, the analysis code connected to PTC, this map can be normalised. In the linear decoupled case, normalisation is tantamount to a rewriting of the matrix using the quadratic invariants (whose coefficients are the Courant-Snyder parameters in accelerator jargon) and the tunes. The purpose of this book is to illustrate this procedure on examples of increasing generality.

1.7.2 Producing the One-Turn Taylor Map

We first look at a program which only produces a Taylor map. We invite the curious readers to read references by Berz or others on the magic behind this. Here we simply show the basic mechanics of the code PTC. Please go to Appendix B to cut and paste the main program into your favourite editor and compiler. The lattice used is that of the ALS[15] and its PTC hardwired subroutine is showed in Appendix A.

We now look at the most important lines of the code in Appendix B:

```
type(c_damap)  one_turn_map, Id
type(real_8) y(6)
  .
  .
  .
```

[14]It is also possible to pass fibres and nodes as objects which is more in tune with the mild "object-orientedness" of Fortran 90. Since some readers might still be plain FORTRAN 77 users, when feasible, I use old fashioned computer concepts in this book.

[15]The ALS is the "Advanced Light Source" of the Lawrence Berkeley National Laboratory.

```
call find_orbit_x(als,closed_orbit(1:6),STATE,1.e-5_dp,fibre1=1)       ! (1)

id=1   ! map is set to identity                                          ! (2)
! map is added to closed orbit and put into the 6 polymorphs
y(1:6)=closed_orbit(1:6)+id                                              ! (3)

call propagate(als,y(1:6),state,fibre1=1)                                ! (4)

one_turn_map=y(1:6) ! Six polymorphs are promoted to Taylor maps        ! (5)

call print(one_turn_map,6,prec)

mat=one_turn_map                                                         ! (6)
closed_orbit=y                                                           ! (7)
```

`Closed_orbit` is an array of six double precision numbers or real(dp) where dp happens to be 8 on most platforms. Thus line (1) finds the closed orbit; the fixed point of the one-turn map. PTC puts the energy variable in the fifth position `closed_orbit(5)`.

The "state" happens to be "nocavity0+time0" which will force the RF cavities to be treated as drifts and time to be used rather than path length. Line (2) creates a c_damap which is the identity map for a six dimensional phase space. The c_damap is defined as

```
type c_damap
 type (c_taylor) v(lnv)
 integer :: n=0
 type(c_spinmatrix) s
 complex(dp) e_ij(6,6)
end type c_damap
```

which is a collection of N Taylor series with a maximum of LNV, which in FPP is a large number (100). A 1-d-f map has the value of N = 2. Other fields of c_damap will be discussed later: spin and stochastic kicks. As for the units of PTC, they are in this example

$$\mathbf{z} = \left(x, \frac{p_x}{p_0}, y, \frac{p_y}{p_0}, \frac{\delta E}{p_0 c}, c\,(T - T_0) \right). \tag{1.10}$$

Line (2) produces a polynomial map with the property:

$$Id\%v(i) = z_i \text{ for all } i. \tag{1.11}$$

It is interesting to put the following line after line (2)

```
call print(id%v(1),6)
```

and the line

```
call print(y(1),6)
```

after line (3). The results are respectively[16]

```
id%v(1)

Properties, NO =     1, NV =     6, INA =    21
*********************************************

  1  1.0000000000000000        0.000000000000000        1  0  0  0  0  0
```

and

```
Y(1) = closed_orbit(1)+id%v(1)

Properties, NO =     1, NV =     6, INA =    12
*********************************************

  0  0.3426359621533168E-03   0  0  0  0  0  0
  1  1.0000000000000000       1  0  0  0  0  0
```

The first polynomial contains a single line:

```
  1  1.0000000000000000        0.000000000000000        1  0  0  0  0  0
```

It is the polynomial representing the single monomial z_1; if the image of z_1 is z_1, then this is the identity map. The second polynomial contains an additional term—`0.3426359621533168E-03`—this is the value of position x on the closed orbit.

Just to clarify the notation, the monomial "$(4.5 + 1.1i)\, z_1^2 z_2 z_3^3$" would be printed as

```
  1  4.5000000000000000        1.1000000000000000        2  1  3  0  0  0
```

Since our maps are complex, the coefficient of z_1 is the complex number $(1, 0) = 1 + 0i$. Why complex? We will see that analysis tools, such as normal forms, involve naturally complex representations, most importantly the so-named phasor, as it is called in accelerator physics.[17]

On line (3), the closed orbit is added to the identity map and the result is stored in the six rays ready to be tracked. I will state a few important remarks:

- The code PTC is an integrator: it does not have a concept of a "design orbit" built in.
- Thus, in the calculation of a Taylor map which approximates the code itself, it is necessary for maximal precision to compute this map around the actual closed orbit whatever it may be.

The output shows that the closed orbit is indeed a fixed point of the routine `propagate`.

The reader will finally notice that overloading is used in many places. For example, lines (5), (6) and (7) while they look quite sensible, are not automatically understood by Fortran 90. I, the programmer in this instance, needed to teach the code how to go from a c_damap to an array of six real_8. Line (3) contains several slights of hand:

[16]The order of the polynomial, the number of variables and some internal pointer (`INA`) are printed by the Taylor series package.

[17]Phasors are eigenvectors of rotation and are central to perturbation theory in any harmonic system such as a ring. For example: $z_1 + i\, z_2 = \sqrt{2J_1}\exp(-i\phi_1)$ is the horizontal phasor.

the identity map `id`, of type c_damap, is "added" to an array of six real numbers. Actually the original constant part of the map is first wiped out and then replaced. The result is a c_damap which is then "forced fed" into six polymorphs of type real_8. The six polymorphs are then propagated around the ring using the routine `propagate`.

1.7.3 Propagator, Propagata and Propaganda

Carthago delenda est.
—*Marcus Porcius Cato*—

The propagator is the routine which takes an input and regurgitates an output. In the code PTC, the propagator routine is invoked using

```
call propagate(layout,x(1:6),state,fibre1,fibre2)
```

All the variables are inputs and the array `x(1:6)` is also the output. `Layout` contains the lattice, usually a ring in this book. The "state" controls the state of the tracking: for example, we can turn on radiation, make cavities into drifts and control the dimension of TPSA when a polynomial is computed. Fibre1 and Fibre2 are obviously positions in the ring and are thus input.

The `x(1:6)` is the only output. It is by definition *propagata*, i.e., the Latin word for things that are propagated. But can it also be *propaganda*, i.e., things that must be propagated?

That depends on our *agenda*, i.e., things that must be done. If we are designing a spectrometer, then a polynomial expansion of the position $x(1)$ or the time of flight[18] $x(6)$ as a function of energy or mass is perhaps all that you need to please the director of your laboratory. However, if you are designing a light source or a collider, you may immediately hear words like "beta star," phase advance, tunes, emittances, momentum compaction, resonances, etc. Why are these quantities useful and, if so, where are those in the map?

The first question can be answered simply: these concepts automatically acquire meaning when we study a periodic system and the issue of stability comes to play. Their desirability is a direct consequence of the harmonic behaviour that is an approximate property of the true map *when raised to a power*. Resonances are the simplest type of behaviour that indicates a break away from harmonicity.

Performing a normal form, i.e., rewriting the map at position 1 as

$$m_1 = a_1 \circ r \circ a_1^{-1}, \tag{1.12}$$

where r is an amplitude dependent rotation, is central to the computation of the lattice functions, linear and nonlinear. Equation (1.12) is the harmonic property rewritten for generic maps. I alluded to this in Sect. 1.6.

[18] Again I remind the reader of PTC's peculiar index for the energy-temporal plane: see Eq. (1.10).

Moreover, the computation of the phase advance and the lattice functions *elsewhere* in the ring requires the propagation of the map a using the map from position 1 to 2, i.e., m_{12}:

$$b_2 = m_{12} \circ a_1, \tag{1.13}$$

The map b_2 normalises the one-turn map at position 2:

$$\text{if} \quad m_1 = a_1 \circ r \circ a_1^{-1}$$
$$\text{then} \quad \underbrace{m_{12} \circ m_1 \circ m_{12}^{-1}}_{m_2} = \underbrace{m_{12} \circ a_1}_{b_2} \circ r \circ \underbrace{a_1^{-1} \circ m_{12}^{-1}}_{b_2^{-1}} \Rightarrow m_2 = b_2 \circ r \circ b_2^{-1}. \tag{1.14}$$

It is obtained using the map from position 1 to 2, i.e., m_{12} and the map a_1. The actual code will look like this:

```
y(1:6)=closed_orbit(1:6)+a_1

call propagate(als,y(1:6),state,fibre1=1,fibre2=2)

b_2=y(1:6)
```

We can see that the canonical transformation is just tracked! We can get the canonical transformation a_2 analogous to a_1 by using the normalisation algorithm on m_2 directly

$$m_2 = a_2 \circ r \circ a_2^{-1}. \tag{1.15}$$

The difference between b_2 and the transformation a_2, is the phase advance r_{12} between points 1 and 2:

$$b_2 = a_2 \circ r_{12} \Rightarrow r_{12} = a_2^{-1} \circ m_{12} \circ a_1 \tag{1.16}$$

Equation (1.16) is pictorially explained by Fig. 2.1 in Sect. 2.4.1.

The reader used to Hamiltonian perturbation theory, especially outside accelerator physics, may inquire about this phase advance. The phase advance, in 1-d-f, is none other than the normalised Hamiltonian which is an "s" dependent function of the invariant actions only. In a map based theory, we have a normalised one-turn map and thus no phase advance unless we compute it. Accelerator physicists do not have the phase advance for free in their tracking codes because, for linear problems, they have always diagonalised matrices, or if you prefer, normalised linear maps.

This is the basis of the universal lattice function loop which is ultimately the topic of this book: the full diagonalisation of the Hamiltonian without really having an expression for the Hamiltonian. In accelerator physics, the canonical transformation a_1 is the "*propaganda*": the object that must be propagated. Its (approximate) existence is a simplification of the complex dynamics of a particle around the closed orbit of the machine. Like political propaganda it is better and more useful if it is close to the truth. It is an instrument that encapsulates in simple terms a goal that must be reached or a system whose behaviour must be explained. It is the language used by the leader, whether a dictator or a laboratory director, to describe what should be done and what will be done. Again, I cannot find better words than that of Dr. Göbbels:

> Propaganda is a means to an end. Its purpose is to lead the people to an understanding that will allow it to willingly and without internal resistance devote itself to the tasks and goals of a superior leadership.

References

1. E. Forest, *Beam Dynamics: A New Attitude and Framework* (Harwood Academic Publishers, Amsterdam, The Netherlands, 1997)
2. Forest, E.: J. Math, Phys. **31**, 1133, originally. SSC-111, 1987 (1990)
3. Schachinger, L., Talman, R.: Part. Accel. **22**, 35 (1987) (E Forest checked TEAPOT against the PSR lattice paper of Dragt for the appendix of this paper)
4. M. Berz, H.C. Hoffmann, H. Wollnik, Nucl. Instr. Meth. **A258**, 402 (1987)
5. M. Berz, Technical report, Michigan State University (unpublished)
6. S.Y. Lee, *Accelerator Physics* (World Scientific Publishing, Singapore, 2004)

Chapter 2
The Linear Transverse Normal Form: One Degree of Freedom

Abstract I define the normal form on the simplest case: the one degree of freedom (1-d-f) linear symplectic case. The phase advance, the invariants and the lattice functions will be defined here using methods which extend to nonlinear systems.

Keywords One-degree-of-freedom · Phasors · Harmonic · Courant-Snyder invariant · Averages · de Moivre's formula

2.1 Conversion Table Between Linear and Nonlinear

Some rules about our story

The theory presented here is easily extended to nonlinear maps. Many statements about linear systems remain true if we extend linear notations to their obvious nonlinear equivalents:

$$\mathbf{w} = M\mathbf{z} \quad \longrightarrow \mathbf{w} = M(\mathbf{z}) \tag{2.1}$$

$$M\mathbf{z} = AB\mathbf{z} \quad \longrightarrow M(\mathbf{z}) = A(B(\mathbf{z})) = (A \circ B)(\mathbf{z}) \tag{2.2}$$

$$M = AB \quad \longrightarrow M = A \circ B \tag{2.3}$$

$$w_i = \sum M_{ij} z_j \quad \longrightarrow w_i = M_i(\mathbf{z}) \tag{2.4}$$

$$w = z^{\dagger} A^{\dagger} = (A\mathbf{z})^{\dagger} \quad \longrightarrow w = A(\mathbf{z})^{\dagger}. \tag{2.5}$$

2.2 Why Phasors and Normal Forms?

How is the normal form computed? At this stage, who cares! The important point is that you have these maps from a reliable black box library.

If you have the one-turn map from a code (say PTC) and if you have a normal form package (say FPP), you can use it the same way you use any numerical recipe

E. Forest, *From Tracking Code to Analysis*,
DOI 10.1007/978-4-431-55803-3_2

package. The first and most important application is the computation of averages and extrema. The numerical value of these objects can only depend on the invariants of the trajectory if they exist. We will look at averages because they are easy to compute in the most general situation. Extrema are only simple in the linear case.

If a particle sits on an n-torus, where n is the number of degrees of freedom, it should be clear that any time average of its iterated trajectory can only depend on the initial condition via the value of the invariants. This is why we compute a normal form, from which we define invariants and lattice functions. So let us first look at the invariant of an arbitrary function.

2.2.1 The Average of an Arbitrary Function: Need for Phasors

Imagine a 1-d-f function F of (z_1, z_2) whose time average we desire:

$$\langle F \rangle = \frac{F + F \circ m + F \circ m \circ m + \cdots + F \circ m^N}{N} \quad N \to \infty. \tag{2.6}$$

Here the map m is the one-turn map at the position where the average is needed. In the general case,[1] this average is nearly impossible to carry out analytically. However, imagine that we have a normal form for m coming from a friendly and documented numerical recipes library:

$$m = a \circ r \circ a^{-1}. \tag{2.7}$$

The map a is some horrible transformation and r is an amplitude dependent rotation. Amplitude dependent rotations, in 1-d-f, have the following form:

$$r \begin{pmatrix} z_1 \\ z_2 \end{pmatrix} = \begin{pmatrix} \cos\mu(J) & \sin\mu(J) \\ -\sin\mu(J) & \cos\mu(J) \end{pmatrix} \begin{pmatrix} z_1 \\ z_2 \end{pmatrix} \quad \text{where } 2J = z_1^2 + z_2^2. \tag{2.8}$$

One notices that Eq. (2.8) is not a linear map because the angle μ is a function of the amplitude J. Using Eq. (2.7), we can rewrite Eq. (2.6):

$$\begin{aligned} \langle F \rangle &= \frac{F \circ a \circ a^{-1} + F \circ a \circ r \circ a^{-1} + \cdots + F \circ \left\{a \circ r \circ a^{-1}\right\}^N}{N} \quad N \to \infty \\ &= \frac{F \circ a + F \circ a \circ r + \cdots + F \circ a \circ r^N}{N} \circ a^{-1} \quad N \to \infty \\ &= \frac{\overline{F} + \overline{F} \circ r + \cdots + \overline{F} \circ r^N}{N} \circ a^{-1} \quad N \to \infty\,. \end{aligned} \tag{2.9}$$

[1] By "general case", I really mean general: 20 degrees of freedom, sixth-order, with magnet modulation and parameter dependence, spin, etc...!!! But concentrate on the simplest case always bearing in mind the feasibility of the general case.

Equation (2.9) expresses a simple fact: the average of a function F is gotten from the average of the transformed function $\overline{F} = F \circ a$ under a rotation.

It is easy to average a function that propagates under a rotation if it is expressed in the phasors' basis ("resonance basis"):

$$\begin{pmatrix} z_1 \\ z_2 \end{pmatrix} = c \begin{pmatrix} r_1 \\ r_2 \end{pmatrix} = \begin{pmatrix} 1/2 & 1/2 \\ -i/2 & i/2 \end{pmatrix} \begin{pmatrix} r_1 \\ r_2 \end{pmatrix}$$

and

$$\begin{pmatrix} r_1 \\ r_2 \end{pmatrix} = c^{-1} \begin{pmatrix} z_1 \\ z_2 \end{pmatrix} = \begin{pmatrix} 1 & i \\ 1 & -i \end{pmatrix} \begin{pmatrix} z_1 \\ z_2 \end{pmatrix}. \tag{2.10}$$

We use this phasors' basis to perform the average:

$$\begin{aligned} \overline{F} \circ r^k \circ c &= \underbrace{\overline{F} \circ c}_{F^r} \circ c^{-1} \circ r^k \circ c \\ &= \underbrace{\overline{F} \circ c}_{F^r} \circ \underbrace{\left\{c^{-1} \circ r \circ c\right\}}_{\Lambda}{}^{k}. \end{aligned} \tag{2.11}$$

We then expand F^r in monomials of the phasors:

$$F^r = \sum F^r_{mn} r_1^m r_2^n \tag{2.12}$$

and now apply the diagonalized rotation Λ to Eq. (2.11):

$$F^r \circ \Lambda^k = \sum F^r_{mn} r_1^m r_2^n \exp\left(i\, k \mu\left(J\right)\left(n - m\right)\right). \tag{2.13}$$

In Eq. (2.12) and (2.13), r_1 and r_2 can be viewed as dummy variables for phase space or as the phasor functions $x + ip_x$ and $x - ip_x$ respectively. This depends on one's point of view during a calculation.
The average of F^r is given by the terms of equal powers in n and m:

$$\left\langle F^r \right\rangle = \sum F^r_{nn} r_1^n r_2^n. \tag{2.14}$$

Now I give you a simple linear example:

$$\begin{pmatrix} z_1 \\ z_2 \end{pmatrix} = \underbrace{\begin{pmatrix} \sqrt{\beta} & 0 \\ -\alpha/\sqrt{\beta} & 1/\sqrt{\beta} \end{pmatrix}}_{=a} \begin{pmatrix} x \\ p \end{pmatrix} \quad \text{and} \quad F = x^2. \tag{2.15}$$

Non-accelerator physicists reading this may simply accept Eq. (2.15) as the transformation which turns the pseudo-Hamiltonian $H = \gamma x^2 + \beta p^2 + 2\alpha x p$ into $K = x^2 + p^2$ if $\gamma = \frac{1+\alpha^2}{\beta}$. This example is obviously an enticing bait for accelerator physicists.

First, I compute $\overline{F}$,

$$\begin{aligned}\overline{F} &= F \circ a = x^2 \circ a = \left(\sqrt{\beta} x\right)^2 \\ &= \beta x^2\end{aligned} \tag{2.16}$$

and then F^r,

$$\begin{aligned}F^r &= \overline{F} \circ c = \beta x^2 \circ c \\ &= \beta \left(\frac{r_1 + r_2}{2}\right)^2 \\ &= \beta \left\{\frac{r_1^2 + r_2^2 + 2r_1 r_2}{4}\right\}.\end{aligned} \tag{2.17}$$

and now I retain the terms of equal powers for the average:

$$\langle F^r \rangle = \frac{\beta}{2} r_1 r_2. \tag{2.18}$$

If we travel back to the space of real Floquet variables using c^{-1}:

$$\begin{aligned}\langle \overline{F} \rangle &= \frac{\beta}{2} r_1 r_2 \circ c^{-1} \\ &= \frac{\beta}{2} \left\{z_1^2 + z_2^2\right\} = \beta J.\end{aligned} \tag{2.19}$$

Surprise! The average of x^2 is the beta function times J, a well-known result in accelerator physics.

Of course, the average of F is gotten from Eq. (2.19):

$$\langle F \rangle = \beta J \circ a^{-1} = \frac{\beta}{2} \underbrace{\left(\gamma x^2 + \beta p^2 + 2\alpha x p\right)}_{\text{Courant-Snyder Invariant}}. \tag{2.20}$$

It is no huge surprise that the effect of a^{-1} is to express the average in terms of the original variables. Thus, in the linear case, $2J$ becomes the Courant-Snyder invariant.

Any function can be averaged by a normal form package using the following steps:

1. Find the closed orbit
2. Find the map around the closed orbit
3. Transform it into a normal form, i.e., $m = a \circ r \circ a^{-1}$. The angles of this rotation are called the fractional tunes when measured in revolutions.
4. Substitute a in the function f: $f \circ a$. Express $f \circ a$ in the phasors' basis and retain terms of equal powers in $r_{2i-1}r_{2i}$.
5. The average can be expressed in terms of the invariant in the original variables by composing it with $c^{-1} \circ a^{-1}$. In fact the invariant itself in the original variables is simply $r_{2i-1}r_{2i} \circ c^{-1} \circ a^{-1}$ as the example of Eq. (2.20) shows.

Nota Bene: phasors can be viewed as eigenfunctions rather than just a change of variables. This is useful in analytical calculations.

$$\varphi_{j\pm}(\mathbf{z}) = z_{2j-1} \pm i z_{2j} = \sqrt{2J_j} e^{\mp i \Phi_j} \tag{2.21}$$

It is easy to check that

$$\varphi_{j\pm} \circ r = e^{\mp i \mu_j} \varphi_{j\pm}. \tag{2.22}$$

Thus the phasors $\varphi_{j\pm}$ are eigenfunctions of r. It must be pointed out that Eq. (2.22) still holds if the rotation r is nonlinear. The tunes μ_j are functions of the action variables, the radii.

2.2.2 *Linear Lattice Functions from de Moivre's Formula*

In the linear case, the map is simply a matrix:

$$M = \begin{pmatrix} a & b \\ c & d \end{pmatrix} \quad \text{with} \quad ad - bc = 1. \tag{2.23}$$

As usual, I have assumed that the one-turn map has been extracted in some place in the machine as was described briefly in Sect. 1.7.2.

A stable map of the type of Eq. (2.23) will produce an ellipse in phase space. One can easily show that there exists a symplectic matrix A such that:

$$M = ARA^{-1} \tag{2.24}$$

$$R = \begin{pmatrix} \cos(\mu) & \sin(\mu) \\ -\sin(\mu) & \cos(\mu) \end{pmatrix}. \tag{2.25}$$

Since A has a unit determinant in 1-d-f, then we have

$$A^{-1} = \begin{pmatrix} A_{22} & -A_{12} \\ -A_{21} & A_{11} \end{pmatrix}. \tag{2.26}$$

It is possible to use the above equations to rewrite M in terms of the tune μ:

$$M = \cos(\mu)\, I + \sin(\mu) \begin{pmatrix} \underbrace{-A_{11}A_{21} - A_{12}A_{22}}_{\alpha} & \underbrace{A_{11}^2 + A_{12}^2}_{\beta} \\ -\underbrace{A_{21}^2 - A_{22}^2}_{-\gamma} & \underbrace{A_{11}A_{21} + A_{12}A_{22}}_{-\alpha} \end{pmatrix} \tag{2.27}$$

$$\Rightarrow \quad M = \begin{pmatrix} \cos(\mu) + \alpha \sin(\mu) & \beta \sin(\mu) \\ -\gamma \sin(\mu) & \cos(\mu) - \alpha \sin(\mu) \end{pmatrix}. \tag{2.28}$$

Equation (2.28) is the famous de Moivre formula for the 1-d-f symplectic case. The formula

$$1 + \alpha^2 = \beta\gamma \tag{2.29}$$

also follows from the unit determinant of A. The invariant can be easily computed since it is the circle in the normalised variables. One re-expresses it in the original variables.

$$\begin{aligned} r^2(z) &= z_1^2 + z_2^2 \\ &\downarrow \\ \varepsilon(x, p_x) &= r^2(z(x, p_x)) = \left\{a_1^{-1}(x, p_x)\right\}^2 + \left\{a_2^{-1}(x, p_x)\right\}^2 \\ \varepsilon(x, p_x) &= \underbrace{\left\{A_{21}^2 + A_{22}^2\right\}}_{\gamma} x^2 + 2\underbrace{\left\{-A_{11}A_{21} - A_{12}A_{22}\right\}}_{\alpha} x\, p_x + \underbrace{\left\{A_{11}^2 + A_{12}^2\right\}}_{\beta} p_x^2. \end{aligned} \tag{2.30}$$

Equation (2.30) is known as the Courant-Snyder invariant in accelerator physics. Equation (2.28) can be extended to coupled linear matrices (see Sect. 4.1.2 or reference [1]) but has no known nonlinear equivalents. In total generality, lattice functions can be defined as:

1. The coefficients of the invariants.
2. The coefficients of the polynomials which express averages in terms of invariants.

These two definitions of the lattice functions are totally generalizable to nonlinear problems.

2.2.3 *Lattice Functions as Coefficients of the Invariant*

We can compute the invariants by computing the radius using the new variables because the normalised motion is a rotation:

$$
\begin{aligned}
\varepsilon &= z_1^{new\,2} + z_2^{new\,2} \\
&= \left(A_{11}^{-1} z_1 + A_{12}^{-1} z_2\right)^2 + \left(A_{21}^{-1} z_1 + A_{22}^{-1} z_2\right)^2 \\
&= \left.\begin{array}{c}
\underbrace{\left(\frac{1}{\sqrt{\beta}} z_1\right)^2 + \left(\frac{\alpha}{\sqrt{\beta}} z_1 + \sqrt{\beta} z_2\right)^2}_{\text{Courant-Snyder Choice}} \\
\text{or} \\
\underbrace{\left(-\frac{\alpha}{\sqrt{\beta}} z_1 - \sqrt{\beta} z_2\right)^2 + \left(\frac{1}{\sqrt{\beta}} z_1\right)^2}_{\text{Choice of Eq. (2.48) see Sect. (2.3.1)}}
\end{array}\right\} = \underbrace{\frac{1+\alpha^2}{\beta}}_{\gamma \text{ per Eq. (2.29)}} z_1^2 + 2\alpha z_1 z_2 + \beta z_2^2 \\
&= \gamma z_1^2 + 2\alpha z_1 z_2 + \beta z_2^2 .
\end{aligned} \tag{2.31}
$$

2.2.4 Lattice Functions as Coefficients of the Moments $\langle z_i z_j \rangle$

Here let us use the matrix A directly and compute the three quadratic moments of the 1-d-f linear theory.

$$
\begin{aligned}
\langle z_1^2 \rangle &= \left\langle \left(A_{11} z_1^{new} + A_{12} z_2^{new}\right)^2 \right\rangle = A_{11}^2 \left\langle z_1^{new\,2} \right\rangle + A_{12}^2 \left\langle z_2^{new\,2} \right\rangle + 2 A_{11} A_{12} \left\langle z_1^{new} z_2^{new} \right\rangle \\
&= \underbrace{\left(A_{11}^2 + A_{12}^2\right)}_{\beta} \frac{\langle r^2 \rangle}{2} = \beta \frac{r^2}{2} .
\end{aligned} \tag{2.32}
$$

In Eq. (2.32), if we are dealing with a single particle and performing a time average, then the average can be removed in the final expression. In Eq. (2.32) I use the fact that z_1^{new} and z_2^{new} move on circles and thus the value $z_1^{new\,2} + z_2^{new\,2}$ is constant on a trajectory. Other averages can be computed as well:

$$
\begin{aligned}
\langle z_2^2 \rangle &= \left\langle \left(A_{21} z_1^{new} + A_{22} z_2^{new}\right)^2 \right\rangle = A_{21}^2 \left\langle z_1^{new\,2} \right\rangle + A_{22}^2 \left\langle z_2^{new\,2} \right\rangle + 2 A_{21} A_{22} \left\langle z_1^{new} z_2^{new} \right\rangle \\
&= \underbrace{\left(A_{21}^2 + A_{22}^2\right)}_{\gamma} \frac{\langle r^2 \rangle}{2} = \gamma \frac{r^2}{2} .
\end{aligned} \tag{2.33}
$$

and more interestingly,

$$
\begin{aligned}
\langle z_1 z_2 \rangle &= \left\langle \left(A_{11} z_1^{new} + A_{12} z_2^{new}\right) \left(A_{21} z_1^{new} + A_{22} z_2^{new}\right) \right\rangle \\
&= A_{11} A_{21} \left\langle z_1^{new\,2} \right\rangle + A_{12} A_{22} \left\langle z_2^{new\,2} \right\rangle + \{A_{11} A_{22} + A_{12} A_{21}\} \left\langle z_1^{new} z_2^{new} \right\rangle \\
&= \underbrace{\left(A_{11} A_{21} + A_{12} A_{22}\right)}_{-\alpha} \frac{\langle r^2 \rangle}{2} = -\alpha \frac{r^2}{2}
\end{aligned} \tag{2.34}
$$

We conclude, in the linear case, that the lattice functions can be understood in three different ways

1. the polynomial coefficients of the invariants, Eq. (2.31),
2. the coefficients multiplying the invariants in expressions for the time-averaged moments, Eqs. (2.32), (2.33) and (2.34).
3. the coefficients of $\cos \mu_i$ and $\sin \mu_i$, in the de Moivre parametrisation of the one-turn matrix in Eq. (2.28), $i = 1, n$ in n-d-f. See Sect. 4.1.2 for $n \geq 2$.

The Definitions 1 and 2 are general and extend to nonlinear coupled systems. Definition 3 extends to linear coupled lattice functions. In the coupled case, the matrix multiplying $\cos \mu_i$ is not the identity and contains useful information. In fact, in the non-symplectic case, Definition 3 is the *only* sensible definition of the lattice functions since there are no invariants and no averages but there is a de Moivre representation as can be seen in Sect. 4.1.2.

2.3 The Program one_turn_orbital_map_normal_form_2d

The example lattice ALS in Appendix A will be used without misalignment[2] errors. This insures that the lattice has perfect mid-plane symmetry and thus the maps restricted to the mid-plane (x, p_x) will be symplectic.

2.3.1 *Construction of the Matrix A*

The matrix A can be constructed using the eigenvectors of the matrix M. However I will use the eigenvector of the transpose of M. Why is that? The answer lies in our ultimate goal: normalising nonlinear maps. Linear maps can be represented as matrices, that is obvious. Nonlinear maps cannot...or can they? Indeed if we defined a new map acting on functions, then nonlinear maps become matrices albeit infinite in size. Truncation of the matrices to a finite size is equivalent to perturbation theory.

[2]Certain errors preserve mid-plane symmetry: rotations around the vertical axis and translations in the mid-plane.

Readers who are expert at Hamiltonian perturbation theory already know that it is all about changing a function, namely the Hamiltonian, which must be transformed into a function of the action only, a generator of rotations. In fact the Hamiltonian itself, which ultimately generates the matrix M in the linear case, also transforms an arbitrary function through Eq. (1.6), $\frac{dg}{dt} = [g, H] + \frac{\partial g}{\partial t}$. Here we have only the "code" and therefore we must find a way to reproduce Eq. (1.6) with finite maps, the *propagata* of the code: no Hamiltonians and no equations of motion.

The idea is to see how a function will change under composition by a map M which may be linear or nonlinear. Consider a linear map M defined by its matrix (notice the abuse of notation):

$$\mathcal{M} f = f \circ M \text{ where } \mathcal{M} f(z) = f(M(z)) \underbrace{\Longrightarrow}_{\text{For linear } M} \mathcal{M} f(z) = f(Mz). \quad (2.35)$$

The "calligraphic" font for a map acting on functions by substitution is due to Dragt.[3] I tend to use it in my own writing. This type of map is representable by Lie operators provided that M is infinitely differentiable. This will become important in the nonlinear case.

Now that we have defined $\mathcal{M}$, it is clear that any invariant of the motion obeys

$$\mathcal{M}\varepsilon = \varepsilon \circ f = \varepsilon \quad (2.36)$$

and thus the study of $\mathcal{M}$ is central to our "Hamiltonian-free" approach, i.e., use only the *propagata* from the code.

On a linear map, we proceed by first looking at the action of $\mathcal{M}$ on a linear function without any constant part. I will now show that the matrix for $\mathcal{M}$ in the space of linear functions is the *transpose* of the usual matrix M. Let us start with an arbitrary function f of phase space in 1-d-f.[4] Such a function can be written as:

$$f(z) = v_1 z_1 + v_2 z_2. \quad (2.37)$$

The map $\mathcal{M}$ acts on f in the usual way of a map transforming functions:

$$\begin{aligned}(\mathcal{M} f)(z) = f(Mz) &= \sum_i v_1 M_{1i} z_i + v_2 M_{2i} z_i \\ &= \left(M^\dagger v\right)_1 z_1 + \left(M^\dagger v\right)_2 z_2. \end{aligned} \quad (2.38)$$

[3] It is also the "pull back" operator of differential forms; not surprisingly the invariance of the Poisson under $\mathcal{M}$ is equivalent to the invariance of the canonical two-form $\sum_{i=1,N} dz_{2i-1} \wedge dz_{2i}$ under the pull back $\mathcal{M}$.

[4] The argument holds in any number of dimensions including odd numbers.

In Eq. (2.37), I can think of the function f as the sum of two projection functions. It then follows that the array v are the components of f in this basis. These components according to Eq. (2.38), change under the action of $\mathcal{M}$ with the transpose of M:

$$\mathcal{M}(v) = M^{\dagger} v. \tag{2.39}$$

Thus the search for the linear eigenfunctions of $\mathcal{M}$ is equivalent to the search of the eigenvectors of $M^{\dagger}$. The reader must keep in mind that this change of perspective, in a linear context, looks like an attempt at complicating an equivalent process, the diagonalisation of M. Please be patient and remember that we are aiming at the full nonlinear map.

Now, let us assume that our favourite numerical recipe package returns the eigenvalues and eigenvectors of the matrix $M^{\dagger}$:

$$M^{\dagger} w = \lambda w \quad \text{where} \quad w = w_r + i\, w_i . \tag{2.40}$$

In Eq. (2.40), w_r is the real part of the eigenvector and w_i is the imaginary part. We can construct the following "normalised" functions

$$\begin{aligned} g_1(z) &= \frac{w_r \cdot z}{\sigma |[w_r \cdot z, w_i \cdot z]|^{1/2}} \\ g_2(z) &= \frac{w_i \cdot z}{|[w_r \cdot z, w_i \cdot z]|^{1/2}} \\ \text{where } \sigma &= \text{sign}\, [w_r \cdot z, w_i \cdot z] \end{aligned} \tag{2.41}$$

where $[f, g]$ refers to the Poisson bracket as defined in Eq. (1.1). Without loss of generality, I can take $\sigma = 1$ since $\sigma = -1$ amounts to the exchange of the eigenvectors and eigenvalues $\{\lambda, w\}$ and $\{\lambda^*, w^*\}$. The eigenfunctions of $\mathcal{M}$ can be easily constructed:

$$\begin{aligned} f_{\pm}(z) &= g_1(z) \pm i g_2(z) \\ \mathcal{M} f_{\pm} &= \exp(\mp i \mu_0)\, f_{\pm}. \end{aligned} \tag{2.42}$$

Consider the following transformation:

$$\begin{aligned} z_1^{new} &= g_1(z) \\ z_2^{new} &= g_2(z). \end{aligned} \tag{2.43}$$

Theorem 2.1 *The variables* z_1 *and* z_2 *move on a circle and therefore* (g_1, g_2) *defines the matrix* A^{-1},

This is trivial to prove. We notice that

$$z_1^{new2} + z_2^{new2} = (f_+ f_-)(z) \tag{2.44}$$

but we can evaluate this function one turn later:

$$z_1^{new2} + z_2^{new2} = (f_+ f_-)(Mz) \tag{2.45}$$

which by the definition of the map $\mathcal{M}$ (Eq. (2.35)) is just

$$\begin{aligned}(f_+ f_-)(Mz) &= (\mathcal{M} f_+ f_-)(z) \\ &= (\{\mathcal{M} f_+\}\{\mathcal{M} f_-\})(z) \\ &= \left(\lambda f_+ \lambda^* f_-\right)(z) = (f_+ f_-)(z) \quad \text{since } \lambda\lambda^* = 1.\end{aligned} \tag{2.46}$$

Thus, the linear map defining z^{new} defines A^{-1} per Eq. (2.24).

Just for fun, I can get an eigenvector using Eq. (2.28) where the matrix is already in terms of "proper" linear lattice functions. For example, I get the following result:

$$w = \begin{pmatrix} \frac{i-\alpha}{\sqrt{\beta}} \\ -\sqrt{\beta} \end{pmatrix} \quad \text{and} \quad M^\dagger w = e^{-i\mu} w \tag{2.47}$$

where in Eq. (2.47) I have already "normalised" the eigenvector to ensure a unit Poisson bracket. Using Eq. (2.41), I construct the new variables and get A^{-1}:

$$\begin{aligned} z_1^{new} &= g_1(z) = -\frac{\alpha}{\sqrt{\beta}} z_1 - \sqrt{\beta} z_2 \\ z_2^{new} &= g_2(z) = \frac{1}{\sqrt{\beta}} z_1 . \end{aligned} \tag{2.48}$$

Thus the matrix[5] for A and A^{-1} are:

$$A = \begin{pmatrix} 0 & \sqrt{\beta} \\ -\frac{1}{\sqrt{\beta}} & -\frac{\alpha}{\sqrt{\beta}} \end{pmatrix} \quad \text{and} \quad A^{-1} = \begin{pmatrix} -\frac{\alpha}{\sqrt{\beta}} & -\sqrt{\beta} \\ \frac{1}{\sqrt{\beta}} & 0 \end{pmatrix}. \tag{2.49}$$

Theorem 2.2 *In an n-d-f symplectic system, where $n \geq 2$, if all the tunes are on the unit circle and distinct (harmonic case), then the construction of* Eq. (2.41) *guaranties that the Poisson bracket of variables belonging to different planes is zero:*

$$\text{if } i \neq j \text{ then } \left[z_{2i-1}^{new}, z_{2j-1}^{new}\right] = \left[z_{2i-1}^{new}, z_{2j}^{new}\right] = \left[z_{2i}^{new}, z_{2j}^{new}\right] = 0. \tag{2.50}$$

[5]Note to accelerator physicists: this is not the usual Courant-Snyder transformation. See the next "grey box."

The accelerator physics readers may be puzzled by Eq. (2.49): this is not the so-called Courant-Snyder transformation. I wanted to emphasize that a normal form package will give you eigenvectors based on some scheme which pays no respect to any of our cultural biases. Indeed accelerator physicists will find it necessary to "canonise" the transformation of Eq. (2.49) for reasons explained later in Sect. 7.4. This amounts to a phase in the definition of the eigenvectors. In the nonlinear case, an arbitrary amplitude dependent rotation can be added in the definition of A. Thus we have

$$\underbrace{\begin{pmatrix} \sqrt{\beta} & 0 \\ -\frac{\alpha}{\sqrt{\beta}} & \frac{1}{\sqrt{\beta}} \end{pmatrix}}_{\text{Courant–Snyder Choice}} = \begin{pmatrix} 0 & \sqrt{\beta} \\ -\frac{1}{\sqrt{\beta}} & -\frac{\alpha}{\sqrt{\beta}} \end{pmatrix} \underbrace{\begin{pmatrix} 0 & -1 \\ 1 & 0 \end{pmatrix}}_{\text{rotation of } -\pi/2} . \tag{2.51}$$

Other readers, astronomers for example, may wonder what this is all about: why do we care about the canonical transformation? If it makes circles, then what is the big deal? One first notices that the particular phase advance derived from Courant-Snyder, even in the 2-d-f coupled case, has a special empirical meaning which is discussed in Sect. 7.4. Moreover, since it modifies position (i.e. z_1) minimally, it is ideally suited for analytical calculations since most perturbations in accelerator physics involve an electro-magnetic potential A_z which depends only on position.

The departure from the Courant-Snyder choice was rather trivial here: a difference of $\frac{-\pi}{2}$! But it is also possible to choose A with a non-trivial difference:

$$A = \begin{pmatrix} 1/\sqrt{\gamma} & -\alpha/\sqrt{\gamma} \\ 0 & \sqrt{\gamma} \end{pmatrix}, \tag{2.52}$$

in which case many properties associated with the phase advance would have to be rewritten since the difference between the Courant-Snyder choice and Eq. (2.52) is a rotation which depends on position, on α to be precise. I like to refer, somewhat facetiously, to the choice of Eq. (2.52) as the anti-Courant-Snyder choice.

2.3.2 *The Computation of the Map*

The computation proceeds following the explanation in Sect. 1.7.2.

```
state=only_2d0

map_order=2
call init_all(state,map_order,0)

call alloc(one_turn_map,id)
call alloc(y)
call alloc(normal_form)
call alloc(e,r2,z1,z2)

call find_orbit_x(als,closed_orbit(1:6),STATE,1.e-5_dp,fibre1=1)      ! (1)

id=1   ! map is set to identity                                         ! (2)

! map is added to closed orbit and put into the 6 polymorphs
y(1:6)=closed_orbit(1:6)+id                                              ! (3)

call propagate(als,y(1:6),state,fibre1=1)                                ! (4)

one_turn_map=y(1:6) ! Six polymorphs are promoted to Taylor maps        ! (5)
closed_orbit=y                                                           ! (6)

one_turn_map=one_turn_map.sub.1                                          ! (7a)
call print(one_turn_map,6,prec)                                          ! (7b)

call  c_normal(one_turn_map,normal_form)                                 ! (8)
```

The only remarkable feature of this piece of code is the `state=only_2d0`. Although PTC is a serious code, this lines allows the user to forget the dimensions of phase space besides the horizontal plane: $(z_1, z_2) = (x, p_x)$.

For the benefit of all readers, I point out again that for "ideal planar rings," the restriction to the horizontal plane is a *bona fide* nonlinear symplectic map. The literature is full of introductory treatments that concentrate heavily on the horizontal plane (see [2–4]). In describing my techniques, particularly their relationship to the code, I could start with a general discussion. But I follow here the gradual approach. The reader will notice that the software tools and the approach will **not** change as we move towards a more general situation: coupled, nonlinear and with spin.

The particular analysis program I wrote, FPP, requires us to set the maximum degree needed at the onset. Therefore our quest for quadratic invariants of linear maps forces us to set the order of the Taylor series to 2. Later, in line (7a), I truncate the map to first-order so as to retain solely the matrix part. The result is printed by the program at line (7b): it is properly truncated to first-order.

2.3.3 Numerical Computation of the Canonical Transformation

Now we come to some new features: a normal form is performed. On line (8), the map `one_turn_map` is normalised. The results are stored in an object called `normal_form`. In particular the transformation a of Eq. (2.7) is stored in `normal_form%a_t`. The routine `c_normal` is very general. In particular it can and will perform a nonlinear normal form if demanded by the user. This is discussed in 1-d-f in Chap. 3. The result of this routine, made to match Eq. (2.49), is printed by line (8d):

```
a=normal_form%a_t
write(6,*)"";write(6,*) " Canonical transformation A";
do i=1,c_%nd2
 write(6,'(a5,i1,a5,6(1x,g12.5))') "row",i,"-->",a(i,1:c_%nd2)   ! (8d)
enddo
```

The output is just

```
   Canonical transformation A
 row 1 -->     0.0000        3.3404
 row 2 -->   -0.29936       0.13421E-02
```

This transformation was selected to match the case of Eq. (2.49). To get it, I expressed the normalised map of Eq. (2.7) in phasors' basis using Eq. (2.10):

$$
\begin{aligned}
m &= a \circ r \circ a^{-1} \\
&= a \circ c \circ \Lambda \circ c^{-1} \circ a^{-1} \text{ where } \Lambda = \begin{pmatrix} e^{i\mu} & 0 \\ 0 & e^{-i\mu} \end{pmatrix} \\
&\Downarrow \\
m^{\dagger} &= \underbrace{a^{-1^{\dagger}} \circ c^{-1^{\dagger}}}_{ac} \circ \Lambda \circ c^{\dagger} \circ a^{\dagger}.
\end{aligned}
\tag{2.53}
$$

The eigenvector of Eq. (2.47) should be the column vector of the matrix $ac = a^{-1^{\dagger}} \circ c^{-1}$. In the program of Appendix C, this is checked by the lines

```
write(6,*) "";write(6,*) "w_1 =",(i_-alpha)/sqrt(beta)          ! (12a)
               write(6,*) "w_2 =",-sqrt(beta)                     ! (12b)


ac=from_phasor(-1) * normal_form%a_t**(-1)                        ! (13a)
ac=transpose(ac)                                                  ! (13b)

write(6,*) "";write(6,*) " Complex Canonical transformation A";write(6,*) "";
write(6,'(a8,17x,a1,18x,a1,1/)') "column","1","2"
do i=1,c_%nd2
 write(6,'(a5,i1,a5,6(1x,g12.5,1x,g12.5))') "row",i,"-->",ac(i,1:c_%nd2)
enddo
```

Lines (12a) and (12b) construct the eigenvector by direct evaluation of Eq. (2.47). Lines (13a) and (13b) is the FPP implementation[6] of Eq. (2.53). The first row of the matrix should be the same as the output of lines (12a, b). The result of this program is:

```
  w_1 = ( 1.34213518775613719E-003, 0.29936426981922848      )
  w_2 =  -3.3404120024205004

Complex Canonical transformation A

column                    1                  2

row 1 -->   0.13421E-02  0.29936       0.13421E-02 -0.29936
row 2 -->   -3.3404        0.0000      -3.3404        0.0000
```

2.4 The Phase Advance and the Invariant

The code must be able to simulate the motion from one point in the machine to another. In a real accelerator we might be content to observe a beam at various beam position monitors (BPM). Most tracking codes should give us access to all positions: the individual integration steps. Therefore it is reasonable to enquire the value of the canonical transformation a at each position s around the ring, a discrete integer in integrators such as PTC. This will lead us to the phase advance which is the code equivalent of the normalised Hamiltonian.

2.4.1 Phase Advance in a Code: Finite Map Theory

Important! The reader should look at Fig. 2.1: it contains mathematics applicable to nonlinear coupled spin-orbit maps.[7] The red dot represents a ray moving in the "real world." The blue dots (pale and dark) show the corresponding ray, transformed by a_s, moving in the world of circles. In the absence of proof to the contrary, the motion in the circle space, i.e., the angle μ_{12}, must depend on the choice of a_s. This is indeed the case in an "s"-dependent system.

The transformations a_1 and a_2 are simply the transformations one obtains by normalising the one-turn map at position $s = 1$ and $s = 2$. These canonical transformations are the output of the normalising software (FPP in my case). Generally they can be chosen arbitrarily, however in the linear case, the Courant-Snyder-Teng-Edwards choice is preferable for experimental and also analytical reasons. I will come back to this topic later (see Sect. 7.4). Now suffices to say that we have a

---

[6] The line `c_mess_up_vector=.true.; b_mess=-1.0_dp;` at the beginning of the program insures the choice of this section for the matrix A. The reader is invited to comment it out.

[7] As a matter of fact, it even applies to a damped map, i.e., a non-symplectic map such as the map of a classical electron in a radiation source. In that case one must define the normal form to be a "affine dilation," a damped rotation whose complex eigenvalues are slightly under the unit circle.

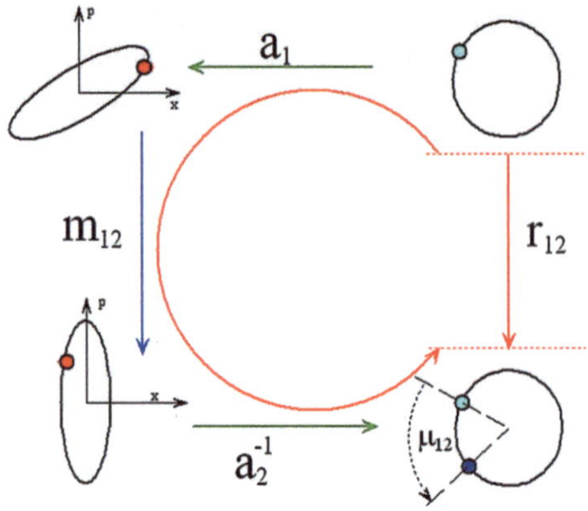

Fig. 2.1 Pictorial view of the phase advance

recipe for the (non)linear calculation of the transformation a_s that depends only on the one-turn map m_s at location s.

It can be shown in the sympletic case, quite generally, that since the normal form is made of commuting rotations, then the map r_{12} on Fig. 2.1 must a rotation.[8] The formula for r_{12} is trivially derived from Fig. 2.1.

$$r_{12} = a_2^{-1} \circ m_{12} \circ a_1 . \tag{2.54}$$

Imagine that in Eq. (2.54), the map can be diagonalised by the same transformation: $a_2 = a_1$. In that case we say that the positions 1 and 2 are matched. They have the same invariants, i.e., the same lattice functions. Moreover, it is possible to prove that the map r_{12} does not depend on the method we used to diagonalise the one-turn map provided that this method depends only on the one-turn map alone and not on extraneous factors. Incidentally this is not the case of Hamiltonian perturbation theory since the map a_s depends on the Hamiltonian, local derivatives of the map with respect to "s". See Chap. 8.

Equation (2.54) defines the phase advance μ_{12} associated to the canonical transformation a_s. This definition works in a code and it works perfectly well in analytical perturbation theory. As I pointed out previously, this is a mute point in astronomy because the normalisation of the Hamiltonian (rather than a map) immediately produces the phase advance, nothing more than the normalised Hamiltonian.

[8]This statement applies to spin as well since the normal form is a rotation around a fixed axis, say the y-axis, for all trajectories in phase space. This set of rotations forms a commutative subgroup of SO(3).

It is instructive, from a tracking code point of view, to rewrite Eq. (2.54):

$$a_2 \circ r_{12} = \underbrace{m_{12} \circ a_1}_{b_2} \Rightarrow a_2 = b_2 \circ r_{12}^{-1}. \tag{2.55}$$

The map b_2 has three remarkable properties:

1. First, it diagonalise the map at position $s = 2$ just like the map a_2. Therefore the invariant ε_2 can be computed using b_2:

$$\varepsilon_2 = r^2 \circ a_2^{-1} = r^2 \circ b_2^{-1}. \tag{2.56}$$

2. Second, b_2 can be tracked by the code using a_1 as the input. Therefore it does not require any normal form calculation.
3. Third, the difference between a_2 and b_2 is the phase advance r_{12}.

Property 1 simply comes from the invariance of a radius under the rotation r_{12}:

$$r^2 \circ b_2^{-1} = r^2 \circ r_{12}^{-1} \circ a_2^{-1} = r^2 \circ a_2^{-1} = \varepsilon_2. \tag{2.57}$$

Proposition 2 is crucial. The canonical transformation a_s can be tracked. In Eq. (2.55), the map for b_2, is made of two parts m_{12} and a_1. If this was the map for a part of the ring, one would conclude that the particle travels first through a segment described by the map a_1 and then through a segment described by m_{12}. Thus if the code that computes m_{12} is initialised by a_1 rather than by the identity, then b_2 will be the output of the propagator. In Sect. 1.7.2, for example, line (3) of the PTC code is replaced by

```
y(1:6)=closed_orbit(1:6)+a_1              ! (3)
```

Proposition 3 gives the most efficient wait to compute the phase advance: one needs only to rotate the map b_2 into the chosen form for the canonical transformation a_2. This is really different from Hamiltonian perturbation theory where, by the very definition of the normalisation process, the phase advance is the transformed Hamiltonian. In accelerator physics, we have only the code, we do **not** have this transformed Hamiltonian: we have the information at chosen surfaces of section.

Let us look for the time being at a piece of code which produces all the information of items 1, 2 and 3.

2.4.2 *Numerical Computation of the Invariant and the Phase Advance*

The program in Appendix D, `one_turn_orbital_map_normal_form_2d.f90`, computes the phase advance for the two simple choices of canonical transformation shown of Sect. 2.2.3, namely the Eqs. (2.51) and (2.52) which I rewrite here for convenience:

$$\underbrace{\begin{pmatrix} 1/\sqrt{\gamma} & -\alpha/\sqrt{\gamma} \\ 0 & \sqrt{\gamma} \end{pmatrix}}_{Anti-Courant-Snyder} \quad \text{and} \quad \underbrace{\begin{pmatrix} \sqrt{\beta} & 0 \\ -\alpha/\sqrt{\beta} & 1/\sqrt{\beta} \end{pmatrix}}_{Courant-Snyder}. \tag{2.58}$$

The program asks the user to choose between these two definitions. For example I can select between the two transformations by typing either `t` or `f`:

```
Write't' for Courant-Snyder
Write'f' for Anti-Courant-Snyder
t
```

In the example code, the phase advance through the first drift is computed. Thus the map m_{01} is that of a drift. The matrix for a drift of length L is just:

$$m_{01} = \begin{pmatrix} 1 & L \\ 0 & 1 \end{pmatrix}. \tag{2.59}$$

Let us start with the Courant-Snyder definition and the second part of Eq. (2.55) to find the phase advance:

$$a_2 = \underbrace{\begin{pmatrix} 1 & L \\ 0 & 1 \end{pmatrix}}_{m_{12}} \underbrace{\begin{pmatrix} \sqrt{\beta} & 0 \\ -\alpha/\sqrt{\beta} & 1/\sqrt{\beta} \end{pmatrix}}_{a_1} \underbrace{\begin{pmatrix} \cos\left(\mu_{12}^{cs}\right) & -\sin\left(\mu_{12}^{cs}\right) \\ \sin\left(\mu_{12}^{cs}\right) & \cos\left(\mu_{12}^{cs}\right) \end{pmatrix}}_{r_{12}^{-1}} \tag{2.60}$$

which, expanded, is given by:

$$a_2 = \begin{bmatrix} \left(\sqrt{\beta} - \frac{\alpha L}{\sqrt{\beta}}\right)\cos\left(\mu_{12}^{cs}\right) + \frac{L\sin\left(\mu_{12}^{cs}\right)}{\sqrt{\beta}} & -\left(\sqrt{\beta} - \frac{\alpha L}{\sqrt{\beta}}\right)\sin\left(\mu_{12}^{cs}\right) + \frac{L\cos\left(\mu_{12}^{cs}\right)}{\sqrt{\beta}} \\ -\frac{\alpha\cos\left(\mu_{12}^{cs}\right)}{\sqrt{\beta}} + \frac{\sin\left(\mu_{12}^{cs}\right)}{\sqrt{\beta}} & \frac{\alpha\sin\left(\mu_{12}^{cs}\right)}{\sqrt{\beta}} + \frac{\cos\left(\mu_{12}^{cs}\right)}{\sqrt{\beta}} \end{bmatrix}. \tag{2.61}$$

The Courant-Snyder choice forces us to select the phase advance so that the A_{12} part of the matrix is zero. Therefore the equation for the phase advance is:

$$A_{12} = -\left(\sqrt{\beta} - \frac{\alpha L}{\sqrt{\beta}}\right)\sin\left(\mu_{12}^{cs}\right) + \frac{L\cos\left(\mu_{12}^{cs}\right)}{\sqrt{\beta}} = 0 \tag{2.62}$$

from which we deduce that

$$\mu_{12}^{cs} = \tan^{-1}\left(\frac{L}{\beta - L\alpha}\right). \tag{2.63}$$

The same thing can be done with the "anti-Courant-Snyder" definition:

$$a_2 = \underbrace{\begin{pmatrix}1 & L\\ 0 & 1\end{pmatrix}}_{m_{12}} \underbrace{\begin{pmatrix}1/\sqrt{\gamma} & -\alpha/\sqrt{\gamma}\\ 0 & \sqrt{\gamma}\end{pmatrix}}_{a_1} \underbrace{\begin{pmatrix}\cos\left(\mu_{12}^{acs}\right) & -\sin\left(\mu_{12}^{acs}\right)\\ \sin\left(\mu_{12}^{acs}\right) & \cos\left(\mu_{12}^{acs}\right)\end{pmatrix}}_{r_{12}^{-1}} \tag{2.64}$$

which expanded out is just

$$a_2 = \begin{bmatrix} \frac{\cos(\mu_{12}^{acs})}{\sqrt{\gamma}} + \left(-\frac{\alpha}{\sqrt{\gamma}} + L\sqrt{\gamma}\right)\sin\left(\mu_{12}^{acs}\right) & -\frac{\sin(\mu_{12}^{acs})}{\sqrt{\gamma}} + \left(-\frac{\alpha}{\sqrt{\gamma}} + L\sqrt{\gamma}\right)\cos\left(\mu_{12}^{acs}\right) \\ \sqrt{\gamma}\sin\left(\mu_{12}^{acs}\right) & \sqrt{\gamma}\cos\left(\mu_{12}^{acs}\right) \end{bmatrix}. \tag{2.65}$$

The anti-Courant-Snyder choice forces us to select the phase advance so that the A_{21} part of the matrix is zero. Therefore the equation for this phase advance is:

$$A_{21} = 0 \Rightarrow \sqrt{\gamma}\sin\left(\mu_{12}^{acs}\right) = 0 \Longrightarrow \mu_{12}^{acs} = 0. \tag{2.66}$$

We look now at the actual code concentrating first on the one-turn map at $s = 1$

```
z1=1.e0_dp.cmono.'10'                                                  ! (1a)
z2=1.e0_dp.cmono.'01'                                                  ! (1b)
r2=z1**2+z2**2                                                         ! (1c)

call find_orbit_x(als,closed_orbit(1:6),STATE,1.e-5_dp,fibre1=1)       ! (2)

id=1   ! map is set to identity                                        ! (3)

! map is added to closed orbit and put into the 6 polymorphs
y(1:6)=closed_orbit(1:6)+id                                            ! (4)

call propagate(als,y(1:6),state,fibre1=1)                              ! (5)

one_turn_map=y(1:6) ! Six polymorphs are promoted to Taylor maps       ! (6)
closed_orbit=y                                                         ! (7)

call  c_normal(one_turn_map,normal_form)                               ! (8a)

write(6,'(1/,a50,1/)')"Canonical Transformation coming from Normal Form"
call print(normal_form%a_t,6)                                          ! (8b)

call c_canonise(normal_form%a_t,a_1,phase=phase);phase(1)=0.0_dp;      ! (9a)

if(courant_snyder_teng_edwards) then
 write(6,'(1/,a50,1/)') "Courant-Snyder Canonical Transformation "
  else
 write(6,'(1/,a50,1/)') "Anti-Courant-Snyder Canonical Transformation "
endif
call print(a_1,6,prec)                                                 ! (9b)
```

Line (5) is the usual propagation that gives the one-turn map. Lines (8a,b) is the result of the normal form: it provides a certain form for a_s "randomly" chosen by the programmer of the normal form routine. It is printed by (9a). In this case, the result is neither the Courant-Snyder nor this bizarre anti-Courant-Snyder transformation but something else. A normal form, like a diagonalisation routine for eigenvalues and eigenvectors, makes no special[9] guaranty: it diagonalises with a transformation of its own.

The next step consist in invoking the routine "`c_canonise`." We owe the word "canon" to the Church, it refers to a set of rules[10] and regulations imposed by the Church as in "Canon Law." Here it also refers to a certain set of rules imposed by accelerator physicists on the transformation a_s. They are somewhat justified but not truly indispensable. The routine `c_canonise` imposes either (2.62) or (2.66) and additional conditions in the nonlinear case. Let us look at the code.

```
y(1:6)=closed_orbit(1:6)+a_1                                           ! (10)

call propagate(als,y(1:6),state,fibre1=1,fibre2=pos)                   ! (11)

b_2=y(1:6)                                                             ! (12a)
call c_canonise(b_2,a_2,phase=phase)                                   ! (12b)
!!!!!!!!!!!!!!!!!!!!!!!!   One turn map at pos = 2 !!!!!!!!!!!!!!!!!!!!
call find_orbit_x(als,closed_orbit(1:6),STATE,1.e-5_dp,fibre1=pos)    ! (13)

y(1:6)=closed_orbit(1:6)+id                                            ! (14)

call propagate(als,y(1:6),state,fibre1=pos)                            ! (15)

one_turn_map=y(1:6)                                                    ! (16)

call  c_normal(one_turn_map,normal_form)                               ! (17)
call  c_canonise(normal_form%a_t,a_2);                                 ! (18)

!!!!!!!!!!!!!!!!!!   Computes invariants at 1 and 2 !!!!!!!!!!!!!!!!!

e1 = r2*a_1**(-1)                                                      ! (19a)
e2t = r2*b_2**(-1)                                                     ! (19b)
e2 = r2*a_2**(-1)                                                      ! (19c)

write(6,'(1/,a54,1/)')"Invariant at position = 1 computed from one turn map"
call print(e1,6)                                                       ! (20a)
write(6,'(1/,a54,1/)') "Invariant at position = 2  tracked from position = 1"
call print(e2t,6)                                                      ! (20b)
write(6,'(1/,a54,1/)') "Invariant at position = 2  computed from one turn map"
call print(e2,6)                                                       ! (20c)


write(6,'(1/,a54,1/)') "Phase advance from position = 1 to position = 2      "
!!!!!!!!!!!!!!!!!!!!!!!!!!!!   Analytic results !!!!!!!!!!!!!!!!!!!!!!
```

---

[9]My normalisation algorithm, in the absence of parameter dependence, accidentally picks up A to be Courant-Snyder. So I have some parameters, namely `c_mess_up_vector=.true.;` `b_mess=-1.0_dp;` to insure a real messed up choice for pedagogical reasons. You are invited to comment that out if you want.

[10] We also owe the present usage of the word "propaganda" to the Church: the necessary dogma that must be propagated!

```
if(pos==2.and.courant_snyder_teng_edwards) then
 L=2.832695d0 ! Length of first element L1
 beta=e1.sub.'02'
 alpha=(e1.sub.'11')/2

 write(6,"(a48,G20.13)") "Based on theory, the phase advance should be =", &
                                                    atan(L/(beta-L*alpha))/twopi
else
 write(6,*) "Based on theory, the phase advance should be zero"
endif
!!!!!!!!!!!!!!!!!!!!!!!!!! The code's results !!!!!!!!!!!!!!!!!!!!!!!!!!

call print(phase(1),6,prec)                                              ! (21)
```

Line (10), line (11) and (12a, b) correspond to Eq. (2.55). The routine `c_canonise` returns a polynomial for the angle of the phase advance in units of revolutions which is customary in accelerator physics. This is printed on line (21) and compared to the analytical results of Eq. (2.63). The agreement in this case is perfect because the ring of this example is "ideal." The reader must accept that the result of line (12b) would remain correct even if nonlinearities and coupling were present. This is the virtue of a code with self-consistent tools.

Lines (19a, b, c) serve two purposes. First of all, line (19a) gives us the invariant at position 1. From this we extract the lattice functions "beta" and "alpha" needed for the analytical formula of Eq. (2.63). Lines (19b, c) are there to reassure us that the invariant at position 2 can be computed from the tracked transformation b_2. Here is the output in abbreviated form:

```
Invariant at position = 1 computed from one turn map


Properties, NO =     2, NV =     2, INA =  148
*********************************************

  2  0.8962076737126187E-01   0.000000000000000        2  0
  2 -0.8966568980107243E-02   0.000000000000000        1  1
  2    11.15835234591495      0.000000000000000        0  2

Invariant at position = 2  tracked from position = 1


Properties, NO =     2, NV =     2, INA =  147
*********************************************

  2  0.8962076737126187E-01   0.000000000000000        2  0
  2 -0.5167031682375807       0.000000000000000        1  1
  2    11.90288336404888      0.000000000000000        0  2

Invariant at position = 2  computed from one turn map


Properties, NO =     2, NV =     2, INA =  141
*********************************************

  2  0.8962076737126205E-01   0.000000000000000        2  0
  2 -0.5167031682375732       0.000000000000000        1  1
  2    11.90288336404884      0.000000000000000        0  2
```

Finally, the phase advance is printed. If we select the Courant-Snyder phase advance then the result is:

```
Phase advance from position = 1 to position = 2

Based on theory, the phase advance should be =  0.3952456186311E-01

Properties, NO =     2, NV =     2, INA =    62
*********************************************

   0  0.3952456186310726E-01   0.000000000000000        0  0
```

More bizarrely, at least for accelerator physicists, the phase advance for the "anti-Courant-Snyder" choice of Eq. (2.66) is indeed zero!

```
Phase advance from position = 1 to position = 2

 Based on theory, the phase advance should be zero

Properties, NO =     2, NV =     2, INA =    62
*********************************************

 Complex Polynomial is zero
```

2.4.3 Computation of the Phase: Some Theory

The code, equipped with TPSA, is always right as I pompously proclaimed in Sect. 1.3. Therefore the call to `c_canonise` in Sect. 2.4.1 works in all cases. Nevertheless it is very instructive, using matrices only, to derive a formula for the phase advance using the most arbitrary 1-d-f Hamiltonian. First of all, I write the most arbitrary quadratic Hamiltonian in 1-d-f:

$$H = \frac{1}{2}\left\{k_{11}z_1^2 + 2k_{12}z_1z_2 + k_{22}z_2^2\right\}. \tag{2.67}$$

The matrix for an infinitesimal propagation due to the Hamiltonian of Eq. (2.67) is

$$m_{ds} = \begin{pmatrix} 1 + k_{12}ds & k_{22}ds \\ -k_{11}ds & 1 - k_{12}ds \end{pmatrix} \tag{2.68}$$

and the matrix for an infinitesimal rotation is

$$r_{ds}^{-1} = \begin{pmatrix} 1 & -d\mu_{ds}^{cs} \\ d\mu_{ds}^{cs} & 1 \end{pmatrix}. \tag{2.69}$$

Then, using the analogue of Eq. (2.60) and imposing $A_{12} = 0$, I get

$$d\mu_{ds}^{cs} = \frac{k_{22}}{\beta} ds. \tag{2.70}$$

And, it can been shown easily by symmetry that

$$d\mu_{ds}^{acs} = \frac{k_{11}}{\gamma} ds. \tag{2.71}$$

In the case of accelerator physics, we have $k_{11} = k(s)$, $k_{12} = 0$ and $k_{22} = 1$ and therefore we get the usual result:

$$d\mu_{ds}^{cs} = \frac{1}{\beta} ds. \tag{2.72}$$

Equation (2.72) is completely consistent with Eq. (2.63) if we simply let L tend towards zero.

2.4.4 Tracking of the Invariant: Some Hamiltonian Theory

We can first start with equation Eq. (2.55):

$$\varepsilon_2 = r^2 \circ b_2^{-1} = r^2 \circ a_1^{-1} \circ m_{12}^{-1} = \varepsilon_1 \circ m_{12}^{-1}. \tag{2.73}$$

Equation (2.73) expresses a simple fact: the invariant at position 2 is gotten from the known invariant at position 1 by simply expressing it in terms of the variables at position 2 using the map from 1 to 2. This is completely general inside the code and applies to nonlinear invariants in many degrees of freedom.

However, I can apply this to the general 1-d-f quadratic Hamiltonian of Eq. (2.67) and its related infinitesimal map (Eq. 2.68):

$$\begin{aligned} \varepsilon_{s+ds} &= \varepsilon_s \circ m_{ds}^{-1} \\ &= \gamma\{(1 - k_{12}ds)\, z_1 - k_{22}ds z_2\}^2 \\ &+ 2\alpha\, \{(1 - k_{12}ds)\, z_1 - k_{22}ds z_2\}\{-k_{11}ds z_1 + \{1 - k_{12}ds\}\, z_2\} \\ &+ \beta\, \{-k_{11}ds z_1 + \{1 - k_{12}ds\}\, z_2\}^2. \end{aligned} \tag{2.74}$$

I then deduce a differential equation for the lattice functions:

$$\begin{aligned} \frac{d\varepsilon_s}{ds} &= 2(\alpha\, k_{11} - \gamma\, k_{12})x^2 + 2(\beta k_{11} - \gamma k_{22})xp \\ &+ 2(\beta k_{12} - \alpha k_{22})p^2 \end{aligned}$$

$$\Downarrow$$

$$\frac{d\gamma}{ds} = 2(\alpha\, k_{11} - \gamma\, k_{12}) \tag{2.75}$$

$$\frac{d\alpha}{ds} = \quad \beta k_{11} - \gamma k_{22} \tag{2.76}$$

$$\frac{d\beta}{ds} = 2(\beta k_{12} - \alpha k_{22}). \tag{2.77}$$

In accelerator physics, this set of equations reduces to:

$$\begin{aligned} \frac{d\gamma}{ds} &= 2\alpha k_{11} \\ \frac{d\alpha}{ds} &= \quad \beta k_{11} - \gamma \\ \frac{d\beta}{ds} &= -2\alpha. \end{aligned} \tag{2.78}$$

It is interesting to relate the Courant-Snyder phase advance of Eq. (2.70) with the anti-Courant-Snyder one of Eq. (2.71). This can be done using the equation for α (Eq. 2.76):

$$\begin{aligned} \frac{d\mu^{cs}_{ds}}{ds} &= \frac{d\mu^{acs}_{ds}}{ds} - \frac{\frac{d\alpha}{ds}}{\beta\gamma} \\ &= \frac{d\mu^{acs}_{ds}}{ds} - \frac{\frac{d\alpha}{ds}}{1+\alpha^2} \quad \text{(according to Eq. (2.29))}. \end{aligned} \tag{2.79}$$

We can check explicitly the statement near Eq. (2.54) on the case of the two phases used in Eq. (2.79): between matched points, the phase advance does not depend on the method used to normalise the map. To do this we integrate Eq. (2.79) between two position 1 and 2:

$$\mu^{cs}_{12} - \mu^{acs}_{12} = -\int_{\alpha_1}^{\alpha_2} \frac{d\alpha}{1+\alpha^2} = \tan^{-1}(\alpha_1) - \tan^{-1}(\alpha_2)\,. \tag{2.80}$$

Equation (2.80) tells us that if the parameters α are equal, then the phase advances can only differ by an irrelevant multiple of 2π. It is certainly true looking at Eq. (2.80), that if position 1 and 2 are matched, then the phase advances are equal modulo 2π: with two specific choices for a_s the condition is found to be weaker.

References

1. P. Nishikawa, JINST **7** (2012)
2. D.A. Edwards, M.J. Syphers, *An Introduction to the Physics of High Energy Accelerators, Wiley Series in Beam Physics and Accelerator Technology* (Wiley, New York, 1993)
3. E. Wilson, Technical Report No. CERN 84–19, CERN (Unpublished)
4. S.Y. Lee, *Accelerator Physics* (World Scientific Publishing, Singapore, 2004)

Chapter 3
The Nonlinear Transverse Normal Form: One Degree of Freedom

Abstract I extend in this chapter the concept of a normal form to nonlinear systems in 1-d-f. I try to contrast Hamiltonian methods and one-turn map methods. The pendulum is solved by "undergraduate tools" and the calculation is corroborated by the perturbation tools of this book. The standard map, which is just the map for an RF-cavity restricted to the longitudinal plane, is introduced as the archetype of the trouble maker: it can display resonances and even chaos if "pushed" too hard.

Keywords One-degree-of-freedom · Nonlinear · Pendulum · Standard map · Poisson brackets · Vector fields

3.1 The Pendulum with Exact Methods

3.1.1 Preliminary

The world on 1-d-f nonlinear systems relevant to accelerator physics consist mainly of maps that are stable around the closed orbit of the ring which they purport to describe. Circular accelerators are usually[1] designed with a stable linear part.

These maps can be divided into two classes: integrable and non-integrable. The pendulum and the standard map are closely related examples of these two types of maps. While we would like to have integrable maps, in reality, most maps are not integrable.

In this section I discuss the pendulum since it is the archetype of the integrable map in one degree of freedom. It has an obvious invariant: the Hamiltonian itself.

Moreover, using mathematics of the most elementary kind, we can get the period of oscillation as a function of the initial amplitude. This period, contrary to the impression given in some high school textbook, depends on the initial amplitude. In fact it increases from the famous $T = 2\pi\sqrt{\frac{L}{g}}$ all the way to infinity at the separatrix.

---

[1] It is possible to design pure nonlinear systems which are stable but this is not commonly considered in particle accelerators.

E. Forest, *From Tracking Code to Analysis*,
DOI 10.1007/978-4-431-55803-3_3

Indeed, a rigid pendulum, standing precariously upside down will stay there forever. Moreover, trajectories in phase space are not ellipses except very near the origin.

So I can summarize a few generic points of integrable maps near a stable fixed point:

1. The period of oscillation depends on the initial amplitude.
2. The trajectories are not ellipses but can be distorted into conics using symplectic transformations. In fact they can be distorted into circles.
3. These two results extend to integrable Hamiltonians in several degrees of freedom. However most Hamiltonians taken out of a hat[2] are not integrable.

The standard map, which is a close relative of the pendulum, is not integrable. The standard map emerges from the pendulum when we apply numerical integration to the pendulum. For example, if I use a first- or second-order symplectic integrator on the pendulum, I produce the standard map. As long at the time step is small, the map will appear integrable: it better be since it approximates the pendulum!

In Sect. 3.1.2, I look at the pendulum with elementary methods appropriate to simple time independent Hamiltonians whose linearised motion is that of a harmonic oscillator. The purpose of this section is to provide the solution of a problem that the reader could have gotten without using the tools of perturbation theory.

Then, in Sect. 3.2, I produce the map with the package FPP. This map will be analysed with Hamiltonian methods and the result reproduced by a little algorithm in the program `pendulum` of Appendix E.

Then, in Sect. 3.3, I will finally look at the standard map which is representative of an accelerator: the Hamiltonian is intrinsically s-dependent and thus new tools must be introduced if we are to stay "faithful" to the Code!

3.1.2 The Hamiltonian of the Pendulum

The Hamiltonian of the pendulum—its energy function—is

$$H = \frac{p^2}{2ml^2} + mgl\,(1 - \cos\theta) = \frac{p^2}{2ml^2} + 2mgl\,\sin^2\frac{\theta}{2} \tag{3.1}$$

where l is the length of a massless rigid rod, m is the mass of a bob at the end of that rod and g is the acceleration of gravity. This system has a constant of the motion which is the energy function H. The trajectories in the phase space (θ, p) are the solutions of the algebraic equation

$$H(\theta, p) = E_0. \tag{3.2}$$

[2] To my knowledge, there is only one Hamiltonian taken out of a hat which was surprisingly found to be integrable despite having no apparent symmetries: the Toda lattice.

These trajectories foliate phase space. Moreover the rotation rate around these trajectories depends on the size of the trajectory itself. These are common properties of integrable systems.

To simplify the coming algebra, I change the definition of time t into a new time τ defined as $t = ml^2\tau$. The resulting Hamiltonian is:

$$\begin{aligned} K &= \frac{p^2}{2} + 2gm^2l^3 \sin^2\frac{\theta}{2} \\ &= \frac{p^2}{2} + 2(2\pi f)^2 \sin^2\frac{\theta}{2}. \end{aligned} \tag{3.3}$$

The quantity f is the frequency of the linearised system in the new time units. Indeed, if I expand K to second-order, I get:

$$K_2 = \frac{p^2}{2} + (2\pi f)^2 \frac{\theta^2}{2}. \tag{3.4}$$

The solution of Eq. (3.4) can be expressed easily in terms of sines and cosines:

$$\begin{pmatrix} \theta_t \\ p_t \end{pmatrix} = \begin{pmatrix} \cos(2\pi f t) & \frac{1}{2\pi f}\sin(2\pi f t) \\ -2\pi f \sin(2\pi f t) & \cos(2\pi f t) \end{pmatrix} \begin{pmatrix} \theta_0 \\ p_0 \end{pmatrix}. \tag{3.5}$$

I can also perform a simple canonical transformation on K_2:

$$\theta^{new} = \sqrt{2\pi f}\theta = \sqrt{2J}\cos(\Phi) \quad p^{new} = \frac{1}{\sqrt{2\pi f}}p = -\sqrt{2J}\sin(\Phi). \tag{3.6}$$

Then the new Hamiltonian is

$$K_2^{new}\left(\theta^{new}, p^{new}\right) = (2\pi f)\left\{\frac{p^{new\,2}}{2} + \frac{\theta^{new\,2}}{2}\right\} = 2\pi f\ J. \tag{3.7}$$

Equations (3.5) and (3.7) display the famous results of the linearised pendulum: the period, namely $1/f$, is obviously not a function of amplitude. Equation (3.7) expresses explicitly this fact: the new Hamiltonian is linear in the action-angle variable J and therefore the period of oscillation is a constant.

3.1.3 The Exact Period of the Pendulum

This section involves only elementary classical physics. I start the pendulum at rest with an angle $\theta = \sqrt{d}$. Taking the root anticipates the fact that the action variable J, while still unknown at this stage, is more closely related to the square of the position than to the position itself as Eq. (3.7) suggests. On this trajectory, the value of the

Hamiltonian $\mathscr{K}$ is:

$$\mathscr{K}_d = 2(2\pi f)^2 \sin^2 \frac{\sqrt{d}}{2}. \tag{3.8}$$

Thus for an arbitrary trajectory I have the relation

$$2(2\pi f)^2 \sin^2 \frac{\sqrt{d}}{2} = \frac{p^2}{2} + 2(2\pi f)^2 \sin^2 \frac{\theta}{2}, \tag{3.9}$$

which in turns implies that

$$dt = \frac{d\theta}{4\pi f \sqrt{\sin^2 \frac{\sqrt{d}}{2} - \sin^2 \frac{\theta}{2}}}. \tag{3.10}$$

This equation can be integrated from 0 to $\sqrt{d}$ to obtain a quarter period. Thus the full period is:

$$T = \int_0^{\sqrt{d}} \frac{d\theta}{\pi f \sqrt{\sin^2 \frac{\sqrt{d}}{2} - \sin^2 \frac{\theta}{2}}}. \tag{3.11}$$

To turn Eq. (3.11) into a form that lends itself to an expansion in d, I perform a change of variables

$$x = \frac{\sin \frac{\theta}{2}}{\sin \frac{\sqrt{d}}{2}}. \tag{3.12}$$

leading to

$$T = \frac{2}{\pi f} \int_0^1 \frac{dx}{\sqrt{1 - \sin^2 \frac{\sqrt{d}}{2} x^2}\ \sqrt{1 - x^2}}. \tag{3.13}$$

If I expand Eq. (3.13), say to third-order, I get

$$T = \frac{1}{f} \left(1 + \frac{d}{16} + \frac{11\, d^2}{3072} + \frac{173\, d^3}{737280} + \cdots \right). \tag{3.14}$$

The contribution to the period are all positive and diverge at the separatrix where the period, correctly, becomes infinite. This corresponds to a pendulum starting in the upside down position. The closer it is to "twelve o'clock", the longer it takes to execute a full oscillation.

The new frequency is the inverse of Eq. (3.14):

$$f(d) = f\left(1 - \frac{d}{16} + \frac{d^2}{3072} - \frac{23\,d^3}{737280}\cdots\right). \tag{3.15}$$

3.1.4 The Frequency and the Hamiltonian as a Function of J

The result of Eq. (3.14) is expressed in terms of a "physical" quantity namely—$d = \theta^2_{\rm max}$—the maximum amplitude reached during one period.

As we have seen, this result is immediately available for simple time-independent Hamitonian systems in 1-d-f. However, when we perform perturbative calculations, we first get the period or frequencies in terms of the invariant action J.

Therefore, for completeness, it is amusing to compute J exactly in terms of a quadrature:

$$J = \frac{1}{2\pi}\int_0^{2\pi} J_\tau d\phi = \frac{J_\tau}{2\pi}\int_0^{2\pi} d\phi = J_\tau \tag{3.16}$$

since J_τ is actually a constant on the trajectory and does not depend on the time τ. Now I use two important facts:

1. Equation (3.16) can be turned into a surface integral in $dJd\phi$ over the surface enclosed by the trajectory.
2. A surface integral in $dJd\phi$ bounded by the trajectory is equal to a surface integral in $d\theta dp$ bounded by the trajectory in the original variables because the two sets of variables are connected through a symplectic (area preserving) transformation.

Using the above facts, I first compute J by integrating $pd\theta$ on a trajectory starting at $\theta = \sqrt{d}$:

$$J = 4 \times \frac{1}{2\pi}\int_0^{\sqrt{d}} pd\theta = 8f\int_0^{\sqrt{d}} \sqrt{\sin^2\frac{\sqrt{d}}{2} - \sin^2\frac{\theta}{2}}\, d\theta. \tag{3.17}$$

Once more I perform the change of variables of Eq. (3.12):

$$J = 16f\sin^2\frac{\sqrt{d}}{2}\int_0^1 \frac{\sqrt{1-x^2}dx}{\sqrt{1-\sin^2\frac{\sqrt{d}}{2}x^2}}. \tag{3.18}$$

To third-order the result is:

$$J = \pi f\left(d - \frac{5\,d^2}{96} + \frac{23\,d^3}{46080} + \cdots\right). \tag{3.19}$$

I can invert Eq. (3.19) and express d in terms of J:

$$d = \frac{J}{\pi f} + \frac{5J^2}{96\pi^2 f^2} + \frac{227J^3}{46080\pi^3 f^3} + \cdots. \tag{3.20}$$

Finally I substitute in Eq. (3.14) into Eq. (3.15):

$$f(J) = f - \frac{J}{16\pi} - \frac{3J^2}{1024\pi^2 f} - \frac{5J^3}{16384\pi^3 f^2} + \cdots. \tag{3.21}$$

We can guess in 1-d-f, the final form of the Hamiltonian: it must be the integral of Eq. (3.21) times 2π:

$$K_\infty(J) = 2\pi f J - \frac{J^2}{16} - \frac{J^3}{512\pi f} + \cdots. \tag{3.22}$$

3.1.5 The Action J as a Function of θ and p

I can relate the maximum amplitude $\sqrt{d}$ to the values of the Hamiltonian using Eq. (3.9). After solving for d, I substitute into Eq. (3.19):

$$I = J(\theta, p) = \frac{4\pi^2 f^2\theta^2 + p^2}{4\pi f} - \frac{80\pi^4 f^4\theta^4 - 24\pi^2 f^2 p^2\theta^2 - 3p^4}{1536\pi^3 f^3} + \frac{1472\pi^6 f^6\theta^6 + 2640\pi^4 f^4 p^2\theta^4 + 1620\pi^2 f^2 p^4\theta^2 + 135p^6}{2949120\pi^5 f^5} + \cdots. \tag{3.23}$$

Of course I can substitute Eq. (3.23) into Eq. (3.22):

$$H(\theta, p) = K_\infty(J(\theta, p)) = \frac{p^2}{2} + 2(2\pi f)^2 \underbrace{\left\{\frac{\theta^2}{4} - \frac{\theta^4}{48} + \frac{\theta^6}{1440} + \cdots\right\}}_{\sin^2(\theta/2)} + \cdots. \tag{3.24}$$

And thus I regain the original Hamiltonian.

I can express Eq. (3.23) in terms of the action-angle variables of Eq. (3.6) which fully normalise the quadratic Hamiltonian as in Eq. (3.7). The result is

$$I(\Phi, J) = J + \frac{\left(1440 - 3840\cos^4(\Phi)\right)\pi f J^2 + \left(135 - 240\cos^4(\Phi) - 128\cos^6(\Phi)\right)J^3}{46080\pi^2 f^2} + \cdots \tag{3.25}$$

In parallel to Eqs. (3.23) or (3.25), there must exist a phase $\Psi(\Phi, J)$ whose Poisson bracket with $I(\Phi, J)$ is one. Using Eq. (3.25), it is a simple exercise to compute a Lie operator which reproduces it to order J^3. From this operator, a formula for $\Psi(\Phi, J)$ can be computed. This formula, consistent with our discussion on the phase advance, can be modified by an arbitrary phase.

Conclusion on high school pendulum: The tools used here were all special.

- They worked because the Hamiltonian of the pendulum is a constant.
- They worked because it is a one degree of freedom system.
- They failed to give us an easy way to get the angle $\Psi = \Phi(\theta, p)$ canonically conjugate to $I = J(\theta, p)$. Actually, as we already know, there is a (nonlinear) phase freedom in choosing $\Phi(\theta, p)$. So it is not surprising that there is no definite formula for Φ.

3.2 The Pendulum with the Numerical Methods of Cosy-Infinity

The "Code" is equipped with a package that has the ability to analyse and manipulate various maps. In particular, I can "exponentiate" vector fields, whether Hamiltonian or not. This technique was first introduced in "matrix" codes by Martin Berz with his code [1] "Cosy-Infinity."[3] I will apply it here to the pendulum.

3.2.1 Phase Space Maps and Lie Maps

I will quote myself (see Sect. 3.2.3):

Normal form is "easy" if the Hamiltonian is time independent. The equivalent of Eq. (2.7) is simply:

$$K^{new} = K \circ a \tag{3.26}$$

The rules equivalent to Eq. (3.26) for a time independent vector field are:

$$F_k^{new} = \left(F \cdot \nabla a_k^{-1}\right) \circ a \tag{3.27}$$

[3]Cosy-Infinity is the ultimate matrix code equipped with TPSA tools beyond the scope of our discussion, interval arithmetic for example.

The transformation a is a change of coordinates: it acts on phase space. For example the Courant-Snyder transformation of Eq. (2.52) transforms the variables (z_1, z_2) into a new set undergoing circular motion if all is done properly, at least in linear approximation.

In analytic calculations we do not transform the tracking data: we transform the map (matrix for example) that generates the tracking data. Sometimes we must transform the Hamiltonian K or the vector field F which generates the map. We will see later that it is convenient to imagine that a itself is generated by a Hamiltonian operator or a vector field. Let us call this vector field A. Then to say that a is generated by A means that

$$a = \exp\left(A \cdot \nabla\right) I. \tag{3.28}$$

The map I is the identity map of phase space: it is made of two trivial projection functions in 1-d-f, in other words

$$I\left(\mathbf{z}\right) = \begin{pmatrix} z_1 \\ z_2 \end{pmatrix}. \tag{3.29}$$

If the vector field A is symplectic (or Hamiltonian), it means that a Hamiltonian H_a can achieve the same result:

$$A \cdot \nabla = \frac{\partial H_a}{\partial z_1}\frac{\partial}{\partial z_2} - \frac{\partial H_a}{\partial z_2}\frac{\partial}{\partial z_1} =: H_a : . \tag{3.30}$$

It can be shown easily, in the Hamiltonian case, that Eq. (3.26) can be written as:

$$K^{new} = K \circ a = \exp\left(: H_a :\right) K. \tag{3.31}$$

Equation (3.31) is useful in perturbative calculations: to increase the order of the calculation one needs to take an additional derivative that is to say an additional term in the expansion of the exponential.

It is important to realise that maps such as $\exp\left(: H_a :\right)$ act on *functions*: indeed Hamiltonian perturbation theory involves transformations of the Hamiltonian (itself a function) into a simpler function depending only on radii, i.e., on actions. In my writings, following Dragt, I like to use calligraphic fonts to denote these operators:

$$K^{new} = K \circ a = \exp\left(: H_a :\right) K = \mathscr{A} K. \tag{3.32}$$

At Equation (2.35), I introduced this calligraphic notation in anticipation of the nonlinear case.

For a vector field, one could have naïvely hoped that

$$\xcancel{F^{new} = F \circ a = \exp(A \cdot \nabla)\, F.} \tag{3.33}$$

It is absolutely incorrect.
As Eq. (3.27) shows, it is a little more complex: the argument of the force must be transformed as well as the "orientation" of the force. In the Hamiltonian case, these two actions are lumped into the change of a single scalar function: the Hamiltonian. Nevertheless we ought to investigate, using a correct approach, if there is something like a Lie operator transformation which can save Eq. (3.33).

One starts by properly defining how a Lie operator, any operator, transforms. It must undergo a similarity transformation:

$$\begin{aligned}
\exp\left(F^{new} \cdot \nabla\right) &= \exp(A \cdot \nabla) \exp(F \cdot \nabla) \exp(-A \cdot \nabla) && (3.34)\\
&= \exp(\exp(A \cdot \nabla)\, F \cdot \nabla \exp(-A \cdot \nabla)) && (3.35)\\
&= \exp\left(\exp\left({}^{\#}_{\#}A \cdot \nabla^{\#}_{\#}\right) F \cdot \nabla\right). && (3.36)
\end{aligned}$$

The notation ${}^{\#}_{\#}\widehat{O}^{\#}_{\#}$ denotes the adjoint representation: a commutator must be taken. This notation is again due to Dragt.

$${}^{\#}_{\#}\widehat{O}^{\#}_{\#}\widehat{P} = \widehat{O}\widehat{P} - \widehat{P}\widehat{O}. \tag{3.37}$$

It must be pointed out that passage from Eqs. (3.34), (3.35) and finally to (3.36) are all very general and apply to *any* linear operators, in particular to matrices. Commutators of vector field operators need to be evaluated if we want an expression germane to our kinds of operators, i.e., operators defined by substitution. The result is:

$$\begin{aligned}
{}^{\#}_{\#}A \cdot \nabla^{\#}_{\#} F \cdot \nabla &= [A \cdot \nabla, F \cdot \nabla] = \langle A, F\rangle \cdot \nabla \\
\text{where } \langle A, F\rangle_a &= A_b \frac{\partial F_a}{\partial z_b} - F_b \frac{\partial A_a}{\partial z_b}. && (3.38)
\end{aligned}$$

In the light of Eq. (3.38), I can define a new Lie operator which acts on the vectors the same way the operator $: K :$ acts on functions:

$$: A : F = \langle A, F\rangle = A_b \frac{\partial F}{\partial z_b} - F_b \frac{\partial A}{\partial z_b}. \tag{3.39}$$

In Eq. (3.39), I "overloaded" Dragt's notation for the Poisson bracket operator. There is no confusion since in Eq. (3.39) the colons surround a vector function not a scalar function. Then Eq. (3.36) can be rewritten as

$$\begin{aligned}\exp\left(F^{new}\cdot\nabla\right) &= \exp\left(\exp\left(\overset{\#}{\underset{\#}{}}A\cdot\nabla^{\#}_{\#}\right)F\cdot\nabla\right)\\ &= \exp\left(\{\exp\left(:A:\right)F\}\cdot\nabla\right)\;\rightarrow\; F^{new}=\exp\left(:A:\right)F.\end{aligned}\tag{3.40}$$

Before I close this short summary on Lie operators, it is worth mentioning an important "homomorphism." In the case of a symplectic operator of the kind $:f:$ where f is indeed a Hamiltonian-like operator, I can derive the corresponding vector field:

$$:f:=\underbrace{-S\nabla f}_{F}\cdot\nabla \;\Longrightarrow F_\alpha=-S_{\alpha\beta}\frac{\partial f}{\partial z_\beta}.\tag{3.41}$$

The first important property is that the commutator of two Poisson bracket operators is the operator of their Poisson brackets:

$$[:f:,:g:]=:[f,g]:.\tag{3.42}$$

Equation (3.42) simply stipulates that one can either deal with an operator or with the function that defines the said operator. Equation (3.41) has a very important practical application for FPP code writers given that the following equation is true:

$$\begin{aligned}&\text{if }:f:=\underbrace{-S\nabla f}_{F}\cdot\nabla\text{ and }:g:=\underbrace{-S\nabla g}_{G}\cdot\nabla\\ &\Longrightarrow:[f,g]:=\;-S\nabla\left[f,g\right]\cdot\nabla=\langle F,G\rangle\cdot\nabla.\end{aligned}\tag{3.43}$$

Equation (3.43) implies that, in the symplectic case, a homomorphism exists between general Lie brackets and Poisson backets. Thus we can write a normalising library which is very general using vector fields: if applied on a symplectic map (or force) the result will be the same as if we had used Poisson brackets (Hamiltonians) all along. If the maps are nearly symplectic, say radiation is present, a code using vector fields will produce correct results and all the transformations will be near the symplectic group to the extent that radiation is a small effect in accelerators.

I can illustrate this homomorphism with an example. Consider two functions f and g:

$$f=\frac{x^2+p^2}{2}\quad\text{and}\quad g=x^2p.\tag{3.44}$$

I can compute the vector fields F and H associated to these two functions:

$$F\cdot\nabla=[f,x]\frac{\partial}{\partial x}+[f,p]\frac{\partial}{\partial p}=-\frac{\partial f}{\partial p}\frac{\partial}{\partial x}+\frac{\partial f}{\partial x}\frac{\partial}{\partial p}=-p\frac{\partial}{\partial x}+x\frac{\partial}{\partial p}\tag{3.45}$$

$$G\cdot\nabla=[g,x]\frac{\partial}{\partial x}+[g,p]\frac{\partial}{\partial p}=-\frac{\partial g}{\partial p}\frac{\partial}{\partial x}+\frac{\partial g}{\partial x}\frac{\partial}{\partial p}=-x^2\frac{\partial}{\partial x}+2xp\frac{\partial}{\partial p}.\tag{3.46}$$

First I can compute the Lie bracket of F and G using Eq. (3.38) which is valid for any vector field:

$$\begin{aligned} H \cdot \nabla &= \langle F, G \rangle \cdot \nabla \\ &= \left(\left(-p \frac{\partial}{\partial x} + x \frac{\partial}{\partial p} \right) (-x^2) - \left(-x^2 \frac{\partial}{\partial x} + 2xp \frac{\partial}{\partial p} \right) (-p) \right) \frac{\partial}{\partial x} \\ &+ \left(\left(-p \frac{\partial}{\partial x} + x \frac{\partial}{\partial p} \right) (2xp) - \left(-x^2 \frac{\partial}{\partial x} + 2xp \frac{\partial}{\partial p} \right) (x) \right) \frac{\partial}{\partial p} \\ &= 4xp \frac{\partial}{\partial x} + \left(3x^2 - 2p^2 \right) \frac{\partial}{\partial p} \\ &\Downarrow \\ H &= \left(4xp, 3x^2 - 2p^2 \right). \end{aligned} \tag{3.47}$$

Secondly I compute the Poisson bracket of the function of f and g:

$$h = [f, g] = \left[\frac{x^2 + p^2}{2}, x^2 p \right] = x^3 - 2xp^2. \tag{3.48}$$

Finally it suffices to check that the vector field generated by the function h agrees with Eq. (3.47):

$$[h, x] \frac{\partial}{\partial x} + [h, p] \frac{\partial}{\partial p} = 4xp \frac{\partial}{\partial x} + \left(3x^2 - 2p^2 \right) \frac{\partial}{\partial p} \quad \text{Q.E.D} \tag{3.49}$$

We are now ready to proceed with our pendulum example using a vector field to compute its map numerically and also to normalise it.

3.2.2 *Vector Field of the Pendulum*

The equation of motion for the pendulum of Eq. (3.3) is:

$$\frac{df}{d\tau} = [-K, f] \tag{3.50}$$

$$= \left\{ \frac{\partial K}{\partial p} \frac{\partial}{\partial \theta} - \frac{\partial K}{\partial \theta} \frac{\partial}{\partial p} \right\} f. \tag{3.51}$$

Equation (3.51) does not assume a Hamiltonian form. Indeed, it is of the form:

$$\frac{df}{d\tau} = F \cdot \nabla f \quad \text{where} \quad F = \begin{pmatrix} \frac{\partial K}{\partial p} \\ -\frac{\partial K}{\partial \theta} \end{pmatrix}. \tag{3.52}$$

In the case when the Hamiltonian is time-independent, I can write a formal solution for Eq. (3.52):

$$f_{\Delta\tau} = \exp(\Delta\tau \; F \cdot \nabla) \, f_0. \tag{3.53}$$

I can apply Eq. (3.53) to the case when f_0 is an arbitrary vector function, a Taylor map for example:

$$m_{\Delta\tau} = \exp(\Delta\tau \; F \cdot \nabla) \, m_0. \tag{3.54}$$

Equation (3.54) expresses the propagation of the map m_0 for a time $\Delta\tau$. If m_0 is the identity, i.e., $m_0 = I$, then Eq. (3.54) gives us the map for a time $\Delta\tau$. In the case of the pendulum, the map for a time $\Delta\tau$ is thus given by:

$$m_{\Delta\tau} = \exp(\Delta\tau \; F \cdot \nabla) \, I = \sum_{n=0}^{\infty} \frac{\Delta\tau^n F \cdot \nabla^n}{n!} I \tag{3.55}$$

$$\text{where} \quad F = \left(p, -\,(2\pi f)^2 \sin(\theta)\right). \tag{3.56}$$

In passing it is worth re-emphasising that in my example the vector field F is derived from a Hamiltonian. Therefore, the operation of $F \cdot \nabla$ on an arbitrary g, $F \cdot \nabla g$, can be written as:

$$F \cdot \nabla g = [-K, g] =: -K : g. \tag{3.57}$$

With this notation, initially introduced by Dragt, Eq. (3.53) takes the explicitly Hamiltonian form

$$m_{\Delta\tau} = \exp(\Delta\tau \; : -K :) \, m_0. \tag{3.58}$$

Equation (3.58) gives us a prescription for generating Hamiltonian maps which are often useful in perturbation theory. The other method is that of mixed generating functions which appears more naturally in the discussion on canonical 1-form. I do not use mixed generating functions in purely analytical discussions; furthermore they do not extend to non-symplectic maps while Eq. (3.54) applies to any type of vector field F including non-Hamiltonian fields.

The program we will use for the pendulum is in Appendix E. This program initialises the code PTC (which I do **not** use[4] in this example!) with a 1-d-f freedom phase space. The important initial steps are as follows:

```
f=1/2.d0 !  This pendulum has a period of tau=2
dt=5.d-2
```

[4]FPP can be used alone. But instead of introducing more mysterious calls, I initialised PTC in 1-d-f which then ripples down to FPP. As I said, I never use the tracking code PTC in this pendulum program.

```
theta=1.d0.cmono.1                              !(1a)
p=1.d0.cmono.2                                  !(1b)

K=p**2/2+2*(2*pi*f)**2*sin(theta/2.d0)**2   !(2)

!!! vf is the force field of the pendulum
vf%v(1)=dt*(theta.pb.K)                         !(3a)
vf%v(2)=dt*(p.pb.K)                             !(3b)

id=1                         !  (4a)
m=exp(vf,id)                 !  (4b)
```

On lines (1a,b) the variables (θ, p) are defined as the monomial vector (z_1, z_2) by the `cmono` operator. On line (2) the Hamiltonian K of the pendulum is written into a complex Taylor series. Lines (3a, b) are the two components of the vector fields[5] which enter into Eq. (3.54). In fact, the Fortran 90 operator `.pb.` takes a Poisson bracket.

The next step is the computation of the map for a time $\Delta\tau = 0.05$ as in Eq. (3.56). It is given by lines (4a,b). This technique was first pioneered by Berz [1, 2] and implemented in his code "Cosy-Infinity".

The Taylor map `m` of line (4b) requires an infinite number of terms in Eq. (3.55); in practice the sum is carried out until convergence is reached.

3.2.3 Hamiltonian Normal Form of the Pendulum

I will use this section to review, in its simplest incarnation, perturbation theory on a time independent Hamiltonian. It is *not* what is used by the "Code": accelerators are strongly dependent on the time like variable "s". The Hamiltonian of an accelerator is actually a horribly complicated beast. In a code capable of misaligning magnets, of putting fringe fields, etc..., the equations of motion are made of force fields with complicated and discontinuous s-dependence. However the one-turn map, like the one-turn matrix, can be normalised by using a simple canonical transformation at the point of observation *irrespective* of the internal complexity of the "Code." This will be examine in Sect. 3.3.

Normal form is "easy" if the Hamiltonian is time independent. The equivalent of Eq. (2.7) is simply:

$$K^{new} = K \circ a. \tag{3.59}$$

[5]I have routines that go back and forth between Hamiltonians and vector fields, but here I wanted to be explicit for this simple example.

The rules equivalent to Eq. (3.59) for a time independent vector field are:

$$F_k^{new} = \left(F \cdot \nabla a_k^{-1}\right) \circ a. \tag{3.60}$$

Here the Hamiltonian is simple and τ independent. I can expand it to the sixth degree:

$$K = \frac{p^2 + 4\pi^2 f^2 \theta^2}{2} - \frac{\pi^2 f^2 \theta^4}{6} + \frac{\pi^2 f^2 \theta^6}{180} + \cdots. \tag{3.61}$$

I can then normalise it at the linear level using Eqs. (3.6) and (3.59):

$$\begin{aligned} K_2^{new} &= \pi f \left(\theta^{new\ 2} + p^{new\ 2}\right) - \frac{1}{24}\theta^{new\ 4} + \frac{1}{1440\pi f}\theta^{new\ 6} + \cdots \\ &= 2\pi f J - \frac{\cos^4(\Phi)}{6} J^2 + \frac{\cos^6(\Phi)}{180\pi f} J^3 + \cdots. \end{aligned} \tag{3.62}$$

To proceed further I need to find a transformation a, such that

$$K_4^{new} = K_2^{new} \circ a = 2\pi f J + \alpha J^2 + \beta J^3 + \cdots. \tag{3.63}$$

Hopefully the result will be that of Eq. (3.22), namely

$$\alpha = -\frac{1}{16} \quad \text{and} \quad \beta = -\frac{1}{512\pi f}. \tag{3.64}$$

In this section, I will use purely Hamiltonian methods. Therefore, the action of the map a can be written as:

$$K^{new} = K \circ a = \cdots \exp\left(: B(\Phi) J^3 :\right) \exp\left(: A(\Phi) J^2 :\right) K. \tag{3.65}$$

The map $\exp\left(: A(\Phi) J^2 :\right)$ acts on functions, such as the Hamiltonian K. If I want the map a, I need to let it act on the identity map itself. To second order, it is just:

$$\begin{aligned} \Phi^{new} &= \exp\left(: A(\Phi) J^2 :\right) \Phi = \Phi - 2AJ + AA' J^2 + \cdots \\ J^{new} &= \exp\left(: A(\Phi) J^2 :\right) J = J + A' J^2 + \left(A'^2 - AA''\right) J^3 + \cdots \end{aligned} \tag{3.66}$$

One could also use a mixed generating function which is to leading order identical:

$$\begin{aligned} \Phi^{new} &= \Phi - 2A(\Phi)\, J^{new} \\ J^{new} &= J + A'(\Phi)\; J^{new2}. \end{aligned} \tag{3.67}$$

The advantage of the Lie operator is to produce explicit formulae that depend only on derivatives. It also keeps the entire theory analytic. The reader can check that the mixed variables generating function will not be a polynomial beyond leading order unless it is expressed in terms of (θ, p).

I expand Eq. (3.65) to second-order first taking into account only $\exp\left(: A(\Phi)J^2 :\right)$:

$$\begin{aligned} \exp\left(: A(\Phi)J^2 :\right) K =& 2\pi f J - \frac{1}{6}\cos^4(\Phi)\, J^2 + \frac{1}{180\pi f}\cos^6(\Phi)\, J^3 \\ &+ \left[AJ^2, 2\pi f J\right] + \frac{1}{2}\left[AJ^2, \left[AJ^2, 2\pi f J\right]\right] \\ &- \frac{1}{6}\left[AJ^2, \cos^4(\Phi)\, J^2\right]. \end{aligned} \tag{3.68}$$

The fourth term must be selected to remove the part of the second term which is not a function of J alone. This is done by expanding $\cos^4(\Phi)$ in terms of phasor eigenfunctions $\exp(\pm i\Phi)$:

$$\begin{aligned} -\frac{1}{6}\cos^4(\Phi)\, J^2 + \left[AJ^2, 2\pi f J\right] =& -\frac{1}{96}\left\{e^{i4\Phi} + e^{-i4\Phi} + 4e^{i2\Phi} + 4e^{-i2\Phi} + 6\right\} J^2 \\ &+ 2\pi f A' J^2. \end{aligned} \tag{3.69}$$

I can select A' as follows:

$$A' = \frac{1}{192\pi f}\left\{e^{i4\Phi} + e^{-i4\Phi} + 4e^{i2\Phi} + 4e^{-i2\Phi}\right\}. \tag{3.70}$$

With this particular choice, the Hamiltonian reduces to

$$\begin{aligned} \exp\left(: A(\Phi)J^2 :\right) K =& 2\pi f J - \frac{1}{16}J^2 + \frac{1}{180\pi f}\cos^6(\Phi)\, J^3 \\ &+ 2\pi f\left\{A'^2 - AA''\right\} J^3 - \frac{1}{6}\left[AJ^2, \cos^4(\Phi)\, J^2\right]. \end{aligned} \tag{3.71}$$

The tune shift term proportional to J^3 originates from the last 3 terms in Eq. (3.71). The reader can check that the term $A'^2 - AA''$ contributes a tune shift term given by:

$$\frac{17}{4608\pi f} J^3,$$

the last terms $-\frac{1}{6}\left[AJ^2, \cos^4(\Phi)\, J^2\right]$ gives us

$$-\frac{17}{2304\pi f}J^3,$$

and finally, the direct term from the expansion of the Hamiltonian $\frac{1}{180\pi f}\cos^6(\Phi)\, J^3$ gives

$$\frac{1}{576\pi f}J^3.$$

All these terms add up to to $\beta = -\frac{1}{512\pi f}$ as in Eq. (3.64).

The map $\exp\left(: B(\Phi)J^3 :\right)$, which I do not bother to compute, is necessary to actually remove the terms in J^3 which are not "rotational."

The reader must notice that I did not deal with polynomial functions in this example up to this point. However the connection to polynomials is simple. I can use the phasors defined in Eq. (2.21) and rewrite the Hamiltonian:

$$K_2^{new}(\varphi_+, \varphi_-) = \pi f \varphi_+\varphi_- - \frac{(\varphi_+ + \varphi_-)^4}{2^7\, 3} + \frac{(\varphi_+ + \varphi_-)^6}{2^{11}\, 45\pi f} + \cdots. \tag{3.72}$$

Using phasors, the Lie polynomial AJ^2 of Eq. (3.65) is just:

$$A\, J^2 = \frac{i}{2^2\, 192\pi f}\left\{\frac{\varphi_+^4}{4} - \frac{\varphi_-^4}{4} + 2\varphi_+^3\varphi_- - 2\varphi_+\varphi_-^3\right\} \tag{3.73}$$

All the operations of this section can be carried out again using phasors. It suffices to know that the Poisson bracket operator must be redefined for a function of phasors.

$$\underset{\star}{\overset{\star}{:}} f \underset{\star}{\overset{\star}{:}} g = -2i : f : g = -2i \sum_{j=1,N} \frac{\partial f}{\partial \varphi_{j+}}\frac{\partial g}{\partial \varphi_{j-}} - \frac{\partial f}{\partial \varphi_{j-}}\frac{\partial g}{\partial \varphi_{j+}}. \tag{3.74}$$

The reader may ask the following pertinent question: why can't Forest put a normalising factor in his definition of the phasors and remove the annoying factor of $-2i$ in Eq. (3.74) since this would restore the usual Hamiltonian framework (see Eq. 10.1). In a nutshell, my definition of the phasors guaranties that the usual linear property distinguishing tunes from damping persists in the nonlinear case, namely tunes are complex vector field operators while damping decrements appear real. I illustrate this in supplemental Chap. 10.

3.2.4 *Illustration of Normal Form Theory with FPP Tools*

As I explained in the introduction of this book, the Code, namely PTC in my case, must have two disconnected parts. First, since it is "the Code" it must have tracking routines of its own. These routines must be able to track rays in a self-consistent way: following Talman's idea, I strongly favour (symplectic) integrators.

Secondly the Code must have tools to analyse the Taylor maps it produces. Analysis of Taylor maps entails more that just a "black box" normal form algorithm. If this tool box, which I called FPP, is properly written, it will give access to various internal tools: Lie operators, exponential of Lie operators, "logarithm" of maps, etc.

Therefore a twisted individual, a "thought criminal," like myself, could write his own little matrix code, say a small Cosy-Infinity [1]. He could also write his own normal form for time-independent Hamiltonians or even maps. In Sect. 3.3, using the standard map, I will truly become a "serial thought criminal" when I will display four different normal form algorithms...

Here I examine this type of criminal behaviour, where one dares to go outside the world of the integrator, using our canonical example: the pendulum. This is the program of Appendix E. We will write our own little normal form algorithm as well as using the "canned" map based algorithm of FPP.

I start by writing the Hamiltonian K of Eq. (3.3) as a Taylor series in the variables (θ, p). Lines (1a,b) insures that the two variables are indeed the first and second variable of the Taylor series for K, i.e., line (1c) aka Eq. (3.3).

```
f=1/2.d0 !  This pendulum has a period of tau=2
dt=5.d-2
theta=1.d0.cmono.1                           !(1a)
p=1.d0.cmono.2                               !(1b)

K=p**2/2+2*(2*pi*f)**2*sin(theta/2.d0)**2   !(1c)
```

As in any procedure, whether on maps or Hamitonians, we must first normalise the linear motion. Therefore I must make the quadratic part of line (1c) into a radius. This is achieved by the transformation of Eq. (3.6) and the following lines of code:

```
a1%v(1)=theta/sqrt(2*pi*f)           !(2a)
a1%v(2)=p*sqrt(2*pi*f)               !(2b)

write(6,*);write(6,*)" This is Hamiltonian in linear phasors "; write(6,*);

K=K*a1*from_phasor()                 !(2c)
```

Lines (2a, b) construct the canonical transformation of Eq. (3.6). Line (2c) applies this transformation to K followed by a transformation into the phasors basis. The end result is Eq. (3.72) which is printed by the program:

```
 This is Hamiltonian in linear phasors


Properties, NO =    6, NV =    2, INA =   57
*********************************************
```

```
2   1.570796326794897          0.000000000000000       1  1
4 -0.2604166666666668E-02      0.000000000000000       4  0
4 -0.1041666666666667E-01      0.000000000000000       3  1
4 -0.1562500000000001E-01      0.000000000000000       2  2
4 -0.1041666666666667E-01      0.000000000000000       1  3
4 -0.2604166666666668E-02      0.000000000000000       0  4
6  0.6907766627252409E-05      0.000000000000000       6  0
6  0.4144659976351445E-04      0.000000000000000       5  1
6  0.1036164994087861E-03      0.000000000000000       4  2
6  0.1381553325450482E-03      0.000000000000000       3  3
6  0.1036164994087861E-03      0.000000000000000       2  4
6  0.4144659976351445E-04      0.000000000000000       1  5
6  0.6907766627252409E-05      0.000000000000000       0  6
```

The reader can check these coefficients numerically using Eq. (3.72), for example `0.6907766627252409E-05`, the coefficient of φ_+^6, is just $\frac{1}{2^{11}\,45\pi f}$.

Now comes the normal form. My task is to write a loop which implements the steps from Eq. (3.65) to (3.71) and put it in a recursive algorithm.

```
      tpf=K.sub.'11'

       do io=3,map_order
      a_op=0.0_dp
       k_io=K.sub.io
       call c_taylor_cycle(K_io,size=ns)   !(3a)

!!!   Cycling over all monomials
    do km=1,ns
       call c_taylor_cycle(K_io,ii=km,value=v,j=ji) !(3b)
       if(ji(1)/=ji(2)) then                          !(3c)
         v=v/((ji(1)-ji(2))*(2*i_)*tpf)     !(3d)
         a_op=a_op+(v.cmono.ji)                 !(3e)
       endif
    enddo
    if(io==4) then
       write(6,*); write(6,*)" Eq.\,3.73 of the book is printed "
       call print(a_op,6)
    endif
       vf=cgetvectorfield( a_op )                     !(4a)
       K=exp(vf,K)                                    !(4b)

       enddo
```

Please notice that I print Eq. (3.73) on the screen during the normalising loop. Line `tpf=K.sub.'11'` extracts the quantity πf, the first term of Eq. (3.72). The operator `.sub.` takes a string for fast and dirty work: it contains the exponents of the monomial whose coefficient is desired. Obviously it can only work for exponents less or equal to 9. Actually for more solid implementation, it can take an integer array.

I must now write the equivalent of Eq. (3.69) for a recursive process. If I denote the Hamiltonian at step `io`, K_{io}, then I have:

$$K_{io} = \pi f \varphi_+\varphi_- + T_{<io}(\varphi_+\varphi_-) + R_{io}(\varphi_+, \varphi_-) + R_{>io}(\varphi_+, \varphi_-). \quad (3.75)$$

The term $T_{<io}$ containts all the contribution to the tune (so-called tune shifts) so far computed. The term R_{io} contains the terms of order `io` which require “cleansing.”

This is done with the Lie transform a_op which I denote as A_{io}. The equation to cleanse is similar to Eq. (3.69):

$$K_{io+1} = \exp\left(: A_{io} :\right) K_{io} = \pi f \varphi_+ \varphi_- + T_{<io} + \left[A_{io}, \pi f \varphi_+ \varphi_-\right] + R_{io} + \cdots . \tag{3.76}$$

The operator associated to A_{io} is selected so as to remove all terms with unequal power in the phasors φ_+ and φ_-. Thus if we have

$$R_{io} = \sum_{j_1+j_2=io} v_j \varphi_+^{j_1} \varphi_-^{j_2} \tag{3.77}$$

and the ansatz

$$A_{io} = \sum_{j_1+j_2=io} A_j \varphi_+^{j_1} \varphi_-^{j_2}, \tag{3.78}$$

then Eq. (3.76) becomes:

$$K_{io+1} = \pi f \varphi_+ \varphi_- + T_{<io} + \sum_{j_1+j_2=io} \left\{ v_j - i 2\pi f \left(j_1 - j_2\right) A_j \right\} \varphi_+^{j_1} \varphi_-^{j_2} + \cdots . \tag{3.79}$$

I can clean away most terms in the summation in Eq. (3.79) by choosing

$$A_j = \frac{v_j}{i 2\pi f \left(j_1 - j_2\right)} \quad \text{if} \quad j_1 \neq j_2 . \tag{3.80}$$

Going back to the code, first the part of the K of order io is extracted by the line k_io=K.sub.io. Then the routine c_taylor_cycle, as the name indicates, cycles through each monomial. It is first called at line (3a) to find the number of non-zero monomial in K_{io} namely ns. It is then called again ns times, at line (3b) to retrieve the value v_j and the array j of Eq. (3.79) which are then stored in the variables v and ji respectively.

At line (3c) I check that the coefficient is *not* a tune shift and then the coefficient A_j is computed at line (3d). This is the implementation of Eq. (3.80). Line (3e) constructs monomial by monomial the Lie transformation A_{io}.

Once this is done, Eq. (3.76) is evaluated to the maximum order selected by the user at line (4b). At that point, the variable K of the program is K_{io+1}. The iteration continues until io reaches map_order.

Line (4a), vf=cgetvectorfield(a_op), is needed because my FPP complex package deals with general vector fields and not Poisson bracket operators. In fact I could have written the normalization loop using vector fields in which case it could have applied to a non-symplectic force or a force expressed in non-canonical variables.

Finally the lines (5) to (14) show the normalisation of the pendulum using only the tools specifically developed for the maps produced by PTC. Specifically on line (9), the subroutine `c_normal` of FPP is called. I will explain this routine in 1-d-f in Sect. 3.3.2. In fact I recreate this routine using a simple 1-d-f implementation in the program of Appendix F.

3.3 Normal Form of the Standard Map: The Algorithm for Maps

Before I start this section, I want to tell the reader that four different algorithms are displayed in the code of Appendix F. The codes for the map algorithms are explained in Sects. 3.4.2 and 3.4.3. The algorithms on the "logarithm" of the map are described in Sect. 3.4.4.

The standard map arises naturally if one integrates the pendulum numerically for a time dt. For example, we can approximate the true map of the pendulum by the following second-order integrator.

```
f=1/2.d0 !  This pendulum has a period of tau=2
dt=5.d-2
theta=1.d0.cmono.1                          !(1a)
p=1.d0.cmono.2                              !(1b)

! One quadratic step of integration
!  K=p**2/2+2*(2*pi*f)**2*sin(theta/2.d0)**2

 theta = theta + (dt/2) * p                                       ! (2a)
 p = p - dt * 2*(2*pi*f)**2* sin(theta/2.d0)*cos(theta/2.d0)    ! (2b)
 theta = theta + (dt/2) * p                                       ! (2c)

M%v(1)=theta     ! (3a)
M%v(2)=p         ! (3b)
```

This integrator is an explicit second-order symplectic integrator and is part of the program `standard_map` found in Appendix F. This means that the error for one integration time step is of order dt^3.

This map is also, modulo some scaling, a good model for the "longitudinal" motion due to a single radio-frequency (RF) cavity in a circular accelerator. The map for drifting through half a ring of length L can be approximately written as

$$\tau^f = \tau + \tfrac{L}{2}\alpha\delta. \tag{3.81}$$

The coefficient α of Eq. (3.81) can even contain information pertinent to the coupling between the transverse and longitudinal plane. It is related to the so-called "momentum compaction or time slip." I will show you in Sect. 4.2 that it falls out trivially from a special normal form. The variable τ is the relative time of arrival multiplied by the speed of light c. For an ultra-relativistic particle it is almost the path length. δ is the energy scaled by the momentum $p_0 c$.

The map for a thin RF-cavity is given by

$$\begin{aligned} \delta^f &= \delta + \frac{VL}{p_0 c} \sin\left(\frac{\omega}{c}\tau\right) \\ &= \delta + 2\frac{VL}{p_0 c} \sin\left(\frac{\omega}{c}\frac{\tau}{2}\right) \cos\left(\frac{\omega}{c}\frac{\tau}{2}\right) \end{aligned} \tag{3.82}$$

where V is the voltage and ω is the frequency.

With a simple scaling of time, this is the same as lines (2a, b, c) in the above code fragment. The cavity map has, for most realistic parameters, a very small tune which is equivalent to a small dt. Therefore it can be approximated by a pendulum.

3.3.1 *Linear Part*

We have the Taylor map m constructed at lines (3a, b). The first step of any normalization is the computation of the linear part of the transformation a of Eq. (2.7). I first compute the linear part of m which I called l. The matrix of l is simply:

$$l = \begin{pmatrix} 1 - 2(\pi f\, dt)^2 & dt\left(1 - (\pi f\, dt)^2\right) \\ -4dt\,(\pi f)^2 & 1 - 2(\pi f\, dt)^2 \end{pmatrix}. \tag{3.83}$$

I can compute the transformation a_1 as defined in Eq. (2.51). The result is:

$$\beta = \frac{\left\{1 - (\pi f\, dt)^2\right\}^{1/2}}{2\pi f} \quad \text{and} \quad \alpha = 0. \tag{3.84}$$

This result agrees with Eq. (3.6) if dt is set to zero. Indeed if an integration step is infinitely small, then it must reproduce the exact solution for the pendulum.

The normalised matrix $r_1 = a_1^{-1} l a_1$ is just:

$$r_1 = \begin{pmatrix} 1 - 2(dt\pi f)^2 & 2dt\pi f\sqrt{1 - 2(dt\pi f)^2} \\ -2dt\pi f\sqrt{1 - 2(dt\pi f)^2} & 1 - 2(dt\pi f)^2 \end{pmatrix}. \tag{3.85}$$

The linear tune ν_0 is

$$\mu_0 = 2\pi\nu_0 = 2\sin^{-1}(\pi f\, dt)\,. \tag{3.86}$$

This result is consistent with Eq. (3.21) where $\frac{d\nu_0}{dt}$ is the leading term.

3.3.2 *The Nonlinear Algorithm: Theory*

The idea behind the numerical algorithm is to factor the full canonical a as

$$a = a_1 \circ c \circ a_2 \circ a_3 \cdots \tag{3.87}$$

and to compute term-by-term all the a_i's. As we will see, the resulting normalised map will also be in factored form even though the factors commute in a normal form whose goal is the computation of an amplitude dependent rotation.

I can use Eq. (3.87) on the full map m:

$$a^{-1} \circ m \circ a = \cdots a_3^{-1} \circ \underbrace{a_2^{-1} \circ \underbrace{c^{-1} \circ a_1 \circ m \circ a_1^{-1} \circ c}_{m_1} \circ a_2}_{m_2} \circ a_3 \cdots . \tag{3.88}$$

The map c in Eq. (3.88), defined by Eq. (2.10), insures that the calculation of the nonlinear terms, i.e. beyond m_1, will take place in the phasors' basis of Eq. (2.21). The nonlinear part of the canonical transformation a, in phasors basis, is assumed to have a factored[6] Lie representation:

$$a_2 \circ a_3 \cdots = \cdots \exp(F_3 \cdot \nabla) \exp(F_2 \cdot \nabla) I. \tag{3.89}$$

I denote by m_n the map at the nth step in the iteration:

$$m_n = a_n^{-1} \circ \cdots \circ a_2^{-1} \circ c^{-1} \circ a_1 \circ m \circ a_1 \circ c \circ a_2 \circ \cdots \circ a_n. \tag{3.90}$$

In this iteration procedure, it is assumed that the map m_n has been "cleansed" to order n. Therefore m_n can be factorised as a rotation to "order" n and a map n_{n+1} that contains all the "remaining ideological filth" of higher order that must be purged!

$$m_n = n_{n+1} \circ \rho_n. \tag{3.91}$$

The map ρ_n is a rotation computed so far to order n. Therefore it can have the following Lie representation:

$$\rho_n = \exp(T_1 \cdot \nabla) \cdots \exp(T_n \cdot \nabla) I. \tag{3.92}$$

The operator $T_1 \cdot \nabla$ is already known to us through Eqs. (3.85) and (3.21). It must have the following Lie representation:

[6]Notice the order which is reversed from the usual Taylor map: remember the transpose matrix $M^{\dagger}$ of Sect. 2.3.1!

$$T_1 \cdot \nabla = \begin{pmatrix} -i\mu_0\varphi_+ \\ i\mu_0\varphi_- \end{pmatrix} \cdot \nabla = -i\mu_0\varphi_+ \frac{\partial}{\partial\varphi_+} + i\mu_0\varphi_- \frac{\partial}{\partial\varphi_-} \tag{3.93}$$

$$\Longrightarrow T_1 \cdot \nabla = {}^{\star}_{\star} - t_1{}^{\star}_{\star} = -\frac{\mu_0}{2}{}^{\star}_{\star}\varphi_+\varphi_-{}^{\star}_{\star}. \tag{3.94}$$

In Eq. (3.94), one should not forget the factor of $-2i$ that enters into the operator ${}^{\star}_{\star}\varphi_+\varphi_-{}^{\star}_{\star}$ as explained in Eq. (3.74). Let us apply to the map m_n the transformation a_{n+1}, yet unknown, which will push Eq. (3.91) one order higher. Namely

$$\begin{aligned} m_{n+1} &= a_{n+1}^{-1} \circ m_n \circ a_{n+1} \\ &= a_{n+1}^{-1} \circ n_{n+1} \circ \rho_n \circ a_{n+1}. \end{aligned} \tag{3.95}$$

In the Lie operator language, this equation becomes:

$$\begin{aligned} m_{n+1} &= a_{n+1}^{-1} \circ n_{n+1} \circ \rho_n \circ a_{n+1} \\ &= \underbrace{\exp\left(F_{n+1} \cdot \nabla\right) \mathscr{R}_n \mathscr{N}_{n+1} \exp\left(-F_{n+1} \cdot \nabla\right)}_{\mathscr{M}_{n+1}} I. \end{aligned} \tag{3.96}$$

The first step is to extract the map n_{n+1} from m_n.

$$n_{n+1} = m_n \circ \rho_n^{-1} \tag{3.97}$$

$$= \exp\left(-T_n \cdot \nabla\right) \exp\left(-T_{n-1} \cdot \nabla\right) \cdots \exp\left(-T_2 \cdot \nabla\right) m_n \circ \rho_1^{-1}. \tag{3.98}$$

Equation (3.98) is part of the recursive algorithm and therefore all the operators T_i's and F_i's are all known up to order n. In Eq. (3.98) the Lie map $\exp\left(-T_1 \cdot \nabla\right)$ is left as the linear phase space map ρ_1^{-1} simply because this is how it appears in the numerical algorithm: it came from a linear (matrix) calculation.

To compute F_{n+1} I reorder the various factors in Eq. (3.96):

$$\begin{aligned} \mathscr{M}_{n+1} &= \exp\left(F_{n+1} \cdot \nabla\right) \mathscr{R}_n \mathscr{N}_{n+1} \exp\left(-F_{n+1} \cdot \nabla\right) \\ &= \mathscr{R}_n \mathscr{R}_n^{-1} \exp\left(F_{n+1} \cdot \nabla\right) \mathscr{R}_n \mathscr{N}_{n+1} \exp\left(-F_{n+1} \cdot \nabla\right) \\ &= \mathscr{R}_n \exp\left(\mathscr{R}_n^{-1} F_{n+1} \cdot \nabla \mathscr{R}_n\right) \mathscr{N}_{n+1} \exp\left(-F_{n+1} \cdot \nabla\right) \\ &\approx \mathscr{R}_n \exp\left(\mathscr{R}_1^{-1} F_{n+1} \cdot \nabla \mathscr{R}_1 - F_{n+1} \cdot \nabla + G_{n+1} \cdot \nabla + \cdots\right). \end{aligned} \tag{3.99}$$

In Eq. (3.99), the map $\mathscr{N}_{n+1}$ has been approximated to order $n+1$ as $\exp\left(G_{n+1} \cdot \nabla\right)$. Then the equation for the $n+1$ order part is simply:

$$\begin{aligned} \mathscr{R}_1^{-1} F_{n+1} \cdot \nabla \mathscr{R}_1 - F_{n+1} \cdot \nabla + G_{n+1} \cdot \nabla &= \exp\left({}^{\#}_{\#} - T_1 \cdot \nabla{}^{\#}_{\#}\right) F_{n+1} \cdot \nabla - F_{n+1} \cdot \nabla + G_{n+1} \cdot \nabla \\ &= \left\{\exp\left(:-T_1:\right) F_{n+1} - F_{n+1} + G_{n+1}\right\} \cdot \nabla. \end{aligned} \tag{3.100}$$

And thus we get the equation for the normal form:

$$\exp(:-T_1:)\,F_{n+1} - F_{n+1} + G_{n+1} = \text{Tune shifts}. \tag{3.101}$$

In Eq. (3.101) the vector operator F_{n+1} removes all the resonances. On occasion one can also leave a resonance in the map. This will be discussed later in Sect. 5.4. To see how the calculation of the vector F_{n+1} proceeds, it is necessary to expand it.

$$F^a_{n+1} = \sum_m F^a_{n+1;m}\varphi_+^{m_1}\varphi_-^{m_2} \quad \text{with} \quad F_{n+1} = \left(F^1_{n+1}, F^2_{n+1}\right). \tag{3.102}$$

I can look at a single "vector" monomial of Eq. (3.101). First, one writes the vector $-T_1$ correspounding to the unperturbed tune; this is given by Eq. (3.93) which I rewrite here:

$$-T_1 = \begin{pmatrix} i\mu_0\varphi_+ \\ -i\mu_0\varphi_- \end{pmatrix}. \tag{3.103}$$

Then I let $\exp(:-T_1:)$ act on a single vector monomial belonging to the ath component of the vector field F:

$$\begin{aligned} :-T_1: F^a_{n+1;m}\varphi_+^{m_1}\varphi_-^{m_2} &= -\sum_k T_1^k \frac{\partial}{\partial\varphi_k} F^a_{n+1;\,m}\varphi_+^{m_1}\varphi_-^{m_2} - F^a_{n+1;m}\varphi_+^{m_1}\varphi_-^{m_2}\frac{\partial}{\partial\varphi_a}T_1^a \\ &= i\mu_0\left(m_1 - m_2 + (-1)^a\right) F^a_{n+1;m}\varphi_+^{m_1}\varphi_-^{m_2}. \end{aligned} \tag{3.104}$$

In Eq. (3.104) I used the formula for the Lie bracket remembering that $F^a_{n+1;m}\varphi_+^{m_1}\varphi_-^{m_2}$ is actually a vecfor field with a single non-zero component namely the ath component. Remarkably this vector monomial is an eigen-monomial of the Lie bracket operator of the unperturbed tune. Thus I can solve Eq. (3.101) by selecting F_{n+1} as follows:

$$m_1 - m_2 + (-1)^a \neq 0 \quad \rightarrow \quad F^a_{n+1;m} = \frac{G^a_{n+1;m}}{1 - \exp\left(i\mu_0\left(m_1 - m_2 + (-1)^a\right)\right)}. \tag{3.105}$$

The tune shift terms are given by

$$\text{Tune shifts} = \sum_{m_1 - m_2 + (-1)^a = 0} G^a_{n+1;m}\varphi_+^{m_1}\varphi_-^{m_2}. \tag{3.106}$$

The algorithm on Lie polynomials

In the symplectic case, particularly in analytic calculations, one may elect to work on Hamiltonian operators. In that case, Eq. (3.101) is a polynomial equation in the phasors' basis.

$$\exp\left({}^{\star}_{\star} - t_1 {}^{\star}_{\star}\right) f_{n+1} - f_{n+1} + f_{n+1} = \text{Tune shifts as in Eq.(3.102).} \quad (3.107)$$

The polynomial t_1 is given by Eq. (3.94). The polynomial f_{n+1} is also expanded in phasors:

$$f_{n+1} = \sum_k f_{n+1;k} \varphi_+^{k_1} \varphi_-^{k_2} \text{ as in Eq. (3.102)} \quad (3.108)$$

I then proceed with the action of $-t_1$ on the polynomial f_{n+1}:

$$\begin{aligned} {}^{\star}_{\star} - t_1 {}^{\star}_{\star} f_{n+1;k} \varphi_+^{k_1} \varphi_-^{k_2} &= -i\mu_0 f_{n+1;k} \left[\varphi_+ \varphi_-, \varphi_+^{k_1} \varphi_-^{k_2}\right] \\ &= i\mu_0 f_{n+1;k} (k_1 - k_2) \varphi_+^{k_1} \varphi_-^{k_2} \text{ as in Eq. (3.104)} \end{aligned} \quad (3.109)$$

The final result for f_{n+1} is:

$$k_1 - k_2 \neq 0 \rightarrow f_{n+1;k} = \frac{g_{n+1;k}}{1 - \exp\left(i\mu_0 (k_1 - k_2)\right)} \text{ as in Eq. (3.105)} \quad (3.110)$$

Of course, Eq. (3.110) must reproduce Eq. (3.105). This is done by computing the vector field associated with f_{n+1}. Let us look at the effect on φ_+:

$${}^{\star}_{\star} \varphi_+^{k_1} \varphi_-^{k_2} {}^{\star}_{\star} \varphi_+ = 2i \varphi_+^{k_1} \varphi_-^{k_2 - 1} \quad (3.111)$$

Thus one concludes from Eq. (3.111) that $F^1_{n+1;m}$ is

$$F^1_{n+1;m} = 2i f_{n+1;k} \text{ with } \begin{matrix} m_1 = k_1 \\ m_2 = k_2 - 1 \end{matrix} . \quad (3.112)$$

Eq. (3.112) tells us that the condition $k_1 - k_2 = 0$ for the tune shift is indeed equivalent to the condition for the vector field in Eq. (3.105).

The normal form of the standard map will be performed *five* different ways[7] and therefore this necessitates a section of its own i.e., Sect. 3.4.

[7] Well...four different ways but I also include the call to FPP which is actually a generalised version of the application of vector fields to one-turn maps.

3.4 Standard Map: Normalisation in Software

This section is devoted to the discussion of the code `standard_map` found in Appendix F.

First I will invoke the general normal form of the complex FPP library. This is tantamount to a manual for the code FPP. This tasteless exposition does have the virtue of emphasizing that normal form should often be viewed, by the average user, Orwell's Outer Party Member, as a black box whose purpose is to "diagonalise the map." No question asked; no dissidence tolerated!

Secondly, I do examine the process by writing a small 1-d-f normal form loop in the main program rather than opening the routine `c_normal` of FPP which is too general for this introduction. This small loop uses vector fields like `c_normal`.

Thirdly, I repeat the calculation with Poisson bracket operators rather vector fields. The results should be the same.

Finally in a fourth and fifth stage I compute the "logarithm" of the map and normalise directly this operator. I do it again with the Poisson bracket (pseudo-Hamiltonian) and the vector field (pseudo-force).

3.4.1 *Creation of Standard Map and Using a "Canned" Normal Form*

We start again with the code fragment of Sect. 3.3 for the standard map.

```
map_order=8
call init_all(state,map_order,0)
          .
          .
          .
f=1/2.d0 !  This pendulum has a period of tau=2
dt=5.d-2
theta=1.d0.cmono.1                          !(1a)
p=1.d0.cmono.2                              !(1b)

! One quadratic step of integration
!  K=p**2/2+2*(2*pi*f)**2*sin(theta/2.d0)**2

 theta = theta + (dt/2) * p                                      ! (2a)
 p = p - dt * 2*(2*pi*f)**2* sin(theta/2.d0)*cos(theta/2.d0)    ! (2b)
 theta = theta + (dt/2) * p                                      ! (2c)

m%v(1)=theta     ! (3a)
m%v(2)=p         ! (3b)

write(mf,*);
write(mf,*)" Normalising the standard map using FPP software ";

call  c_normal(m,n)        ! (4)
```

This fragment of code contains two distinct parts. Lines (1a, b) and (2a, b, c) would be part of the "Code." They simulate the pendulum (or the RF cavity in an accelerator). Because the input, namely lines (1a, b), are Taylor series, the code will return two Taylor series. If the input had been real numbers, the code would have returned two real numbers like an ordinary simulation code.

Lines (3a, b) and (4) are not part of the "code" but really part of the analysis of the code's result. In lines (3a, b) the output of the code is fed into a Taylor map, an object of type c_damap in my own FPP.

It is important to realise that this action could be utterly meaningless. Whoever writes lines (3, 4) must know that the object exiting the tracking loop contains Taylor series which can be *reinterpreted* as an approximate map of the 1-d-f phase space. For example, lines (3a, b) would produce a meaningless map if the output was made of real numbers or even Taylor series expansions depending only on system parameters such as the frequency `f` of the pendulum. Lines (1a, b) are critical: the first and second Taylor parameters will be the two phase space variables (θ, p). Thus at the exit, they can be fed in the object `m` of type c_damap and produce an approximate representation of the standard map. This representation is an eight order Taylor series in (θ, p) as specified by the parameter `map_order`.

Line (4) calls a routine of FPP which normalises the map `m`. Here I am regrettably in a manual mode. It is nevertheless instructive to describe the normal form `n`. It is given in part by the structure:

```
type c_normal_form

 type(c_damap) a1 ! brings to fix point at least linear
 type(c_damap) a2 ! linear normal form
 type(c_factored_lie) g !nonlinear part of a in phasors
 type(c_factored_lie) ker ! kernel i.e. normal form in phasors
 type(c_damap) a_t ! transformation a (m=a n a^-1)
 type(c_damap) n !transformation n (m=a n a^-1)
 type(c_damap) as !for spin (m = as a n a^-1 as^-1)
! resonances to be left in the map, including spin (ms)
 integer nres,m(ndim2t/2,nreso),ms(nreso)
! stores simple tune information
 real(dp) tune(ndim2t/2),damping(ndim2t/2),spin_tune
 logical positive ! forces positive tunes (close to 1 if <0)
!!!envelope radiation stuff to normalise radiation (sand's like theory)
 complex(dp) s_ij0(6,6) !equilibrium beam sizes
 complex(dp) s_ijr(6,6) ! equilibrium beam sizes in resonance basis
! equilibrium emittances as defined by chao (computed from s_ijr(2*i-1,2*i)
i=1,2,3)
real(dp) emittance(3)

end type c_normal_form
```

According to the definition of the c_normal_form type, the orignal map `m` of the code at lines (3a,b) can be expressed as

```
    m= n%a_t * n%n * n%a_t**(-1)
```

where "*" represent the concatenation of two Taylor maps in FPP.

The other object `n%ker` is the factored Lie map representing `n%n`. This object is of type `c_factored_lie` which is defined as:

```
type c_factored_lie
  integer :: n= 0
```

```
  integer :: dir= 0
  type (c_vector_field), allocatable :: f(:)
end type c_factored_lie

---------------------------------------------
type c_vector_field !
  integer :: n=0,nrmax ! n dimension used v(1:n) (nd2 by default)
! nrmax some big integer if eps<1
  real(dp) eps ! if eps=-integer then |eps| lie brackets are taken
! otherwise eps=eps_tpsalie=10^-9 used for convergence test
  type (c_taylor) v(lnv)
end type c_vector_field
```

An object of type c_factored_lie is defined as a product of Lie operators:

$$\begin{aligned} \mathscr{F} &= \exp\left(F_1 \cdot \nabla\right) \cdots \exp\left(F_N \cdot \nabla\right) \quad \text{if} \quad dir = 1, \\ \mathscr{F} &= \exp\left(F_N \cdot \nabla\right) \cdots \exp\left(F_1 \cdot \nabla\right) \quad \text{if} \quad dir = -1. \end{aligned} \tag{3.113}$$

In particular, as is shown in Eq. (3.89), the nonlinear canonical transformation $a_2 \circ \cdots \circ a_n$ appears naturally factored. Notice that the order of the Lie operators is the inverse of that of the normal phase space maps.

For the sake of comparing the results of the FPP algorithm and the small algorithms of the coming sections, I compute the pseudo-Hamiltonian of the normalised map n. This is done by the code fragment[8]

```
        id=exp(n%ker)           ! (5)
        call print(id,mf,prec)

!!!!!!!!!!!!!! Normalised Pseudo-Hamiltonian !!!!!!!!!!!!!!
        vf=0

        do io=1,size(n%ker%f)
         vf%v(1)=vf%v(1)+n%ker%f(io)%v(1)   ! (6a)
         vf%v(2)=vf%v(2)+n%ker%f(io)%v(2)   ! (6b)
        enddo

        k_io=-cgetpb(vf)/dt                 ! (6c)
```

The normalised map n is stored in the Dragt-Finn style object `n%ker` with `n%ker%dir=1`. However, since all the terms that survived cleansing are generators of rotations, I can add them up: they all commute. The sum is written in the vector field `vf` of type c_vector_field through lines (6a, b).

Line (6c) turns a vector field, in phasors' basis, into a Hamiltonian operator. For example, let us start backwards with the pseudo-Hamiltonian K_{io}

$$K_{io} = \sum_k t_k \varphi_+^{k_1} \varphi_-^{k_2}. \tag{3.114}$$

We can compute the vector field `vf` of line (6c), which I denote as V, using the correspondence

[8] FPP has a routine called `flatten_c_factored_lie` that performs this task, but here I prefer to be explicit.

$$n = \exp\left(\underset{\star}{\overset{\star}{:}} - dt\ K_{io}\underset{\star}{\overset{\star}{:}}\right) I \quad \text{where} \quad I = \begin{pmatrix} \theta \\ p \end{pmatrix} \tag{3.115}$$

$$V \cdot \nabla = \underset{\star}{\overset{\star}{:}} - dt\ K_{io}\underset{\star}{\overset{\star}{:}}. \tag{3.116}$$

The vector field V is just

$$V = \underset{\star}{\overset{\star}{:}} - dt\ K_{io}\underset{\star}{\overset{\star}{:}} I = dt\ 2i\ S\nabla K_{io} = dt2i \begin{pmatrix} -\sum_k k_2 t_k \varphi_+^{k_1} \varphi_-^{k_2-1} \\ \sum_k k_1 t_k \varphi_+^{k_1-1} \varphi_-^{k_2} \end{pmatrix} \tag{3.117}$$

Of course, it is possible to go in the reverse direction

$$K_{io} = -\frac{1}{2idt} \int_0^{(\varphi_+,\varphi_-)} SV \cdot d\varphi. \tag{3.118}$$

In Eq. (3.118) the path of integration is arbitrary. In the function `cgetpb`, of line (6c), I selected the diagonal between $(0, 0)$ and (φ_+, φ_-). It can be shown that for a symplectic map the result for K_{io} does not depend on the path of integration; the theory[9] is consistent.

3.4.2 *Normalisation of the Map with Vector Fields*

Here is the code fragment which prepares the nonlinear algorithm:

```
t_op=0 ;f_op=0;

beta=(1-(pi*f*dt)**2)**(1.0_dp/2)/(2*pi*f)        ! (ia)

a1%v(1)=sqrt(beta).cmono.1                         ! (ib)
a1%v(2)=(1.0_dp/sqrt(beta)).cmono.2                ! (ic)

c=from_phasor()                                    ! (iia)
m_n=c**(-1)*a1**(-1)*m*a1*c                        ! (iib)

rho1=m_n.sub.1                    ! (iiia)
expmu=m_n%v(1).sub.'1'            ! (iiib)

t_op%dir=1    ! (iva)
f_op%dir=-1   ! (ivb)
```

[9] If the map is not symplectic due to small computation errors, it pays to choose a path that avoids the region of maximum violation. For example, in the days of the defunct Super Conducting Super Collider, a member of SSC asked me to "symplectify" a map of the Jordan normal form type (see Sect. 4.2) that violated the symplectic condition mostly in the longitudinal direction due to truncation errors. I did not use the diagonal but avoided the longitudinal plane as much as possible in the integration path. This is also useful when a transverse map is known through measurements as a function of energy: one can reconstruct almost completely the longitudinal part of the map by avoiding it in the integration path.

The normalisation of the map starts with the linear part. The linear transformation, denoted by a1, is given by Eq. (3.84) through the β function, line (ia) in the code and subsequently lines (ib, c). Line (iib), i.e., m_n, is exactly the map m_1 of Eq. (3.88). This map is “ready” to be submitted to the nonlinear recursive normal form: let the cleansing start! This is the following code fragment:

```
        t_op%dir=1    ! (iva)
        f_op%dir=-1   ! (ivb)

 do io=2,map_order

         n_np1= m_n*rho1**(-1); n_np1= exp_inv(T_op,n_np1)            ! (v)
         g_io%v(1)=n_np1%v(1).sub.io      ! (via)
         g_io%v(2)=n_np1%v(2).sub.io      ! (vib)
do a=1,2
          call c_taylor_cycle(g_io%v(a),size=ns)

!!!   Cycling over all monomials
    do km=1,ns
       call c_taylor_cycle(g_io%v(a),ii=km,value=v,j=ji)    ! (vii)
       if(ji(1)-ji(2)+(-1)**a/=0) then                     ! (viiia)
         v=v/(1-expmu**(ji(2)-ji(1)-(-1)**a))              ! (viiib)
         f_op%f(io)%v(a)=f_op%f(io)%v(a)+(v.cmono.ji)    ! (viiic)
       else
         t_op%f(io)%v(a)=t_op%f(io)%v(a)+(v.cmono.ji)    ! (viiid)
       endif
    enddo
 enddo
        m_n=exp(-f_op%f(io))*m_n ;m_n=exp(f_op%f(io),m_n) ! (ix)
 enddo

        t_op%f(1)%v(1)= log(expmu).cmono.1       ! (xa)
        t_op%f(1)%v(2)=-log(expmu).cmono.2       ! (xb)
```

Line (v) is very important; often when I write such an algorithm (without say cutting and pasting from somewhere else...) I mess up at line (v). This map n_np1 of line (v) must correspond to the map n_{n+1} of Eq. (3.98). The map n_{n+1} is near the identity with corrections which we have not yet been analysed, so as I said previously, it is of the form

$$\begin{aligned} n_{n+1} &= \exp\left(G_{n+1}\cdot\nabla\right) I \\ &= I + G_{n+1}\cdot\nabla I + \cdots \\ &= I + G_{n+1} + \cdots . \end{aligned} \tag{3.119}$$

In my loop, the vector field G_{n+1} is represented by the variable g_io where io is the index of the recursive loop. Thus io at any given instant is $n+1$ and it starts at io=2. G_{n+1} is extracted at lines (via,b) using the operator .sub. which extracts in this case the ioth order component of a polynomial; these lines are obviously assuming a 1-d-f map.

The outer loop, over io, is clearly over the degree of the Taylor maps; the next loop is over the components of G_{n+1}, G^1_{n+1} and G^2_{n+1}; but the truly interesting loop is the inner loop over the monomials. Lines (viiia, b, c, d) perform the work of Eqs. (3.105) and (3.106) which I rewrite here for convenience:

$$\text{Eq. (3.105)}: m_1 - m_2 + (-1)^a \neq 0 \quad \rightarrow \quad F^a_{n+1;m} = \frac{G^a_{n+1;m}}{1 - \exp\left(i\mu_0 \left(m_1 - m_2 + (-1)^a\right)\right)},$$

$$\text{Eq. (3.106)}: \text{Tune shifts} = \sum_{m_1 - m_2 + (-1)^a = 0} G^a_{n+1;m} \varphi_+^{m_1} \varphi_-^{m_2}$$

Line (vii) fetches a monomial and its properties. Line (viiia) verifies the resonance condition $m_1 - m_2 + (-1)^a \neq 0$. If the monomial does not represent a tune shift, the operator F^a_{n+1} is augmented so that it will have the power to "liquidate" this monomial: line (viiic). Otherwise the monomial is a tune shift and it is stored in T^a_{n+1}: line (viiid).

Concatenation

The next line is crucial. How do we compute the map m_{n+1} of Eq. (3.95)? m_{n+1} is the map transformed by the canonical a_{n+1} which we have just computed through line (viiic).

$$\begin{aligned} \text{Eq. (3.95)}: m_{n+1} &= a_{n+1}^{-1} \circ m_n \circ a_{n+1} \\ &= \underbrace{\exp\left(F_{n+1}.\nabla\right) \underbrace{\left\{ \underbrace{\exp\left(-F_{n+1}.\nabla\right) I}_{a_{n+1}^{-1}} \circ m_n \right\}}_{a_{n+1}^{-1} \circ m_n}}_{a_{n+1}^{-1} \circ \{m_n \circ a_{n+1}\}}. \end{aligned} \tag{3.120}$$

Equation (3.120) is realised by line (ix) of the code. In the library FPP, `exp(-f_op%f(io))` is equivalent `exp(-f_op%f(io),I)`, in other words the identity map can be ommitted in FPP.

The central idea is to avoid using the CBH theorem simply because we do **not** have a Lie representation of our one-turn map. Thus in Eq. (3.120), one concatenation is used and two trivial applications of a Lie map which only require a finite number of Lie brackets, i.e., derivatives.

It is my opinion, held strongly from experience, that the normal form algorithms described by A. Chao for example in his lectures see [3] are pretty much sitting in a "no-man's-land" of applicability. They involve a Lie representation of the one-turn map of the Dragt-Finn type which we do not have coming from an integrator code and the CBH theorem which we do not need. In the case of analytical calculations, this approach also fails to separate clearly the solved problem (on which the numerical algorithm of the code can be applied) from the analytically represented force that is the actual perturbation. Of course if the unperturbed problem is trivial, then the methods in Chao's lectures can be applied: a linear map followed by a beam-beam kick for example.

I can only conclude that the methods utilising a one-turn map factorised as $e^{:h_2:}e^{:h_3:}\cdots$ appear in lectures because, accidentally, this is how the code Marylie of A. Dragt parametrised its maps. It is worthwhile to notice that Cosy-Infinity, the matrix code of Berz [1], must use an algorithm similar to the one I just described: a general concatenator is used and most likely Berz avoids unnecessary Lie maps.

It is possible to avoid the CBH formula if one can get a *single* Lie operator description through the logarithm of supplemental Chap. 11. This is possible numerically (on occasion) but not in purely analytical calculations.

The last step of the normalisation is useful but not necessary. In lines (xa,b) I construct and add the operator T_1 to T. Indeed in any numerical algorithm on the map, the linear part is made of matrices and matrices are used to normalise it. In Sect. 2.3.1, I described that process with nonlinear Lie maps in mind. Nevertheless the reader saw that a linear numerical algorithm with Lie maps in mind simply forces us to consider the *transpose* of the usual phase space matrix rather than the matrix itself. Lie operators for the linear transformation a_1 are not needed and are not necessary. Thus at the end of the exercise we have a matrix for the linear normal form n_1. If we want a Lie operator for the full normalised n, we need to add to the vector field T its linear part T_1. In the code, it is done first by extracting the eigenvalues at line (iiib) and adding them to T at lines (xa, b). This is actually necessary in a one-resonance normal form as described in Sect. 5.4.3 where the "co-moving" map is computed.

This concludes the description of the algorithm. Lines (xiia, b) and (xiii), in Appendix F, reproduce the pseudo-Hamiltonian once more for comparison with the results of Sect. 3.4.1.

3.4.3 Normalisation of the Map with Poisson Brackets

Normalisation with the Poisson bracket is nearly identical to the normalisation over the vector field; it is describe theoretically in the gray box near Eq. (3.107). We can look at the code fragment immediately:

```
do io=2,map_order
        t_op%f(io)=0                       ! (IVa)
        f_op%f(io)=0                       ! (IVb)
        n_np1=m_n*rho1**(-1); n_np1=exp_inv(t_op,n_np1)             ! (V)

        g_io%v(1)=n_np1%v(1).sub.io         ! (VIa)
        g_io%v(2)=n_np1%v(2).sub.io         ! (VIb)

        k_io=cgetpb( g_io )                 ! (VIc)

        f_pb=0.0_dp                        ! (VIIa)
        t_pb=0.0_dp                        ! (VIIb)
```

```
          call c_taylor_cycle(K_io,size=ns)    ! same as above

!!!   Cycling over all monomials
    do km=1,ns
       call c_taylor_cycle(K_io,ii=km,value=v,j=ji)
       if(ji(1)/=ji(2)) then                      !(VIIIa)
         v=v/(1-expmu**(ji(2)-ji(1)))             !(VIIIb)
         f_pb=f_pb+(v.cmono.ji)                   !(VIIIc)
       else
         t_pb=t_pb+(v.cmono.ji)                   !(VIIId)
       endif

    enddo
        t_op%f(io)=cgetvectorfield(t_pb)          ! (IXa)
        f_op%f(io)=cgetvectorfield(f_pb)!         ! (IXb)
        m_n=exp(-f_op%f(io))*m_n; m_n=exp(f_op%f(io),m_n) ! (X)

 enddo
```

The first difference is at line (VIc). The Lie operator `K_io` is extracted using Eq. (3.118) which is implemented in the routine `cgetpb` of FPP.

The next difference is in the cleansing part: lines (VIIIa, b, c, d). Line (VIIIb) performs the computation of the canonical transformation a at order `io` which is now a simple Lie polynomial; this is Eq. (3.110).

There is not much more to say except that this algorithm cannot work for non-symplectic maps nor can it work if non-canonical variables are used.

3.4.4 Using a Pseudo-Hamiltonian: Normalising the Logarithm

Here is a bizarre functionality of FPP which will have applications in Chap. 8 when I discuss "Guignard style" [4] perturbation[10] theory. Imagine that our code is made of integration steps (which it is), then symbolically, tracking through one turn of the machine can be written as a sequence of concatenations:

$$m_j = m_{j+N-1\ \ j+N} \circ \cdots \circ m_{j+1\ \ j+2} \circ m_{j\ \ j+1}. \tag{3.121}$$

Obviously these maps are close to the identity, thus we can surmise that they can be expressed using a single Lie operator:

$$m_{j+k\ \ j+k+1} = \exp\left(ds_k F_k \cdot \nabla\right) I. \tag{3.122}$$

The force field F_k may in fact be the actual force of the magnet in question if the local Hamiltonian is s-independent. It is certainly an accurate representation of the average effect over the short distance ds_k over which this action takes place.

[10]This is "s"-dependent Hamiltonian perturbation theory where H is Fourier-transformed in the time-like variable s often mapped into an angular variable θ as I do in Chap. 8. It is commonly described in many accelerator books though rarely carried through even on simple lattices.

Assuming we can compute F_k, it implies that we have a pseudo-Hamiltonian for the entire ring:

$$H(s) = \sum_{k=1}^{N} u(s, s + ds_k) H_k. \tag{3.123}$$

The function $u(s, s + ds_k)$ is equal to one in the interval $[s, s + ds_k]$ and zero elsewhere. In Chap. 8 I will show how Eq. (3.123) allows us to perform the kind of perturbation theory often described in standard treatments but rarely carried out (see [5] for example).

The length s is any length more or less equal to the length along the closed orbit. In the case of a simple code, the Hamiltonian H_k may indeed be the exact Hamiltonian used by the writer of the code. Otherwise it is a good average Hamiltonian for a single time step.

In the program of Appendix F, the "one-turn map", i.e., the standard map, is near the identity. So its logarithm can be obtained by the iteration procedure of supplemental Chap. 11. It is given by the syntax

```
vf=log(M)                                    ! (1a)
k=-getpb(vf)/dt                              ! (1b)
```

Here is `M` is simply a map of phase space and `vf` is a vector field. `k` is the Hamiltonian extracted from the vector field.

3.4.4.1 Using the Pseudo-Hamiltonian for the Normalisation

From that point on, the algorithm of the normalisation is that of time-independent perturbation theory. I proceed using the techniques discussed in Sect. 3.2.4. First I normalise the quadratic part of `k` using Eq. (3.84) which, in the code, is just:

```
beta=(1-(pi*f*dt)**2)**(1.0_dp/2)/(2*pi*f)   ! (2a)

a1%v(1)=sqrt(beta).cmono.1                   ! (2b)
a1%v(2)=(1.0_dp/sqrt(beta)).cmono.2          ! (2c)
```

The iterative part of the loop is, as before, a cycling over the monomials of the Hamiltonian at a given order:

```
!!!   Cycling over all monomials
    do km=1,ns
       call c_taylor_cycle(K_io,ii=km,value=v,j=ji) ! (5a)
       if(ji(1)/=ji(2)) then                        ! (5b)
         v=v/((ji(1)-ji(2))*(2*i_)*tpf)             ! (5c)
         a_op=a_op+(v.cmono.ji)                     ! (5d)
       endif
    enddo
```

The important removal of the resonant terms is done by line (5c) which implements Eq. (3.80), namely

$$A_j = \frac{v_j}{i2\pi f\,(j_1 - j_2)} \quad \text{if} \quad j_1 \neq j_2 \;\rightarrow\; v = v/((ji(1) - ji(2)) * (2 * i\_) * tpf).$$

The Hamiltonian is then updated at order `io`

```
vf=cgetvectorfield( a_op )                      ! (6a)
K=exp(vf,K)                                     ! (6b)
```

Line (6b) produces the equivalent of $K_{io+1} = \exp(F_{io} \cdot \nabla)\, K_{io}$. Line (6a) is needed because my complex FPP package is ignorant of Poisson bracket operators: I need to produce the equivalent general vector field before I exponentiate.

3.4.4.2 Using the Pseudo-force for the Normalisation

The procedure with the vector field is more of the same: I start with the logarithm of the map at lines (3A, B, C) where the linear part of this map is already in phasors. At line (3C) I extract the "frequency" of the linear map needed in normal form algorithm.

```
g_io=log(M)  !                                  ! (3A)
g_io=-(1.0_dp/dt)*g_io                          ! (3B)
om= g_io%v(1).sub.'1'            ! (3C)
```

The core of the normal form is found at line (5B)

```
v=v/(om*(ji(1)-ji(2)+(-1)**a))                  ! (5B)
```

Line (5B) uses the "force" equivalent of Eq. (3.105) which applies to a map expressed in terms of a vector field. In fact, for small μ_0, Eq. (3.105) becomes

$$m_1 - m_2 + (-1)^a \neq 0 \quad \frac{G^a_{n+1;m}}{1 - \exp\left(i\mu_0 \left(m_1 - m_2 + (-1)^a\right)\right)} \rightarrow \frac{G^a_{n+1;m}}{-i\mu_0 \left(m_1 - m_2 + (-1)^a\right)} \tag{3.124}$$

which agrees with line (5B) if I take into account the minus sign that enters in the definition of the Hamiltonian and replace `om` of line (5B) by μ_0.

The example of the vector field also illustrates the equivalence between formula Eq. (3.27) (i.e., $F^{new}_k = \left(F \cdot \nabla a_k^{-1}\right) \circ a$) and (3.40) (i.e., $F^{new} = \exp(: A :)\, F$). Here is the code fragment from Appendix F:

```
if(t_o) then     ! if true, use (F.dot.a_k^-1) o a
      id=exp(f_op%f(io))      ! (6A)
      a1=id**(-1)             ! (6B)
      r=0

      do ns=1,2
      do a=1,2
       r%v(ns)=g_io%v(a)*(a1%v(ns).d.a)+r%v(ns)  ! (7)
      enddo
      r%v(ns)=r%v(ns)*id                         ! (8)
      enddo

      do ns=1,2
        g_io%v(ns)=r%v(ns)
      enddo
```

```
else          !  use a Lie Bracket operator
      g_io=exp_ad(f_op%f(io),g_io)          ! (6')
endif
```

The vector field being normalised is g_io. In lines (6) to (8), I evaluate the effect of the canonical transformation using Eq. (3.27). First, I compute the transformation a which is stored in id at lines (6A):

$$a = \exp\left(F_{op} \cdot \nabla\right) I \quad (I \text{ is implicit in the code}). \tag{3.125}$$

I then compute its inverse and store it into a1. Lines (7) and (8) completes the process of Eq. (3.27) by multiplying the derivative of a1 with the vector field g_io at line (7) and then substituting the map a (i.e., id) into the result.

The other option is simply to use the vector Lie bracket as in Eq. (3.40): this is done in line (6').

$$F^{new} = \exp\left(: A :\right) F = \sum_{n=0}^{\infty} \frac{1}{N!} \underbrace{\langle A, \cdots \langle A, \langle A, F \rangle\rangle\rangle}_{N \text{ times}} \tag{3.126}$$

Eq. (3.126) is realised in the code at line (6'): the sum is carried over until convergence is achieved. For a purely nonlinear map, it requires a finite number of steps.

Once may notice, looking at the logarithm explained in supplemental Chap. 11 and Eq. (3.126), that a normalisation of the map can be achieved without using concatenation. This is a funny and unexpected result but it hinges on one's ability to take the logarithm of the map by the process of supplemental Chap. 11. This can only be done for a map near the identity. For maps far from the identity, it is possible to modify the algorithm of supplemental Chap. 11 to achieve convergence for certain maps but it does not always work. For this reason, I think that a TPSA package with an efficient concatenator is always necessary for results emanating from a real tracking code.

References

1. M. Berz, Technical report, Michigan State University (unpublished)
2. M. Berz, Part. Accel. **24**, 109 (1989)
3. A. Chao, Technical Report No. SLAC-PUB-9574, Stanford Linear Accelerator (unpublished)
4. G. Guignard, Technical Report No. CERN 78–11, CERN (unpublished)
5. S.Y. Lee, *Accelerator Physics* (World Scientific Publishing, Singapore, 2004)

Chapter 4
Classification of Linear Normal Forms

Abstract A normal form is an idealisation of the nonlinear world based on an extrapolation from the linear domain. This is its central limitation. But if we accept it, with reservations, it entails that it can only be understood if the linear system has first been studied. In this chapter, I examine several linear systems upon which a nonlinear normal form can be defined: the full harmonic sink, the Jordan coasting beam, the radiative stochastic beam envelope and the periodically modulated magnets. Only spin is left out and treated later. In the case of the damped harmonic coupled oscillator, I introduce its de Moivre representation and some consequences.

Keywords Linear · Multidimensional · Harmonic · Sinks · De Moivre · Jordan normal form · Radiation · Beam envelopes

4.1 The Full Harmonically Sinking Phase Space

The previous chapters described a system which was undergoing harmonic oscillations around the origin in one degree of freedom. It turns out that it is very easy to extend all of this to many dimensions.

The hardest part is the linear part! The algorithm for the nonlinear part remains identical and no new concepts are introduced. Of course, when I say "easy" I am assuming that no new topological features need to be explained: neither resonances nor chaos appears in the region of interest. This is obviously the assumption behind the adequacy of a normal form: the phase space is a tranquil extension of the linear regime.

So I will devote some time to the linear regime and introduce some new concepts such as the de Moivre formula in many dimensions and the concept of generalised dispersions. As usual I will pepper this exposé with examples from the "Code," namely my own FPP for analysis and PTC for the tracking code.

This section will be illustrated by the program in Appendix G.

E. Forest, *From Tracking Code to Analysis*,
DOI 10.1007/978-4-431-55803-3_4

4.1.1 Radiation: The Multidimensional Drain

I extend the 1-d-f case, "multiplicatively", to three[1] degrees of freedom.

$$M = A \Lambda R A^{-1} \tag{4.1}$$

$$R = \begin{pmatrix} r_1 & 0 & 0 \\ 0 & r_2 & 0 \\ 0 & 0 & r_3 \end{pmatrix} \qquad r_i = \begin{pmatrix} \cos\,\mu_i & \sin\,\mu_i \\ -\sin\,\mu_i & \cos\,\mu_i \end{pmatrix}$$

$$\Lambda = \begin{pmatrix} \Lambda_1 & 0 & 0 \\ 0 & \Lambda_2 & 0 \\ 0 & 0 & \Lambda_3 \end{pmatrix} \qquad \Lambda_i = \begin{pmatrix} \exp(-\alpha_i) & 0 \\ 0 & \exp(-\alpha_i) \end{pmatrix}. \tag{4.2}$$

The computation of the matrix A follows the steps outlined Sect. 2.3.1. There is no difference between the construction of the symplectic and radiative case. In the symplectic case the eigenvectors belonging to different eigenvalues will automatically have vanishing Poisson brackets: this insures that the matrix A is symplectic. In the non-symplectic case, the Poisson bracket between two distinct eigenplanes is not zero, but we really do not care.

Many papers have been written on analytical descriptions and calculations of the matrix A in the symplectic case. Often they are extensions of the Teng-Edwards method [1] or elucidations of that method. One important extension was developed by Ohmi, Hirata and Oide in Ref. [2]. It is notable because it is the first paper connecting the Chao synchrotron integrals [3] to the more approximate methods [4] of Sands. Perhaps more interestingly, they also point out the existence of a dispersion which is proportional to time rather than energy; they call this transverse vector ζ. The vector ζ appears in all expressions which are "first-order" in the coupling between the longitudinal and transverse planes. This can be derived, more generally, using the de Moivre representation of the 6×6 transfer matrix.

Personally I hold the view that all reference to particular parametrizations and factorizations of A are useless particularly in the context of a code. It is possible to regain the ζ function of Ohmi et al., without appealing to any particular parametrization of A. It is also possible to demonstrate the connection between Chao and Sands without resorting to any special parametrization of A. Moreover the factorization of A in the style proposed in Refs. [1, 2] is not global and fails for certain highly coupled systems. This is not surprising because the symplectic group can only be represented by a minimal set of parameters near the identity due to its non-compactness. The factorization of Teng and Edwards, for example, requires only the minimal 10 parameters of the Lie algebra of Sp(4,$\mathbb{R}$) and is thus bound to fail at large coupling.

However the concept of the function ζ, for example, does not depend on any factorization and indeed any choice for the matrix A. In a code, we compute A

[1] Actually the results of this section apply to N degrees of freedom; but in accelerators $N = 3$ is the maximum size of the symplectic space.

using the eigenvectors of the one-turn matrix M and all properly defined physical concepts can be extracted from A or M in a canonically invariant way as I will now demonstrate.

4.1.2 De Moivre Representation of the One-Turn Matrix

Once we have found the matrix A by any methods whatsoever, we can isolate the contribution to individual planes by rewriting the damped rotation of Eq. (4.1):

$$M = A\left(\sum_i e^{-\alpha_i}\left\{\cos(\mu_i)\, I^i + \sin(\mu_i)\, SI^i\right\}\right)A^{-1} \tag{4.3}$$

$$= \sum_i e^{-\alpha_i}\left\{\cos(\mu_i)\underbrace{AI^iA^{-1}}_{H^i} + \sin(\mu_i)\underbrace{ASI^iA^{-1}}_{B^i}\right\}$$

$$= \sum_i e^{-\alpha_i}\left\{\cos(\mu_i)\, H^i + \sin(\mu_i)\, B^i\right\} \tag{4.4}$$

where I^i is the identity restricted to the ith plane, for example here is I^3:

$$I^3 = \begin{pmatrix} 0&0&0&0&0&0\\ 0&0&0&0&0&0\\ 0&0&0&0&0&0\\ 0&0&0&0&0&0\\ 0&0&0&0&1&0\\ 0&0&0&0&0&1 \end{pmatrix}. \tag{4.5}$$

This trivial rearrangement of the matrix M in terms of A has surprising consequences. But before we list them, here are a few properties of the matrices H^i and B^i. They form orthogonal representations of the complex numbers which they inherit from I^i and $S^i = SI^i$ which are respectively the identity and symplectic matrix S restricted to the ith plane:

$$H^iH^j = \delta_{ij}H^j \tag{4.6}$$

$$H^iB^j = B^jH^i = \delta_{ij}B^j \tag{4.7}$$

$$B^iB^j = -\delta_{ij}H^j. \tag{4.8}$$

The basic feature of this representation is its simple exponentiation properties:

$$M^N = \sum_i e^{-N\alpha_i}\left\{\cos(N\mu_i)\, H^i + \sin(N\mu_i)\, B^i\right\}. \tag{4.9}$$

This can be written in terms of a single Lie exponent:

$$M = \exp\left(\sum_i -\alpha_i H^i + \mu_i B^i\right). \tag{4.10}$$

The reader notices that the matrices H^i multiply the damping decrements while the B^i's are connected to the tunes. So in the symplectic case, the Lie representation which is basically the pseudo-Hamiltonian as we will see in Sect. 4.1.3, does not naturally involve the H^i's.

In one degree of freedom, where de Moivre is well known, H^1 reduces to the identity in the symplectic case and B^1 is made of the usual lattice functions:

$$M = \cos(\mu)\, H + \sin(\mu)\, B \tag{4.11}$$

where

$$H = \begin{pmatrix} 1 & 0 \\ 0 & 1 \end{pmatrix} \text{ and } B = \begin{pmatrix} \alpha & \beta \\ -\gamma & -\alpha \end{pmatrix} \text{ with } \mathrm{B}^2 = -\mathrm{H}. \tag{4.12}$$

In many dimensions, in the symplectic case, the matrices B^i are still connected to the invariants and the matrices H^i are redundant measures of coupling intimately associated with dispersive quantities, for example η and ζ. Dispersive quantities measure the position of (quasi-)fixed points as a function of (quasi-)constant initial conditions.

4.1.3 Ripken Lattice Functions Representation: Invariants and Moments

Now let me concentrate on the symplectic case. In the sympletic case, the damping matrix Λ is identity. Furthermore, I can express A^{-1} in terms of A as shown in Eq. (1.5).

I define two new matrices in each plane in terms of the matrix B^i:

$$B^i = SK^i \tag{4.13}$$
$$B^i = E^i S \tag{4.14}$$
$$K^i = SE^i S^\dagger. \tag{4.15}$$

The matrix K^i is the quadratic form of the coupled Courant-Snyder invariant in the ith plane. The matrix E^i connects moments to the invariants. But first I will prove some trivial properties that K^i and E^i must have if indeed they are connected to invariants and moments. I first rewrite B^i using the symplectic condition Eq. (1.5):

$$B^i = ASI^iA^{-1} = -ASI^iSA^\dagger S$$
$$= AI^iA^\dagger S. \tag{4.16}$$

It follows from Eq. (4.16) that K^i is a symmetric matrix:

$$K^{i\dagger} = \left(-SB^i\right)^\dagger$$
$$= \left(-SAI^iA^\dagger S\right)^\dagger = -S^\dagger AI^iA^\dagger S^\dagger = K^i \text{ (since } S^\dagger = -S). \tag{4.17}$$

It follows trivially, looking at Eq. (4.15), that the moment matrix E^i is symmetric because K^i is symmetric.

4.1.3.1 The Quadratic Invariants

The uncertain reader can review the material in 1-d-f in Sect. 2.2.3; otherwise please keep reading!

The connection between the matrix K^a and the invariant are documented and verified in the program of Appendix G. In the following code fragment, I display two different ways to compute the quadratic invariants. Line (17) works in all instances, linear and nonlinear. Line (18) works for the linear map executing harmonics oscillations in 2,4 or 6 dimensions. In this program the variable `nda` is either 4 or 6.

```
  do i=1,nda
      expo=0
      expo(2*i-1)=2
     e(i)=1.d0.cmono.expo
      expo=0
      expo(2*i)=2
     e(i)=e(i)+(1.d0.cmono.expo)
     e(i)=e(i)*normal_form%a_t**(-1)              ! (17)
  enddo

  do i=1,nda
write(mf,*)" ";
write(mf,*) " Invariant (q^2+p^2) o A^(-1)  in plane ",i;
write(mf,*)" ";
   call print(e(i),mf,prec)
   do j=1,2*nda
    write(mf,'(6(5x,g12.5))') Ka(j,1:6,i)          ! (18)
   enddo
  enddo
```

Let me explain the linear theory. The "Courant-Snyder" or quadratic invariant in the ith plane can be written as

$$\varepsilon_i\left(z\right) = z_a K^i_{ab} z_b \text{ (summed over } a, b). \tag{4.18}$$

I now produce a proof. I start with the symplectic matrix M where I introduce a pseudo-time t:

$$z(t) = M(t)z = \exp\left(t\sum_i \mu_i B^i\right)z. \tag{4.19}$$

I then construct the invariant at time t and take its derivative. Hopefully, its time derivative will be zero!

$$\begin{aligned}\frac{d\varepsilon_i}{dt} &= \frac{d}{dt}\left(z_a K^i_{ab} z_b\right)\\ &= \left(\frac{d}{dt}z_a\right)K^i_{ab}z_b + z_a K^i_{ab}\left(\frac{d}{dt}z_b\right)\\ &= \left\{\sum_j \mu_j B^j_{ac}\right\} z_c K^i_{ab} z_b + z_a K^i_{ab}\left\{\sum_j \mu_j B^j_{bc}\right\} z_c.\end{aligned} \tag{4.20}$$

The product of two B's belonging to different planes vanishes. Therefore I can simplify Eq. (4.20):

$$\begin{aligned}\frac{d\varepsilon_i}{dt} &= \mu_i\left\{B^i_{ac} z_c K^i_{ab} z_b + z_a K^i_{ab} B^i_{bc} z_c\right\}\\ &= \mu_i z\left\{{B^i}^\dagger K^i + K^i B^i\right\} z^\dagger = z\left\{\left(SK^i\right)^\dagger K^i + K^i SK^i\right\} z^\dagger\\ &= z\underbrace{\left\{K^i S^\dagger K^i + K^i SK^i\right\}}_{=0 \text{ since } S^\dagger = -S} z^\dagger = 0\end{aligned} \tag{4.21}$$

Thus ε_i is indeed an invariant. Of course this invariant can be derived from the Lie operator theory. In fact, the Lie map for the matrix M, which I denote $\mathscr{M}$ following Dragt as in Eq. (2.35) in Sect. 2.3.1, is just:

$$\mathscr{M} = \exp\left(: -\frac{1}{2}\sum_{i=1,3} \mu_i z^\dagger K^i z :\right). \tag{4.22}$$

From the Lie map definition, it is clear that its exponents, actually the three polynomials that make it, are invariants. The reader can check that this is the Lie representation of the matrix M by showing that

$$\left[\frac{\mu_i}{2} z^\dagger K^i z, z_a\right] = \mu_i \sum_{b=1,6} B^i_{ab} z_b \tag{4.23}$$

which is in agreement with Eq. (4.19) with $t = 1$.

4.1.3.2 The Quadratic Moments

As for the case of the invariants, the 1-d-f material can be found in Sect. 2.2.4, otherwise keep reading. The code fragment relevant to this calculation is part of the program of Appendix G:

The idea is to perform a time average of the functions $z_i z_j$. As was explained in Sect. 2.2.4, we express the moment in terms of the normalised variables z^{new}:

$$\langle z_i z_j \rangle = \sum_{a,b} A_{ia} A_{jb} \langle z_a^{new} z_b^{new} \rangle . \tag{4.24}$$

The average $\langle z_a^{new} z_b^{new} \rangle$ can be computed for the normalised variables

$$\langle z_a^{new} z_b^{new} \rangle = \delta_{ab} \left\langle \frac{z_a^{new2} + z_{\overline{a}}^{new2}}{2} \right\rangle = \langle J_{p(a)} \rangle , \tag{4.25}$$

where I denote the variable conjugate to a by $\overline{a}$ and the plane of a or $\overline{a}$ by $p(a)$. For example, if $a = 4$ then $\overline{a} = 3$ and $p(3) = p(4) = 2$. Using this notation, I substitute Eq. (4.25) into Eq. (4.24):

$$\langle z_i z_j \rangle = \sum_{p(a)=1,2,3} \underbrace{\left(A_{ia} A_{ja} + A_{i\overline{a}} A_{j\overline{a}} \right)}_{E_{ij}^{p(a)}} \langle J_{p(a)} \rangle . \tag{4.26}$$

The next step consists in noticing that $E_{ij}^{p(a)}$ can be rewritten using the matrix $I^{p(a)}$, the identity which restricts the summation to the plane $p(a)$:

$$\begin{aligned} E^{p(a)} &= A I^{p(a)} A^{\dagger} \\ &= A I^{p(a)} S A^{-1} S^{\dagger} = B^{p(a)} S^{\dagger} \\ &= S K^{p(a)} S^{\dagger} . \end{aligned} \tag{4.27}$$

Equation (4.27) completes the proof: the moments are connected to the invariants by a similarity transformation involving S.

So we get the final result:

$$\langle z_i z_j \rangle = E_{ij}^1 \langle J_1 \rangle + E_{ij}^2 \langle J_2 \rangle + E_{ij}^3 \langle J_3 \rangle . \tag{4.28}$$

4.1.4 De Moivre-Ripken H^a Matrices in the Symplectic Case: The Dispersions η and ζ

In the entire work of Ripken, the matrices H^a $(a = 1, 3)$ never appear. This accident comes from the fact that, in a Hamiltonian treatment, they are "useless" and do not appear naturally. When we look at the Lie exponent of Eq. (4.22), we see that it contains only the B^a's through the matrices K^a. Only a non-symplectic map would naturally involve the H^a's as confirmed by Eq. (4.10).

So we are left with a simple question, beside the introduction of a de Moivre formula, do the H^a's have any particular meaning in the (quasi-)symplectic case?

The answer is that the H^a reduce to dispersive quantities including the usual dispersion η, but unlike the dispersion which is ill-defined in the presence of an RF-cavity, the H^a's are always well defined!

First I will recall for the reader the usual definition of dispersion which will be important in Sect. 4.2. In the *absence* of a cavity, the energy is a constant. Thus it makes sense to talk about the closed orbit[2] as a function of the energy. Particles with different energies will undergo transverse oscillations around a different closed orbit.

Assuming that the "on-energy" particle has a closed orbit, i.e.,

$$m(\underbrace{x_c, p_{xc}, y_c, p_c}_{f}, 0, 0) = (x_c, p_{xc}, y_c, p_c, 0, \tau_c). \tag{4.29}$$

I can ask if the function $f(\delta)$ defined as

$$m\left(f(\delta), \delta, 0\right) = \left(f(\delta), \delta, \tau_c(\delta)\right) \tag{4.30}$$

has a solution in the neighbourhood of $\delta = 0$? The answer is always yes if the linear transverse map for $\delta = 0$ has tunes away from the integer. This is usually true in all accelerator designs for rings. This is a consequence of the implicit function theorem in mathematics. Thus the dispersion is a well defined mathematical object in the absence of a cavity.

In the example code PTC, the energy is the fifth variable, thus the linear dispersion $f = \delta\eta$ obeys the equation:

$$M \begin{pmatrix} f \\ \delta \\ 0 \end{pmatrix} = \underbrace{\begin{pmatrix} N & v & 0 \\ 0 & 1 & 0 \\ w & \alpha & 1 \end{pmatrix}}_{6\times 6 \text{ matrix}} \begin{pmatrix} f \\ \delta \\ 0 \end{pmatrix} = \begin{pmatrix} f \\ \delta \\ \tau \end{pmatrix}. \tag{4.31}$$

[2] The dispersive ζ of the Japanese trio, Ohmi, Hirata and Oide, cannot be deduced from a cavity-less system because nothing depends on the time of arrival in the absence of longitudinal(temporal) focussing. Thus the ζ functions can only emerge from an examination of the full three-dimensional oscillator.

Expanding Eq. (4.31) I get an equation for f:

$$Nf + \delta v = f \Rightarrow f = \delta(1-N)^{-1}v \Longrightarrow \eta = (1-N)^{-1}v. \tag{4.32}$$

The matrix M in Eq. (4.31) represents motion *without* an RF-cavity. Yet we know that the concept of dispersion is used a lot in the design of circular accelerators with longitudinal focussing present. Thus logic dictates two distinct possibilities:

1. Accelerators physicists use an ultimately irrelevant concept while designing their machines.
2. The concept of dispersion retains in some adiabatic sense its meaning when the cavities are turned on, that is to say, when time and energy start undergoing harmonic oscillations.

I must assume that my colleagues, at least some of them, know what they are doing and thus option 2 must be correct. Therefore I look at de Moivre's formula for M in Eq. (4.9) in the adiabatic limit: I take the limit $\nu_3 \ll \nu_1, \nu_2$. I then compute the average position of a ray over a number of turns large compare to $1/\nu_{1,2}$ but small compared to $1/\nu_3$:

$$\begin{aligned} z(N) &= M^N z \\ &= \sum_{i=1,3} \left\{ \cos\left(N2\pi\nu_i\right) H^i + \sin\left(N2\pi\nu_i\right) B^i \right\} z \\ &\approx H^3 z \quad if \quad \frac{1}{\nu_{1,2}} \ll N \ll \frac{1}{\nu_3}. \end{aligned} \tag{4.33}$$

Equation (4.33) gives the average position of a ray which is frozen in "time." In my code PTC, the energy variable is z_5, thus I claim that the dispersion computed in Eq. (4.32) should be very close to

$$f_k = H^3_{k5} z_5 \Rightarrow \eta_k = H^3_{k5}. \tag{4.34}$$

Mathematically this adiabatic process can be viewed has an average over the invariant tori corresponding to the first two planes. It can also be seen as a stroboscopic average where one observes the variable z near a full cycle in the third plane which is nothing more than the cosine Fourier transform. The reader may notice that these two definitions of H^3, one mathematical and one more physical, apply equally well to H^1 and H^2.

Looking at Eq. (4.33), we notice that they are five more dispersion vectors. In particular, we have something called ζ,

$$f_k = H^3_{k5} z_6 \Rightarrow \zeta_k = H^3_{k6} \tag{4.35}$$

which is the dependence of the adiabatic closed orbit on the initial longitudinal position or time.

The dispersive quantity ζ has no equivalent in the framework of Eq. (4.31) for the simple reason that in the absence of temporal focussing, absolutely nothing depends on the time variable.

Nevertheless ζ cannot be neglected in approximate treatments where cavities are present and Ripken lattice functions are re-expressed in terms of dispersion. This was a fact first recognised by the Ohmi, Hirata and Oide in Ref. [2].

More precisely, any Ripken lattice function that couples the transverse plane with the longitudinal, for example $E^3_{15} = \frac{\partial \langle z_1 z_5 \rangle}{\partial I_3} = \frac{\partial \langle x \delta \rangle}{\partial I_3}$ can be expressed in terms of the longitudinal β, α and γ and the six dispersive vectors H^3_{k1}, H^3_{k2}, H^3_{k3}, H^3_{k4}, $H^3_{k5} = \eta$ and $H^3_{k6} = \zeta$. Expressions which are correct only to leading order in the coupling will involve only components of η and ζ. This is explicitly obvious in the treatment of Ref. [2]. Here it depends on the fact that the generalised Ripken lattice functions form a representation of the complex numbers.

In this book, I will illustrate this with a practical example I lifted from the work of Erlichman et al. in Ref. [5]. They defined the crab crossing angle as

$$\theta_{16} = \frac{1}{2} \tan \left(\frac{2\sigma_{16}}{\sigma_{66} - \sigma_{11}} \right). \tag{4.36}$$

This can be immediately expressed in terms of the Ripken lattice functions:

$$\Theta_{16} = \frac{1}{2} \tan^{-1} (\theta_{16}) = \frac{\sum_{a=1,3} E^a_{16} \varepsilon_a}{\sum_{a=1,3} E^a_{66} \varepsilon_a - E^a_{11} \varepsilon_a} = -\frac{\sum_{a=1,3} K^a_{52} \varepsilon_a}{\sum_{a=1,3} K^a_{55} \varepsilon_a - K^a_{22} \varepsilon_a}. \tag{4.37}$$

In Eq. (4.37) I use the usual lattice function of Ripken. The K^a are the coupled Courant-Snyder invariants in the ath plane.

We can simplify this expression if the longitudinal emittance is small. First we can keep terms which are leading order in the coupling. Using the explicit formula of Eq. (4.26) for the moments, we see that a term like E^2_{16} is second-order in the coupling because it involves three separate degrees of freedom. Furthermore E^3_{16} is first-order and E^3_{66} is zeroth-order. Thus to leading order in the coupling, Eq. (4.37) can be written as:

$$\Theta_{16} = \frac{1}{2} \tan^{-1} (\theta_{16}) = \frac{E^3_{16} \varepsilon_3 + E^1_{16} \varepsilon_1}{E^3_{66} \varepsilon_3 - E^1_{11} \varepsilon_1}. \tag{4.38}$$

We can simplify this further, as in Ref. [5], by assuming that the longitudinal emittance is much bigger, in fact we often have $\varepsilon_3 \gg \varepsilon_1 \gg \varepsilon_2$:

$$\Theta_{16} = \frac{1}{2} \tan^{-1} (\theta_{16}) = \frac{E^3_{16}}{E^3_{66}} = -\frac{K^3_{25}}{K^3_{55}} \tag{4.39}$$

All the expressions above involve the usual Ripken lattice functions. Now we use Eqs. (4.7) and (4.13) to relate K^3 to itself and H^3:

$$B^3 = B^3 H^3 \Rightarrow S^{\mathrm{T}} B^3 = S^{\mathrm{T}} B^3 H^3 \Rightarrow K^3 = K^3 H^3. \tag{4.40}$$

I use Eq. (4.40) on $K^3_{25} (= K^3_{52})$:

$$\begin{aligned}
K^3_{52} &= \sum_{i=1,6} K^3_{5i} H^3_{i2} \\
&= K^3_{51} \underbrace{H^3_{12}}_{0} + K^3_{52} H^3_{22} + \underbrace{K^3_{53} H^3_{32} + K^3_{54} H^3_{42}}_{\text{second order coupling}} + K^3_{55} H^3_{52} + K^3_{56} H^3_{62} \\
K^3_{52} &\approx \frac{1}{1 - H^3_{22}} \left\{ K^3_{55} H^3_{52} + K^3_{56} H^3_{62} \right\} \\
&= K^3_{55} H^3_{52} + K^3_{56} H^3_{62} + (\text{second}-\text{order}).
\end{aligned} \tag{4.41}$$

It is easy to show, in the symplectic case, that

$$H^\dagger = S H S^\dagger \quad \text{and} \quad B^\dagger = S B S \tag{4.42}$$

from which one gets that

$$H^3_{12} = 0 \quad H^3_{52} = -H^3_{16} = -\zeta_1 \quad H^3_{62} = H^3_{15} = \eta_1. \tag{4.43}$$

It then follows that

$$\Theta_{16} = \frac{-K^3_{55} H^3_{52} - K^3_{56} H^3_{62}}{K^3_{55}} = \zeta_1 - \frac{K^3_{56}}{K^3_{55}} \eta_1 = \zeta_1 - \frac{\alpha_3}{\beta_3} \eta_1. \tag{4.44}$$

Equation (4.44) relates the approximate expression for the crab crossing angle in terms of purely longitudinal lattice functions and the adiabatic (small ν_3) coupling functions η and ζ. Whether this is a good idea or not remains to be seen since it is equally easy to compute the Ripken lattice function in Eq. (4.39).

Of course, I must repeat again that all the expressions here do not depend on any particular parametrization of the canonical transformation A unlike the work in Ref. [5] which depends heavily on the extension of Teng-Edwards introduced by Ohmi, Hirata and Oide [2]. Here is the code fragment from Appendix H:

```
if(cas/=0) then
 write(mf,'(a56)') " Almost Exact Crab Angle ignoring transverse emittances "
 write(mf,'((5x,g12.5))') -Ka(5,2,3)/Ka(5,5,3)
 write(mf,'(a56)') " Approximate Crab Angle ignoring transverse emittances  "
 write(mf,'((5x,g12.5))')  Ha(1,6,3)-Ka(5,6,3)*Ha(1,5,3)/Ka(5,5,3)
endif
```

4.1.5 *Kinematic Invariants of Linacs Coming Out of H^i and B^i*

Kinematic invariants are a function of the moments which are invariant under any—I repeat *any*—symplectic matrix M. In the previous section, I discussed time (ergodic) averages or equivalently averages of matched distributions. Thus my formulae depend, even in the ergodic case, on the existence of de Moivre's representation and thus on the normal form.

Kinematics invariants do not dependent on anything beside the assumption that the particles undergo linear symplectic motion. The most famous kinematic invariant is the so-called emittance in 1-d-f:

$$\langle \varepsilon \rangle = 2\sqrt{\langle x^2 \rangle \langle p^2 \rangle - \langle xp \rangle^2}. \tag{4.45}$$

Take any distribution, say 5 particles whose coordinates are decided by throws of the dice, then evaluate the quantity $\langle \varepsilon \rangle$ in Eq. (4.45), then transform the coordinates of the 5 particles by *any* linear symplectic map in 1-d-f: the value of $\langle \varepsilon \rangle$ will remain constant!

It is clear that 5 particles randomly selected are not matched to the invariant ellipse of any one-turn map. Since the invariance of Eq. (4.45) is general, it certainly applies to the case of a distribution which sits on an ellipse. Therefore we ought to be able to derive Eq. (4.45) in the case of a distribution matched to a one-turn matrix.

Plausibility Argument in 1-d-f for the kinetic invariant

The simplest pseudo-derivation uses the relationship between the moment matrix E and the invariant matrix K:

$$\langle \varepsilon \rangle = K_{11} \langle x^2 \rangle + 2K_{12} \langle xp \rangle + K_{22} \langle p^2 \rangle . \tag{4.46}$$

Remembering that

$$K = SES^{\mathrm{T}},$$

I can write the moments of Eq. (4.46) in terms of K,

$$\langle x^2 \rangle = K_{22}\frac{\langle \varepsilon \rangle}{2} \qquad \langle p^2 \rangle = K_{11}\frac{\langle \varepsilon \rangle}{2} \qquad \langle xp \rangle = -K_{12}\frac{\langle \varepsilon \rangle}{2} \tag{4.47}$$

which, if substituted in Eq. (4.46) gives

$$\begin{aligned} \langle \varepsilon \rangle &= \frac{2}{\langle \varepsilon \rangle} \langle x^2 \rangle \langle p^2 \rangle - \frac{4}{\langle \varepsilon \rangle} \langle xp \rangle^2 + \frac{2}{\langle \varepsilon \rangle} \langle x^2 \rangle \langle p^2 \rangle \\ \Rightarrow \langle \varepsilon \rangle^2 &= 4 \langle x^2 \rangle \langle p^2 \rangle - 4 \langle xp \rangle^2 \rightarrow \text{Eq. (4.45)}. \end{aligned} \tag{4.48}$$

The kinetic invariant in N-d-f

The slight of hand pulled for the 1-d-f case does not work beyond two dimensional phase space. However I can use de Moivre's matrices to pull out of our hat one class of the general invariants of Dragt et al. [6]. I first write the following identity for a beam matched to the Courant-Snyder invariants of M:

$$\Sigma = \frac{1}{2}\sum_{k} \langle \varepsilon_k \rangle \, E^k \quad \text{where} \quad \Sigma_{ij} = \langle z_i z_j \rangle . \tag{4.49}$$

I then multiply both sides by $-S$ and use Eq. (4.15):

$$-\Sigma S = \frac{1}{2}\sum_{k} \langle \varepsilon_k \rangle \, B^k . \tag{4.50}$$

I then square Eq. (4.50) and use the property that de Moivre matrices form "N" *orthogonal* representations of the complex numbers:

$$(\Sigma S)^2 = \frac{-1}{2^2}\sum_{k} \langle \varepsilon_k \rangle^2 H^k . \tag{4.51}$$

From the definition of the H^k in Eq. (4.4) and the cyclic property of the trace, I can rediscover Dragt's I^2 invariant (see [6]):

$$Tr\,(\Sigma S)^2 = \frac{-1}{2^2}\sum_{k} \langle \varepsilon_k \rangle^2 \underbrace{Tr(H^k)}_{=2}$$

$$\Rightarrow \; I^2 = \sum_{k} \langle \varepsilon_k \rangle^2 = -2Tr\,(\Sigma S)^2 . \tag{4.52}$$

The result in three degrees of freedom is:

$$\begin{aligned} I^2 = & 4 \sum_{i=1,3} \langle q_i^2 \rangle \langle p_i^2 \rangle - \langle q_i p_i \rangle \\ & + 8\left(\langle q_1 q_2 \rangle \langle p_1 p_2 \rangle - \langle q_1 p_2 \rangle \langle p_1 q_2 \rangle \right) \\ & + 8\left(\langle q_1 q_3 \rangle \langle p_1 p_3 \rangle - \langle q_1 p_3 \rangle \langle p_1 q_3 \rangle \right) \\ & + 8\left(\langle q_2 q_3 \rangle \langle p_2 p_3 \rangle - \langle q_2 p_3 \rangle \langle p_2 q_3 \rangle \right) . \end{aligned} \tag{4.53}$$

This invariant is tested in the code fragment from the program of Appendix G:

```
!!!!!          Evaluation of the Kinetic Invariants          !!!!!!
ray(1,1:6)= (/3,4,3,6,1,3/)
ray(2,1:6)= (/3,5,2,6,1,0/)
```

```
ray(3,1:6)= (/2,4,3,6,1,1/)
ray(4,1:6)= (/3,5,3,6,1,2/)
ray(5,1:6)= (/2,3,5,8,1,0/)

 mat=one_turn_map

do t=1,10


 do k=1,5
   ray(k,1:6)=matmul(mat,ray(k,1:6))
 enddo

do i=1,6
do j=i,6
 mom(i,j)=0.d0
do k=1,5
 mom(i,j)=ray(k,i)*ray(k,j)/5+mom(i,j)
enddo
enddo
enddo

inv=0.D0

do i=1,3
inv=4*(mom(2*i-1, 2*i-1)*mom(2*i, 2*i)-mom(2*i-1, 2*i)**2)+inv
enddo
inv1=inv

inv=inv+8*(mom(1, 3)*mom(2, 4)-mom(1, 4)*mom(2, 3)) &
       +8*(mom(1, 5)*mom(2, 6)-mom(1, 6)*mom(2, 5)) &
       +8*(mom(3, 5)*mom(4, 6)-mom(3, 6)*mom(4, 5))

write(mf,'(a19,i2,a2,1x,g12.5,a18,1x,g12.5)') &
" Invariant at turn ",t,  " =",inv,' 1-d-f invariant =',inv1
enddo
```

Indeed, I chose five rays and applied the matrix `mat` for a total of 10 turns in 3 different cases: with radiation, without radiation, and finally, without radiation and without cavity.

In the radiative case, the invariant slowly shrinks at a rate related to the damping decrements. It is important to realise that the initial distribution is certainly not an ellipse: I typed these rays by hitting my keyboard randomly!

There are two important points to make, one mathematical and one which is physical.

Crucial remarks

1. Mathematically, I started with a map which can be put in normal form. From this, I derived de Moivre matrices H^i and B^i from which the invariant of Eq. (4.52) finally emerges. This invariant is a contraction of moments only: it does not depend on the lattice functions. Therefore it is plausibly a kinematic invariant, independent of the map M itself. But that is not proven here. A clever person could fix my argument but it is simply better to read Ref. [6] to get the full picture.

2. I am concerned with rings primarily in this book. The invariants of Eqs. (4.45) and (4.52), as well as the myriad of invariants of Ref. [6] are best suited to the study of single pass systems such as linacs. Even in a very linear ring, the presence of a tiny amount of tune shift with amplitude will produce a large amount of de-coherence that will effectively damp these kinematic invariants. So beware: they emerge from our studies on rings, but are not necessarily useful in rings!

4.2 No Cavities: Jordan Normal Form

4.2.1 The Reasons for a No-Cavity Normal Form

In the absence of a cavity, the energy is constant: the variable z_5 of the example code PTC is constant. Mathematicians say that in such a case the phase space is "foliated" by the variable z_5. This word comes from the Latin *foliatus* that simply means leafy: Western readers can imagine the French pastry *mille-feuille*[3] which is made of alternating layers of cream (*crème pâtissière*) and puff pastry (*pâte feuilletée*). Each layer is independent and they are all stacked on top of each other.

As the reader can tell, I love geometrical pictures and, occasionally, like the great Francis Ford Coppola, I do not mind throwing good food in my professional life. Each leaf is a representation of the transverse phase space for a given $\delta = z_5$. Of course, as you climb up or down in δ, the changes are continuous as we go from one leaf to the other. Let me be honest: this is more like a *terrine* than a mille-feuille but you get the point!

Now, why do we need to consider a frozen δ? There are two reasons: one physical and one mathematical. There are some machines, such as proton storage rings, where protons just wander around without any longitudinal focussing. For such machines δ is a constant and the system is a mille-feuille. Secondly, as I alluded in Sect. 4.1.4, even in the presence of a cavity, the adiabatic motion of the longitudinal plane preserves the importance and the meaning of quantities like the dispersion, which are well defined only in the *absence* of a cavity. Therefore it makes sense to study the "no-cavity" case even for electron machines which are ultimately damped oscillators; resulting formulae might be simpler for analytic understanding. This is the case of the Sands radiation formalism.

The reader will see that I have a morbid fascination for the dispersion and the "Japanese" ζ function as an adiabatic limit of the full three dimensional oscillator: this was the topic of Refs. [7, 8] and it was covered in my book [9].

[3] Sometimes called Napoleon in the USA.

How did I get interested in this topic? As I pointed out just below Eq. (4.32), accelerator physicists who use dispersion in the presence of cavities are either insane or the concept has some residual meaning in the presence of a cavity. If it does have a meaning, then we should be able to extract it in the presence of cavities with or without radiation: this was explained in Sect. 4.1.4.

The reader can again consult the program of Appendix G: the results are printed in the file result_no_cavity.txt.

4.2.2 A Glance at the Nonlinear Jordan Normal Form

The normal form is similar to that of Eq. (1.7):

$$\begin{aligned} H &= H(J_1, J_2, z_5) \text{ where } J_k = \frac{x_k^2 + p_k^2}{2} \\ x_k(t) &= \cos(\omega_k t)\, x_k(0) + \sin(\omega_k t)\, p_k(0) \\ p_k(t) &= \cos(\omega_k t)\, p_k(0) - \sin(\omega_k t)\, x_k(0) \\ \omega_k &= \frac{\partial H}{\partial J_k}. \end{aligned} \tag{4.54}$$

Equation (4.54) is an amplitude dependent rotation in two degrees of freedom foliated by the variable $z_5 = \delta$. Of course the variable z_5 is a canonical variable and thus we can get the motion of the time z_6 in the usual way:

$$z_6(t) = z_6(0) - \left(\frac{\partial H}{\partial z_5}\right) t. \tag{4.55}$$

4.2.3 The Linear Part with No Cavity

The Jordan normal form is characterised by the presence of a "drift" in the temporal plane.

$$M = ARA^{-1} \tag{4.56}$$

$$R = \begin{pmatrix} r_1 & 0 & 0 \\ 0 & r_2 & 0 \\ 0 & 0 & D \end{pmatrix} \qquad r_i = \begin{pmatrix} \cos\,\mu_i & \sin\,\mu_i \\ -\sin\,\mu_i & \cos\,\mu_i \end{pmatrix}$$

$$D = \begin{pmatrix} 1 & 0 \\ \alpha_s & 1 \end{pmatrix} \tag{4.57}$$

I repeat the discussion found near Eq. (4.31). The matrix A can be factorised into two parts. First a part F which brings the map to the fixed point and then a transformation T which turns the transverse part into two rotations. Thus, I can write A as:

$$A = FT. \tag{4.58}$$

We can compute F by first computing the fixed point:

$$M \begin{pmatrix} f \\ \delta \\ 0 \end{pmatrix} = \underbrace{\begin{pmatrix} N & v & 0 \\ 0 & 1 & 0 \\ w & \alpha & 1 \end{pmatrix}}_{6\times 6 \text{ matrix}} \begin{pmatrix} f \\ \delta \\ 0 \end{pmatrix} = \begin{pmatrix} f \\ \delta \\ \tau \end{pmatrix}. \tag{4.59}$$

Expanding Eq. (4.31) I get an equation for f:

$$Nf + \delta v = f \Rightarrow f = \delta(1-N)^{-1}v \Longrightarrow \eta = (1-N)^{-1}v. \tag{4.60}$$

It is then possible to construct the transverse part of the transformation F:

$$\overline{z}_i = z_i + \eta_i \delta = z_i + \eta_i z_5 \qquad \text{for } i = 1,4. \tag{4.61}$$

Of course I insist in the Hamiltonian case that the map F be symplectic. If I check the Poisson bracket between the time z_6 and any of the transverse variables $\overline{z}_i$, then I can easily see that the time z_6 must depend on the transverse variables. The easiest way to insure a symplectic map is to use a Lie operator for Eq. (4.61). In the linear case, the answer is

$$\mathscr{F} = \exp\left(: -z_5 \sum_{i,j=1,4} \eta_i S_{ij} z_j : \right). \tag{4.62}$$

The map of Eq. (4.62), when acting on the transverse variables, reproduces Eq. (4.61). Of course it also implies a change of the time variable z_6:

$$\begin{aligned} \overline{z}_6 &= \mathscr{F} z_6 \\ &= \exp\left(: -z_5 \sum_{i,j=1,4} \eta_i S_{ij} z_j : \right) z_6 \\ &= z_6 - \sum_{i,j=1,4} \eta_i S_{ij} z_j = z_6 + \widetilde{\eta} \cdot z. \end{aligned} \tag{4.63}$$

I can compute the effect of F on the matrix M as represented by Eq. (4.59):

$$
\begin{aligned}
F^{-1}MF &= \begin{pmatrix} I & -\eta & 0 \\ 0 & 1 & 0 \\ -\tilde{\eta} & 0 & 1 \end{pmatrix} \begin{pmatrix} N & v & 0 \\ 0 & 1 & 0 \\ w & \alpha & 1 \end{pmatrix} \begin{pmatrix} I & \eta & 0 \\ 0 & 1 & 0 \\ \tilde{\eta} & 0 & 1 \end{pmatrix} \\
&= \begin{pmatrix} N & N\eta - \eta + v & 0 \\ 0 & 1 & 0 \\ \tilde{\eta} - \tilde{\eta}N + w & (w - \tilde{\eta}N)\,\eta + \alpha - \tilde{\eta}v & 1 \end{pmatrix} \\
&= \begin{pmatrix} N & 0 & 0 \\ 0 & 1 & 0 \\ 0 & \alpha_s & 1 \end{pmatrix} \quad \text{where} \quad \alpha_s = (w - \tilde{\eta}N)\,\eta + \alpha - \tilde{\eta}v.
\end{aligned}
\tag{4.64}
$$

The expression for the phase slip α_s can be simplified:

$$
N\eta - \eta + v = 0 \;\Rightarrow\; \alpha_s = w\eta + \alpha. \tag{4.65}
$$

In Eqs. (4.64) and (4.65) I did not use the dot product notation because some "vectors" are row matrices, w for example, and some are column objects such as η. Therefore $w\eta$ is really $w \cdot \eta$ if you think in terms of vectors.

The map $F^{-1}MF$ is the matrix around the z_5-dependent closed orbit. On that orbit, it takes a time $T_0 + \alpha_s z_5$ for a particle of differential energy z_5 to go around the machine. T_0 is the total time around the closed orbit.

The rest of the normal form proceeds as before: the 4×4 matrix N is normalised using the techniques discussed in the previous sections, namely in Sect. 4.1.

The program of Appendix G contains the following code fragment:

```
      if(cas==0) then
      write(mf,'(a44)') " Standard time slip without cavity"
      write(mf,'((5x,g12.5))') real(mc(6,5))
        write(mf,'(a21)') " Standard dispersion "
       do i=1,6
        write(mf,*) i, a(i,5)
       enddo

!!!!  Computation by "hand" of the time slip alpha_s
call c_full_canonise(normal_form%A_t,a_cs,a0=a0)             ! (0)
       mat=one_turn_map
       f=a0                                                  ! (1)


! not trusting f(6,1:4) from normal form
       f(6,1:4)=0.0_dp
       do i=1,4
       do j=1,4
        f(6,j)=f(6,j)-f(i,5)*S(i,j)                           ! (2)
       enddo
       enddo

       m(6,1:4) = (mat(6,1:4)-matmul(f(6,1:4),mat(1:4,1:4)))  ! (3a)
       alpha_s=mat(6,5)
       do i=1,4
```

```
  alpha_s =  m(6,i)* f(i,5) - f(6,i)*mat(i,5) + alpha_s    !(3b)
 enddo
 write(mf,*) " time slip alpha_s from computation and normal form"
 write(mf,*)   alpha_s,real(mc(6,5))

endif
```

At line (0), I factorise the full canonical transformation and extract the map `a0` whose linear matrix is `f` at line (1).

At line (2), I extract the vector column `f(1:4,5)` which is just the variable η. In the normal form package, η was computed by the inversion in Eq. (4.60).

The variable $\widetilde{\eta}$ could be extracted as well as `f(6,1:4)`, but I decide here to check my algebra and actually use Eq. (4.63).

Lines (3a,b) are simply a direct evaluation of the time slip using the expression at the end of Eq. (4.64). This is compared with the result of the normal form, the (6, 5) coefficient of the normalised matrix. The agreement is perfect.

4.2.4 The Slip Factor and the Longitudinal Tune

In Sect. 4.1.4 I showed that de Moivre's representation of a three dimensional oscillator leads to an adiabatic definition of the dispersion function as well as the more mysterious ζ function. The converse is also true: if we put a "weak" cavity in a ring, we ought to be able to produce the longitudinal tune in terms of the dispersion and the phase slip.

To get this result, I expand Eq. (3.82)

$$\begin{aligned} \delta^f &= \delta + \frac{VL}{p_0 c} \sin\left(\frac{\omega}{c}\tau\right) \\ &\approx \delta + \frac{VL\omega}{p_0 c^2}\tau. \end{aligned} \tag{4.66}$$

To compute to leading order the longitudinal tune produced by a thin cavity, I first act with the matrix of Eq. (4.59) and follow this with Eq. (4.66). Let us concentrate on the longitudinal planes:

$$\begin{aligned} \tau^f &= \tau + \alpha\delta + w \cdot z \\ \delta^f &= \delta + \frac{VL\omega}{p_0 c^2}\left(\tau + \alpha\delta + w \cdot z\right). \end{aligned} \tag{4.67}$$

The map of Eq. (4.67) is really a six dimensional map involving the transverse variable through the product $w \cdot z$. Suppose we assume that the motion in the transverse plane happens at a much higher frequency than in the longitudinal plane, we can then average z in the transverse phases. The answer for $\langle z \rangle$ is simply $\eta\delta$. Hence upon averaging I can reduce Eq. (4.67) into a solely longitudinal map uncoupled from the transverse plane.

$$
\begin{aligned}
\tau^f &= \tau + \underbrace{(\alpha + w \cdot \eta)}_{\alpha_s}\delta + \\
\delta^f &= \delta + \frac{VL\omega}{p_0 c^2}(\tau + \alpha_s \delta) \\
&\Downarrow \\
\begin{pmatrix} \tau^f \\ \delta^f \end{pmatrix} &= \begin{pmatrix} 1 + \alpha_s \frac{VL\omega}{p_0 c^2} & \frac{VL\omega}{p_0 c^2} \\ \alpha_s & 1 \end{pmatrix} \begin{pmatrix} \tau \\ \delta \end{pmatrix}.
\end{aligned}
\tag{4.68}
$$

The tune can be extracted from this matrix by taking the trace:

$$
\cos(2\pi\nu) = 1 + \alpha_s \frac{VL\omega}{2p_0 c^2} \Rightarrow \nu \approx \frac{\left(V/\sqrt{|V|}\right)}{2\pi c}\left|\frac{L\omega\alpha_s}{p_0}\right|^{1/2}. \tag{4.69}
$$

In Eq. (4.69) I selected a definition of the tune which can be negative but small by introducing the sign of the voltage V. The phase slip is a global property of the cavity-less ring; the longitudinal (or temporal) tune is a global property of a ring with cavities. In the approximation where dispersion is meaningful, say a small longitudinal tune, we expect that the phase slip should be the only thing that enters in the tune computation (besides the parameters of the cavities), and this is what happens.

For larger voltages, the tune will increase, and the approximation based on the phase slip will deteriorate. In most rings, the cavities are located at points of small dispersion in which case the approximation based on the phase slip can remain good even for larger longitudinal tunes.

I can estimate the phase slip in the cases with a cavity (with or without radiation) by multiplying the de Moivre matrix component B^3_{65} with $\sin(\mu_3)$. The agreement is excellent and can be checked in the output file "result_no_cavity.txt" of the program of Appendix G. Moreover I also compare the tune computed by Eq. (4.69) with the exact tune obtained from the full six dimensional normal form.

The agreement for the tune is "beyond" excellent. Why is that? It is because the cavity is thin and located at a point of small dispersion. For zero dispersion, the expression of Eq. (4.69) would be exact for a thin cavity. The reader can move the cavity near a bend in the lattice description of Appendix A. The results are still very good but not perfect.[4]

In Sect. 3.3, I mentioned that the standard map is a good approximation of the pendulum. There are two interesting levels of approximation:

1. The cavity map, even when there is coupling with the transverse, can be reduced to the standard map. The phase slip, a global parameter of the cavity-free system, is the only quantity[5] which enters in the model.

---

[4] If you do so, make sure for the sake of Chap. 8 that you keep a copy of the original lattice, markers included.

[5] Nonlinear components of the phase slip may enter as well.

2. The phase slip is the only thing entering precisely because the resulting synchrotron tune is small. When it is small, only quantities which are averages of the transverse plane enters in the physics. For example, we have seen in Sect. 4.1.4, Eq. (4.44), that in leading order it is possible to express effects between the transverse and longitudinal planes in terms of the adiabatic η-ζ and the plain uncoupled longitudinal lattice functions.

In summary, in the small temporal tune limit, the motion in the longitudinal plane is a pendulum. It affects the transverse plane, in leading order, through the adiabatic η and ζ.

4.3 Normal Form for the "AC" Fluctuation of a Magnet Property

In an accelerator it is always possible to imagine that some properties of a magnet are changed slowly with time. In LEP at CERN, the electron machine which was in the LHC tunnel, the size of the machine was so big as to be affected by the periodic tidal effect of the Moon!

On a more serious basis, it is common to excite the beam by applying a periodic change to a magnet; for example an alternating current to a dipole. Some work using analytical normal forms has been done and published by Rogelio Tomàs [10] for example. In his paper, Tomàs uses Lie methods to derive expressions for the effects one should expect from an AC dipole.

I am not against analytical calculation and I did sprinkle them throughout this book. But, let us remember that this book is based on the claim that the "code" rules. Hence we must ask two questions in relation to the code:

1. Can the code simulate AC modulations of some ring property?
2. Can this be also part of a Taylor map extracted from the code?

The answer to the first question is a trivial "yes." In fact one needs only to modulate the magnets during the tracking.

The second question is more tricky. We would like to be able to introduce the amplitude of modulation as a parameter which can be part of the Taylor series and thus part of the normalisation procedure. This does not follow trivially from our obvious ability to change ring properties as a function of time. It is in a way, a discussion of PTC, but in general it is a discussion of "how to avoid reading Tomàs' paper" and trivially do his calculations within the code and the code alone.

It is best to state "right off the bat" that I will use a pseudo-time: a common average time shared by the particles within a bunch that is related to an ideal or fiducial distance along the ring. In the language of accelerator physicists, I state that I will confuse s and t. Now this could be fixed, but fixes would be inside PTC, a topic for another day…

4.3.1 Adding Pseudo Clocks to Phase Space

It is best to give a simple 1-d-f example to explain the idea. Let us assume that the ring is made of two elements which are a drift and a thin quadrupole. I assume that the quadrupole has a small ripple Δk_L:

$$\begin{aligned} q_1(L) &= q_1(0) + L\ p_1(0) \\ p_1(L) &= p_1(0) - L\{k + \Delta k_L\}\, q_1(L). \end{aligned} \tag{4.70}$$

The simple example of Eq. (4.70) represents a one-turn map. In general the form of Δk_s is arbitrary. In such cases it is best to use tables to represent Δk_s for large values of s, say $0 \leq s \leq NL$ where N is a large integer. These implementations are useful during acceleration or ramping processes where various properties of the lattice are changed dynamically including RF cavities, quadrupoles, bends, etc.

Long-term processes are not trivially amenable to normal form analysis although, thanks to the adiabatic invariance of the actions, one can under certain circumstances relate tracking results to theory.

Here I am more modest: I want to submit the ring to a simple periodic modulation. For example, I can construct a code in which the quadrupole kick represented by Δk_s has a simple sinusoidal shape:

$$\Delta k_s = \kappa_0 \sin(\omega s)\,. \tag{4.71}$$

Besides brute force tracking, is there a way to bring the power of perturbation theory to this problem without exiting the Code? In other words can we produce a Taylor map that has the information of Eq. (4.71) and normalise it without additional theory and effort? Another way to put it, and no offence to Dr. Tomàs, is there a way to avoid reading his paper [10]? The answer is of course yes! Now, you are probably bored to death, looking at the arms of the clock, and saying to yourself: "I have better things to do with my time, I give up!" Well this is exactly what happened to me when I was asked to scan Tomàs' paper. I looked at the clock and suddenly realised that the old clock on the wall was my solution. The "hand" marking the seconds shook me and presented the solution: the modulation amplitude is the projection of the hand of a rotating clock!

To be more precise, consider the following map, which is an extension of Eq. (4.70):

$$\begin{aligned} q_1(L) &= q_1(0) + L\ p_1(0) \\ p_1(L) &= p_1(0) - L\{k + \kappa_0 q_2(L)\}\, q_1(L) \\ \begin{pmatrix} q_2(L) \\ p_2(L) \end{pmatrix} &= \begin{pmatrix} \cos(\omega L) & \sin(\omega L) \\ -\sin(\omega L) & \cos(\omega L) \end{pmatrix} \begin{pmatrix} q_2(0) \\ p_2(0) \end{pmatrix}. \end{aligned} \tag{4.72}$$

If the reader iterates Eq. (4.72) with the initial condition $(q_2(0), p_2(0)) = (0, 1)$, then at some "time" nL, the value of the ripple will be:

$$\Delta k_{nL} = \kappa_0 \sin(\omega n L). \tag{4.73}$$

Equation (4.73) shows us that our little extension of phase space works: it produces the right quadupolar ripple. It is evidently less powerful, simulation wise, than a table of the ripples as a function of s. However, when acceptable to our physical needs, it completely renders the paper of Tomàs useless[6] as far as the "code" is concerned. This was our goal. I will explain why in case that it is not self evident.

The original map of Eq. (4.70) is a 1-d-f map that is symplectic but *not periodic*. The map of Eq. (4.72) is a 2-d-f non-symplectic *periodic* map.

The reader will see that the extended map is immediately normalisable. Since the package in FPP handles non-symplectic maps, such as maps with radiation, it is able to deal with this mildly nonsympletic system. To achieve this, I did complicate the map by adding one dimension. *And this is it!*

4.3.2 A Simple Calculation with FPP

The reader will notice that I am pretty short on theory and immediately jump into an example. The reasons for this are twofold:

1. I am emphasizing the code. The point here is that a simple extension of phase space gives us immediate access to the normal form results.
2. Additionally, the extension of phase phase is a trivial rotation…There is not much to say about it except in the case of the Jordan normal form of Sect. 4.2 where things are a little subtle in the linear normalisation. This issue is addressed in Sect. 4.3.4.

The simple example I took is also nonlinear despite the appearances. Indeed the position variable q_1 multiplies the clock variable q_2 in Eq. (4.72).

Moreover, unless the clock resonates with the plane $q_1 - p_1$, the average effect will be zero in first order perturbation theory. Therefore, no matter how one looks at the example, it is a nonlinear example.

Nevertheless I will quote the analytical result for the tune shift to leading order in the modulation. Notice that the amplitude in the clock space, i.e., the length of the arms, is "one" and thus the strength of the modulation is carried by the quantity κ_0:

$$\Delta\mu = -\frac{(\beta\kappa_0 L)^2}{32}\left(\frac{\sin(2\mu + \omega L)}{1 - \cos(2\mu + \omega L)} + \frac{\sin(2\mu - \omega L)}{1 - \cos(2\mu - \omega L)}\right). \tag{4.74}$$

---

[6]The paper of Tomàs is not useless, I am obviously exaggerating. The code is still controlled by users who desire once in a while some analytic understanding! In fact in Sect. 5.3 I perform a calculation which parallels Tomàs' calculation apparently guilty myself of "uselessness."

This result will be derived in Sect. 5.3 as an example of a second order normal form in two degrees of freedom. The calculation for the tune shift due to a sextupole would be similar though a little more complicated.

The simple map of Eq. (4.72) can be immediately computed by any TPSA library and analysed by a normal form package. This is done in the first part of the program in Appendix I.

```
mu_x=acos(1.d0-Kq*L**2/2)                              !  (2a)
dmu=-2*(sin(2*mu_x+mu_mod)/(1-cos(2*mu_x+mu_mod)) & !  (2b)
+sin(2*mu_x-mu_mod)/(1-cos(2*mu_x-mu_mod)))*4*(-beta*k0*L/16)**2

write(mf,*);
write(mf,*) "  Analytical tune in radians =       ",mu_x
write(mf,*) "  Analytical tune shift in radians = ",dmu
write(mf,*);

drift_map=1
quad_map=1

drift_map%v(1)=q(1)+L*q(2)     ! (3) drift
drift_map%v(2)=q(2)
drift_map%v(3)=cos(mu_mod)*q(3) + sin(mu_mod)*q(4)
drift_map%v(4)=cos(mu_mod)*q(4) - sin(mu_mod)*q(3)

quad_map%v(1)=q(1)
quad_map%v(2)=q(2)-L*(Kq+k0*q(3))*q(1) ! (4) quadrupole

! Map of system is made
one_turn_map=quad_map*drift_map    ! total map    (5)
```

Here `drift_map` and `quad_map` are simple Taylor maps of type c_damap. The array `q(1:4)` contains the four monomials z_1, z_2, z_3 and z_4. The last two components of `q`, `q(3)` and `q(4)` are the hands of the clock.

On line (5), the total map is computed. It is then normalised on line (6) and the tunes are printed on line (7, 8).

```
! Map is normalised
call  c_normal(one_turn_map,normal_form)       !  (6)
write(mf,*);
write(mf,*) " Result from the normal form algorithm for hardwired map ";
write(mf,*);

write(mf,*) " Normal form result for tune in radians =        ", &  !  (7)
 -aimag(normal_form%ker%f(1)%v(1).sub.'1000')
write(mf,*) " Normal form result for tune shift in radians = ", &  !  (8)
 -aimag(normal_form%ker%f(3)%v(1).sub.'1011')
```

As expected, the result of line (8) agrees with line (2b) which is the implementation of Eq. (4.74). These results are found in file "result_with_modulation.txt":

```
Analytical tune in radians =        3.16240943656276169E-002
Analytical tune shift in radians =    5.56450104233673491E-005
```

```
Result from the normal form algorithm for hardwired map

Normal form result for tune in radians =        3.16240943656293585E-002
Normal form result for tune shift in radians =  5.56450104233705543E-005
```

4.3.3 *The Way It Can Be Done in a Code: PTC*

How do I extend the phase space in a code? We return to the definition of `probe_8` in Sect. 4.4.3, but this time I show a little more:

```
type probe_8
 type(real_8) x(6) ! polymorphic orbital ray
 type(rf_phasor_8) ac ! modulation of magnet
 real(dp) e_ij(6,6) ! envelope for stochastic radiation
...
end type probe_8
```

Of course there exists a similar object for real number tracking as I already mentionned in Sect. 4.4.3:

```
type probe
   real(dp) x(6)
   type(rf_phasor) ac
    ...
end type probe
```

The type rf_phasor and its polymorphic equivalent rf_phasor_8 contain all the clock information. It suffices to show rf_phasor_8 since the real version is obtained by replacing real_8 by the double precision real(dp):

```
type rf_phasor_8
   type(real_8)  x(2)  ! The two hands of the clock
   type(real_8) om     ! the omega of the modulation
   real(dp) t          ! the pseudo-time
end type rf_phasor_8
```

Let us look at the example of Appendix I once more starting below line (8). A lattice is created which consists of two elements: a drift L1 followed by a thin quadrupole QF1.

```
! Same calculation from with PTC
! using the following little lattice
!  Qf1 = QUADRUPOLE(" QF1 ",L=0.d0, K1= 0.01d0 ); L1  = drift("L1 ",0.1d0);
!  ALS=L1+QF1;
call build_lattice_als(ALS,mis,exact=.false.,sl=.true.)
```

This "lattice" is set up to reproduce exactly the map of Sect. 4.3.2. The modulation amplitude is assigned to the code by reading a file called "AC_modulation.txt." In my example, the file "AC_modulation.txt" is actually created in the main program and read immediately after its creation as seen here:

```
!!!! AC_modulate.txt sets the magnet QF1 as a modulated magnet !!!!
 call kanalnummer(mf1,file="AC_modulation.txt")
  write(mf1,*) "select layout"
  write(mf1,*) 1
  write(mf1,*) " MODULATE"
  write(mf1,*) " QF1"
  write(mf1,*) "1.d0 0 0        !DC_ac,A_ac,theta_ac"
  write(mf1,*) "1.d0   2        ! D_ac,n_ac  "
  write(mf1,*) "2 0.02d0 0       ! n d_bn(n) d_an(n)   "  ! (A)
  write(mf1,*) "0   0 0 "
  write(mf1,*) " return "
 close(mf1)
 call read_ptc_command77("AC_modulation.txt")
```

Line (A) contains the actual amplitude of the modulation: `d_bn(2)=0.02d0`.

As the object `ray` of type probe_8 propagates through the lattice, its pseudo-time ray%t is advanced according to an ideal time corresponding to some fiducial orbit. Actually this ideal time can be modified by the user/programmer: absolute time, like physical models, is what Big Brother says it is! When the quadrupole kick QF1 is encountered, a kick is performed which depends on the value of the pseudo-time at that position and the frequency of the clock. To be more precise, in the case of the example code PTC, the quantity `d_bn(2)=0.02d0` is multiplied with `ray%ac%x(1)` and added to the static quadrupole component `bn(2)` of QF1.

The initial hands of the clock, located in `ray%ac%x(1:2)`, are set to be the identity map in the third and fourth dimension. In fact, their initial values are printed in line (F2) of Appendix I; the result is just:

```
 Initial value of the clock ( in type probe_8)

 AC INFORMATION: omega, pseudo-time, hands of the clock
  7.7565922617131999
  0.0000000000000000

Properties, NO =     3, NV =     4, INA =    48
*********************************************

  1   1.000000000000000       0  0  1  0

Properties, NO =     3, NV =     4, INA =    49
*********************************************

  1   1.000000000000000       0  0  0  1
```

Needless to say, the results of tracking a clock with our "code" agree to machine precision with the simple map of Eq. (4.72) and are displayed in file "result_with_modulation.txt."

4.3.4 A Linear Complication Due to the Case of Sect. 4.2

The discussion of this section is illustrated in the program of Appendix J which I explain at the end of this section. The entire section explore the problems in the linear normal form that results from joining the Jordan normal form of Sect. 4.2 with magnet modulations.

With Taylor based numerical methods, the nonlinear algorithm is a universal meat grinder, but the linear part needs special attention. This was obvious in the Jordan normal form of Sect. 4.3.4 and it shows its ugly head again when we mix the Jordan case with the non-symplectic clocks used in modulation.

In Sect. 4.2 I discussed the normal form used when the energy variable $z_5 = \delta$ is constant and the time-like variable z_6 is computed in the Taylor map. The normal form in this case is the so-called Jordan normal form. It is a product of two rotations for the transverse plane and a drift-like map in the longitudinal plane. Thus, in the presence of a single clock, the one-turn matrix M of Eq. (4.59) is

$$M = \underbrace{\begin{pmatrix} N & v & 0 & t & d \\ 0 & 1 & 0 & 0 & 0 \\ w & \alpha & 1 & a & b \\ 0 & 0 & 0 & \cos(\omega L) & \sin(\omega L) \\ 0 & 0 & 0 & -\sin(\omega L) & \cos(\omega L) \end{pmatrix}}_{8\times 8 \text{ matrix}} . \tag{4.75}$$

In Eq. (4.75), N is the 4×4 transverse matrix which is normalisable into two rotations. v is a column vector of four components which is responsible for the dispersion (see near Eq. (4.60)). The row vector w quantifies the dependence of time on the transverse variables: it is related to v via the symplectic condition. t and d are column vectors quantifying the effect of the modulation on the transverse phase space. a and b are two entries describing the effect of the modulation on the time variable z_6.

In Eq. (4.75), the normal phase space variables are indexed from 1 to 6. The hands of the modulated clock are at positions 7 and 8. It is mathematically and algorithmically simpler to reorder the planes as follows:

$$M' = \underbrace{\begin{pmatrix} N & t & d & v & 0 \\ 0 & \cos(\omega L) & \sin(\omega L) & 0 & 0 \\ 0 & -\sin(\omega L) & \cos(\omega L) & 0 & 0 \\ 0 & 0 & 0 & 1 & 0 \\ w & a & b & \alpha & 1 \end{pmatrix}}_{8\times 8 \text{ matrix}} . \tag{4.76}$$

In Eq. (4.76) the harmonic planes are indexed from 1 to 6. The Jordan drift-like plane appears last in position 7 and 8. This temporary realignment is actually done in the entrails of the normalising code FPP; it is a sensible thing to do.

As I explained before, the map can be normalised by the product $A' = F'T'$ as in Eq. (4.58). The map F' brings M' to the energy dependent fixed point as in Eq. (4.64):

$$F'^{-1}M'F' = \begin{pmatrix} N & t & d & 0 & 0 \\ 0 & \cos(\omega L) & \sin(\omega L) & 0 & 0 \\ 0 & -\sin(\omega L) & \cos(\omega L) & 0 & 0 \\ 0 & 0 & 0 & 1 & 0 \\ 0 & \widetilde{a} & \widetilde{b} & \alpha_s & 1 \end{pmatrix}. \tag{4.77}$$

The symplectic condition applied to F' insures that the removal of w happens automatically with the removal of v in Eq. (4.76). A matrix T' can be used to diagonalise the 6×6 "harmonic" block. However the time-modulation coupling entries $\widetilde{a}$ and $\widetilde{b}$ *will not vanish automatically*:

$$T'^{-1}F'^{-1}M'F'T' = \begin{pmatrix} R & 0 & 0 & 0 & 0 \\ 0 & \cos(\omega L) & \sin(\omega L) & 0 & 0 \\ 0 & -\sin(\omega L) & \cos(\omega L) & 0 & 0 \\ 0 & 0 & 0 & 1 & 0 \\ 0 & \widetilde{a} & \widetilde{b} & \alpha_s & 1 \end{pmatrix}$$

$$\text{where } R = \begin{pmatrix} \cos(\mu_1) & \sin(\mu_1) & 0 & 0 \\ -\sin(\mu_1) & \cos(\mu_1) & 0 & 0 \\ 0 & 0 & \cos(\mu_2) & \sin(\mu_2) \\ 0 & 0 & -\sin(\mu_2) & \cos(\mu_2) \end{pmatrix}. \tag{4.78}$$

To achieve complete normalisation, we must augment T' by a part whose sole "non-symplectic purpose" is the removal of $\widetilde{a}$ and $\widetilde{b}$. As usual, it is easier to go into the phasors basis as in

$$C'^{-1}T'^{-1}F'^{-1}M'F'T'C' = \begin{pmatrix} \Lambda & 0 & 0 & 0 & 0 \\ 0 & \exp(-\omega L i) & 0 & 0 & 0 \\ 0 & 0 & \exp(\omega L i) & 0 & 0 \\ 0 & 0 & 0 & 1 & 0 \\ 0 & \frac{1}{2}(\widetilde{a} - i\widetilde{b}) & \frac{1}{2}(\widetilde{a} + i\widetilde{b}) & \alpha_s & 1 \end{pmatrix}$$

$$\text{where } \Lambda = \begin{pmatrix} \exp(-i\mu_1) & 0 & 0 & 0 \\ 0 & \exp(i\mu_1) & 0 & 0 \\ 0 & 0 & \exp(-i\mu_2) & 0 \\ 0 & 0 & 0 & \exp(i\mu_2) \end{pmatrix}. \tag{4.79}$$

In this basis, I apply a matrix S' that will remove the offensive terms:

$$\Lambda' = S'^{-1}C'^{\,-1}T'^{-1}F'^{-1}M'F'T'C'S'. \tag{4.80}$$

The matrix S' differs from the identity by only two terms:

$$S'_{85} = \frac{\widetilde{a} - i\widetilde{b}}{\exp(-i\omega L) - 1} \quad \text{and} \quad S'_{86} = \frac{\widetilde{a} + i\widetilde{b}}{\exp(i\omega L) - 1}. \tag{4.81}$$

If I go back to the original indices, the equivalent matrix S has the following non-zero coefficients:

$$S_{67} = S'_{85} \quad \text{and} \quad S_{68} = S'_{86} \tag{4.82}$$

I can flesh out this discussion using the example program of Appendix J. Here is a code fragment:

```
!!!! AC_modulate.txt sets the magnet QF1 as a modulated magnet !!!!
 call kanalnummer(mf1,file="AC_modulation.txt")
  write(mf1,*) "select layout"
  write(mf1,*) 1
  write(mf1,*) " MODULATE"
  write(mf1,*) " BEND1"
  write(mf1,*) "1.d0 0 0         !DC_ac,A_ac,theta_ac"
  write(mf1,*) "1.d0   1         ! D_ac,n_ac  "
  write(mf1,*) "1 0.001d0 0        ! n d_bn(n) d_an(n)   "   ! (A)
  write(mf1,*) "0  0 0 "
  write(mf1,*) " return "
 close(mf1)
 call read_ptc_command77("AC_modulation.txt")
                                .
                                .
                                .
                                .
do_linear_ac_longitudinal=.false.              ! (I)
call  c_normal(one_turn_map,normal_form)       ! (J1)
id=normal_form%a_t*from_phasor()
do_linear_ac_longitudinal=.true.
call  c_normal(one_turn_map,normal_form)       ! (J2)
normal_form%a_t=normal_form%a_t*from_phasor()

write(mfmap,*);write(mfmap,*);
write(mfmap,*) " Correct A (M=ARA^-1) from the algorithm for the code ";
write(mfmap,*);write(mfmap,*);

call  print(normal_form%a_t,mfmap,prec)

diagonal=normal_form%a_t**(-1)*one_turn_map*normal_form%a_t     ! (K1)

quasi_diagonal=id**(-1)*one_turn_map*id                         ! (K2)

g1=quasi_diagonal%v(6).sub.'00000010'                           ! (L1)
g2=quasi_diagonal%v(6).sub.'00000001'                           ! (L2)

al1=-g1/(1.d0-(quasi_diagonal%v(7).sub.'00000010'))             ! (L3)
```

```
al2=-g2/(1.d0-(quasi_diagonal%v(8).sub.'00000001'))          ! (L4)

a_ac=1

a_ac%v(6)=a_ac%v(6)+(al1.cmono.'00000010')+(al2.cmono.'00000001') ! (L5)
```

The first thing to realise is that I modified the input from Sect. 4.3.3 from a quadrupole modulation to a dipole modulation. Indeed in an ideal machine, as I said in Sect. 4.3.2, the modulated quadrupole is a *nonlinear problem* in the clock variables. Therefore the problems with the Jordan normal form will not appear. Here line (A) induces a dipole component modulation on the magnet `QF1`:

The code PTC is equipped with a "pedagogical flag," `do_linear_ac_longitudinal` which is defaulted to true. At line (I) it is set to false and the calculation of the matrix S is disabled, allowing us to do it in the main program. This emasculated normal form is done at line (J1) while the correct normal form is performed at (J2) for comparison.

Lines (L1,2,3,4) correspond to Eq. (4.81) except that the indices are those of Eq. (4.82). As I said previously, in the package FPP, the temporary exchange of planes shown in Eq. (4.76) is internally performed. The library FPP can handle an arbitrary number of clocks however the example code PTC is limited to one clock—one frequency of modulation—clearly not a fundamental restriction.

The results are printed in file `maps.txt`.

4.4 The Stochastic Normal Form: Radiation Theory

N.B. I will not delve into the analytical formulae connected with this section on radiation. The interested reader is invited to look at standard references for example Ref. [11], *Part III, Chap. 8. Here we return to the code as the supreme arbiter of truth and look at methods suited for the code.*

The theory of so-called synchrotron integrals is an attempt to find the equilibrium beam sizes in an electron machine. This effect can be easily understood as an "epic combat" between two competing effects: classical radiation and stochastic photon emission.

Under classical radiation an electron continuously emits a small amount of radiation and thus loses energy. It regains the energy, on average and approximately, at the cavity. On the closed orbit, it regains exactly the lost energy at the various cavities located around the ring.

In the absence of radiation, particles go around and around. We know that in the linear regime this motion can be normalised into three circles (tori) on which the particle orbits with a given tune on each torus. When we turn on classical radiation, does the closed orbit survive? The answer to this is contained in pure mathematics:

the inverse function theorem. The fixed point, that is to say the closed orbit, continues to exist for small finite amount of radiation. Therefore we can ask what happens to the tunes: since the map is no longer symplectic, the tunes are allowed to wander away from the unit circle. In stable accelerators, the tunes move inside the unit circle and thus we have a sink.

In the absence of anything else, all particles should collapse on the closed orbit and this is the end of the game. However, we have quantum mechanics: radiation is emitted in quanta that is to say photons. Thus the classical case is only an average behaviour. Therefore we can view the emission of radiation as a two "terms" process over each step of integration of our tracking code:

$$\frac{dz_5}{ds} = r(z) + \frac{\Delta(z)}{\sqrt{ds}}. \tag{4.83}$$

The first term $r(z)$ produces classical radiation: it depends on all the phase space variables and it is deterministic. The second term is a stochastic noise whose average is zero but its variance is finite. Thus we have

$$\langle \Delta \rangle = 0 \quad \text{and} \quad \left\langle \Delta^2 \right\rangle \neq 0. \tag{4.84}$$

I will now describe, in the linear regime only, the type of calculation we can perform to estimate the effect of Eq. (4.84) on a beam described by its quadratic moments.

4.4.1 The Stochastic Map of Moments

A beam around the closed orbit, in the linear regime, does not have any linear moments: the average phase position $\langle z \rangle$ vanishes. In fact in the linear regime I simply assume that $\langle z \rangle$ obeys the deterministic equation where radiation only enters through the term $r(z)$ of Eq. (4.83).

To proceed further, I must compute the effect of the stochastic part over an arbitrary integration step assuming that the map from some initial $s = 0$ to the present value s is known and linear. Obviously this can be computed for any orbit including the closed orbit precomputed with the deterministic term in the equation of motion.

When a photon is emitted, in the ultra relativistic limit, only the longitudinal momentum is affected. Thus, if, $z_2 = x' = p_x$ and $z_4 = y' = p_y$, then only z_5 is changed (actually not always true, see comments below Eq. (4.87)). Therefore the stochastic changes of the moment matrix $\Sigma_{ij} = \left\langle z_i z_j \right\rangle$ are given by

$$d\Sigma_{ij} = 0 \quad \text{except for} \quad d\Sigma_{55} = \left\langle \Delta^2 \right\rangle ds. \tag{4.85}$$

The stochastic effect is equivalent to a translation[7] of the moment matrix Σ at position s. Let us see the effect of a single stochastic kick on the one-turn map for the quadratic moments:

$$\begin{aligned}\Sigma^f &= M^{sC}\left(\Sigma^s + \Delta\Sigma^s\right)M^{sC^\dagger}\\ &= M^{0C}\left(\Sigma^0 + \underbrace{M^{0s^{-1}}\Delta\Sigma^s M^{0s^{-1^\dagger}}}_{\Delta E^s}\right)M^{0C^\dagger}\\ &= M^{0C}\left(\Sigma^0 + \Delta E^s\right)M^{0C^\dagger}. \end{aligned} \tag{4.86}$$

From the structure of Eq. (4.86), I can deduce that the total map, summing up all the stochastic kicks ΔE^s, is the simple integral of Eq. (4.86).

$$\Sigma^f = M^{0C}\left(\Sigma^0 + E^{0C}\right)M^{0C^\dagger} \quad \text{where} \quad E^{0C} = \oint_0^C M^{0s^{-1}}\Delta\Sigma^s M^{0s^{-1^\dagger}}. \tag{4.87}$$

In Eq. (4.87) it assumed that only the energy z_5 is changed by the classical and the quantum fluctuation. This is true if *non* canonical variables are used. In the example code PTC, canonical variables are used. Therefore, during integration using my polymorphic type, it is natural to extract the matrix M^{0s} connecting initial canonical variables to the canonical variables at position s. Without going into details, I will simply say this: if one expresses the non-canonical set x' and y' at s in terms of the original canonical variables, then one can extract the proper matrix M^{0s} which makes formula (4.87) work in all cases.

4.4.2 *The Normal Form of the Quadratic Moment Map*

The "combat" between classical and quantum effects must lead to an equilibrium set of moments. Like the gladiators Verus and Priscus, during the inaugural games at the Coliseum, both emerge equal[8] and victorious! Dropping unnecessary superscripts in Eq. (4.87), the equilibrium moment matrix is given by

$$\Sigma^f = M\left(\Sigma^0 + E\right)M^\dagger \quad \Rightarrow \quad \Sigma^\infty = M\left(\Sigma^\infty + E\right)M^\dagger. \tag{4.88}$$

Before I solve for the equilibrium moments using the elegance of our normal forms, it is worth pointing out that the numbers can be obtained by raising the map of Eq. (4.87) to a large power. Indeed if I assume that an arbitrary Σ can be written as $\Sigma = \Sigma^\infty + \Delta\Sigma$, then it is easy to show that after n turns, the moments are

[7] Translation, like canonical and propaganda, is also a word we owe to the Church. For example, one talks of the "translation" of the relics of St-Nicholas from Myra to Bari!

[8] *Contigit hoc nullo nisi te sub principe, Caesar: cum duo pugnarent, uictor uterque fuit.* From Martial's Liber de Spectaculis.

$$\Sigma^n = \Sigma^\infty + \Delta\Sigma^n = \Sigma^\infty + M^n \, \Delta\Sigma \, M^{\dagger^n}. \tag{4.89}$$

The equation for $\Delta\Sigma$ only involves the deterministic part. Since the eigenvalues of M are below the unit circle, $\Delta\Sigma^n$ tends towards zero as the number of iterations increases.

Now I will perform the calculation for Σ^∞ using the tools of normal form. I apply the normal form of Eq. (4.2) and proceed further, as in Eq. (2.53), to completely diagonalise M using phasors.

$$\begin{aligned} \Lambda &= B^{-1} M B \\ &\Downarrow \text{ for i} = 1, 3 \\ \Lambda_{2i-1\;2i-1} &= \exp\left(-\alpha_i - i\mu_i\right) \\ \Lambda_{2i\;2i} &= \exp\left(-\alpha_i + i\mu_i\right) \;. \end{aligned} \tag{4.90}$$

The matrix $B = CA$ is the product of the phasors' matrix C which diagonalises a rotation and the matrix A which turns M into a dilation: a rotation which shrinks (see Eq. (4.2)). I then apply B on Eq. (4.88):

$$\begin{aligned} B^{-1}\Sigma^\infty B^{\dagger-1} &= B^{-1} M B \left(B^{-1}\Sigma^\infty B^{-1\dagger} + B^{-1} E B^{-1\dagger}\right) B^\dagger M^\dagger B^{\dagger-1} \\ \sigma^\infty &= \Lambda \left(\sigma^\infty + e\right) \Lambda. \end{aligned} \tag{4.91}$$

Equation (4.91) is diagonal; I can solve it for each moment separately:

$$\sigma_{ij}^\infty = \Lambda_{ii} \left(\sigma_{ij}^\infty + e_{ij}\right) \Lambda_{jj} \;\Rightarrow\; \sigma_{ij}^\infty = \frac{e_{ij}}{1 - \Lambda_{ii}\Lambda_{jj}}. \tag{4.92}$$

The beam sizes in the original space are regained by a simple application of the matrix B:

$$\Sigma^\infty = B \sigma^\infty B^\dagger. \tag{4.93}$$

It is worthwhile to examine three very special denominators in Eq. (4.92) in the limit of small damping decrements. They dominate because they produce huge contributions to the equilibrium moments.

$$\begin{aligned} 1 - \Lambda_{2k-1\;2k-1}\Lambda_{2k\;2k} &= 1 - \exp\left(-\alpha_k - i\mu_k\right) \exp\left(-\alpha_k + i\mu_k\right) \\ &\approx 2\alpha_k + O(\alpha_k^2). \end{aligned} \tag{4.94}$$

From Eq. (4.94) I deduce that the equilibrium values for these three special phasors are approximately:

$$\sigma_{2k-1\,2k}^\infty = \frac{e_{2k-1\,2k}}{1 - \exp\left(-2\alpha_k\right)} \approx \frac{e_{2k-1\,2k}}{2\alpha_k} \qquad k = 1, 2, 3. \tag{4.95}$$

The three quantities $\frac{\sigma^{\infty}_{2k-1\,2k}}{2}$ are the so-called equilibrium emittances. The equilibrium beam sizes can be computed by inserting the emittances of Eq. (4.95) into Eq. (4.26), remembering that $\langle J_k \rangle = \sigma^{\infty}_{2k-1\,2k}/2$.

The theory presented here is an exact theory for the quadratic moments of a linear map. The computation of a stochastic shift of the moments, the matrix E of Eq. (4.88), is necessary for the computation of the equilibrium moments. However, when the damping decrements α_k's are small, as they are in a real accelerator, the matrix E is mostly made of three terms, the diagonal terms $e_{2k-1\,2k}$ of Eq. (4.95). These three terms are, to leading order in α_k, the fluctuation of the symplectic quadratic invariants: the so-called "H" functions.

The theory based directly on the fluctuation of these three quadratic invariants is due to Chao [3]. On the other hand, the most common theory shown to accelerator physicists is due to Sands which, in addition to small damping, assumes a small coupling to the longitudinal plane or, equivalently, a small longitudinal tune. (See Ref. [11] Eqs. (8.52) and (8.53).)

Sands' theory is equivalent to Chao's if one uses the lattice functions of the multidimensional de Moivre formulae of Sect. 4.1.2. In particular this requires a careful definition of the dispersion; one must use a slight modification but physicially sensible modification of Eq. (4.34), namely $\eta_i = \langle z_i \rangle / \langle z_5 \rangle = H^3_{i5}/H^3_{55}$. See Ref. [8] if you are interested in the connection between Sands' and Chao's theories of radiation integrals.

4.4.3 *Numerical Example*

I will now examine the program `radiation_map` of Appendix H, again based on the code PTC. This example loops over two cases represented by the variables `cas =1,2`. The case `cas=1`, a plain "no radiation" situation, will give us the symplectic de Moivre matrices which enter in a Chao-like calculation where only the emittances of Eq. (4.95) are used. Then, with `cas=2`, the program will compute the fluctuation matrix E and find the equilibrium beam sizes as well as the equilibrium emittances of Chao.

The numerical calculation requires the ability to track a more complex object: the ray and the stochastic kick E of Eq. (4.87) from any two points in the ring. As I stated in Sect. 1.7, I would explain new *propagata*, beyond phase space rays when necessary. Well the need arose here!

In PTC, there is an object called a probe_8 which contains the polymorphic rays, the stochastic matrix E and other things of no present relevance (hidden under the ellipsis).

```
type probe_8
 type(real_8) x(6) ! polymorphic orbital ray
 real(dp) e_ij(6,6) ! envelope for stochastic radiation
 ...
end type probe_8
```

The reader will remember that my polymorphic type can change into a Taylor series and this produces Taylor approximations of the ray. Since the stochastic moment calculation requires a matrix M for the rays, I need to track my polymorphs. There is also a type probe which in the present example does not add anything to the ray. In fact it contains the phase space ray as `x(6)` and deals only with classical radiation.

```
type probe
   real(dp) x(6)
    ...
end type probe
```

The code in Appendix H looks very much like a code which propagates a simple phase space ray:

```
ray_closed=closed_orbit       ! (3)
id=1;
! ray= closed orbit + identity map
ray=ray_closed+id;            ! (4)


call propagate(als,RAY,state,fibre1=pos)  ! (5)

! Six polymorphs and the fluctuationsare E_ij
! are promoted to Taylor maps

one_turn_map=ray                          ! (6a)
one_turn_map=one_turn_map.sub.1           ! (6b)

if(cas==2) then
 write(mf,*) " "; write(mf,*) "  The linear map";
 call print(one_turn_map,mf,prec)                     ! (7)
endif

call  c_normal(one_turn_map,normal_form)              ! (8)
```

The differences reside in the `ray_closed` and `ray` which are of type probe and probe_8 respectively. These "containers" allow the code to extend the "size" of phase space to include objects such as the fluctuation matrix E. They also include magnet modulation and spin. These are hidden now under the ellipsis "...".

At line (4), the probe `ray` is ready to be propagated. This is done in the normal fashion explained in previous examples: line (5). Finally, the output of the tracking, namely 6 polynomials for the ray and the 36 fluctuations[9] stored in `ray%E_ij` are transferred to an object of type c_damap that can be normalised.

As I said at the beginning of this section, for the value `cas = 1`, I compute the one-turn map in the presence of a cavity but in the *absence* of radiation. The purpose is to get de Moivre's matrices of Eqs. (4.13), (4.14) and (4.15): lines (13a, b, c, d).

```
if(cas==1) then
!!!!!!! Lattice functions !!!!!!!
!! coefficient of invariant  !!
```

[9]36 fluctuations but only 21 independent ones because $\langle z_i z_j \rangle = \langle z_j z_i \rangle$.

```
do i=1,c_%nd
 Ba(1:6,1:6,i)=matmul(matmul(a,Sa(1:6,1:6,i)),ai)  ! (13a)
 Ha(1:6,1:6,i)=matmul(matmul(a,Ia(1:6,1:6,i)),ai)  ! (13b)
 Ka(1:6,1:6,i)=-matmul(S,Ba(1:6,1:6,i))            ! (13c)
 Ea(1:6,1:6,i)=-matmul(Ba(1:6,1:6,i),S)            ! (13d)
enddo

endif
```

These will be used to compare the approximate synchrotron integrals of Chao with the exact moment theory that uses the full fluctuation matrix E.

For `cas=2`, classical radiation is turned on, i.e., the function $r(z)$ of Eq. (4.83) is present in the code. Moreover the effect of the stochastic kick $\Delta(z)$ of Eq. (4.83) is computed via Eq. (4.87). The resulting map is printed in file `result_with_stochastic_radiation.txt` through line (7).

4.4.3.1 Normalisation of the One-Turn Quadratic Stochastic Map

This is done at line (8). The result is a diagonal form for M and the calculation of the equilibrium moments. The moments are first put into phasors' basis and Eq. (4.92) is solved. The result is displayed using line (14).

```
write(mf,'(16X,a50)') "   Equilibrium moments in Phasors Basis              "
 do i=1,6
 do j=i,6
  if(abs(normal_form%s_ijr(i,j))>1.d-20) then
   write(mf,fmd) " Phasors -> ","<x_",i," x_",j,"> = ",  &   ! (14)
                   c_clean(normal_form%s_ijr(i,j),1.d-20)
  endif
 enddo
 enddo
```

Here I called a function `c_clean` whose purpose is to remove parasitic numbers in the real or imaginary part of the results that might be smaller, norm wise, than 10^{-20}. It is interesting to display the results from the output file when *errors* are absent from the lattice.

```
                    Equilibrium moments in Phasors Basis
Phasors ->  <x_1 x_1> = -0.12186203415D-13 -0.79027734122D-14
Phasors ->  <x_1 x_2> =  0.75935118057D-08  0.00000000000D+00
Phasors ->  <x_1 x_5> = -0.25174428585D-12 -0.91374266132D-12
Phasors ->  <x_1 x_6> = -0.87781801866D-12 -0.33899987271D-12
Phasors ->  <x_2 x_2> = -0.12186203415D-13  0.79027734122D-14
Phasors ->  <x_2 x_5> = -0.87781801866D-12  0.33899987271D-12
Phasors ->  <x_2 x_6> = -0.25174428585D-12  0.91374266132D-12
Phasors ->  <x_5 x_5> = -0.46195527875D-08  0.34258629143D-08
Phasors ->  <x_5 x_6> =  0.59332086927D-05  0.00000000000D+00
Phasors ->  <x_6 x_6> = -0.46195527875D-08 -0.34258629143D-08
```

The first important result, at least if you are an accelerator physicist, is the absence of vertical moments. The example code PTC implements the simplest stochastic kick: the photons are emitted *exactly* in the direction of propagation. In a planar lattice, all the photons are in the horizontal plane and therefore there are no moments involving y or p_y, that is to say, the third and fourth variables of phase space.

There are three moments that do not have any imaginary parts: $\langle z_1 z_2 \rangle$, $\langle z_3 z_4 \rangle$ and $\langle z_5 z_6 \rangle$. These are the Chao equilibrium emittances of Eq. (4.95). As we will see through our example, owing to the small damping decrements, they are usually sufficient for the computation of equilibrium moments.

The main advantage of tracking the full stochastic kick E is ease and conceptual simplicity in a code. The code computes maps which are then normalised. If one implements the Chao synchrotron integrals or worse, the Sands integrals, then the procedure is internally more complex.

The Chao integrals, and particularly the Sands integrals, are useful for their analytical simplicity. Sands integrals can be written in terms of cavity-less lattice functions which is a conceptual simplification for the accelerator designer.

4.4.3.2 Comparing Chao with the Exact Computation

To compare Chao with an exact calculation, I first extract the Chao emittances, more precisely the average action which is half the emittance. This is done at lines (15a, b, c) of the following code fragment:

```
emi(1)=real(normal_form%s_ijr(1,2))/2          ! (15a)
emi(2)=real(normal_form%s_ijr(3,4))/2          ! (15b)
emi(3)=real(normal_form%s_ijr(5,6))/2          ! (15c)


write(mf,'(16X,a54)') "     Exact                Chao            (Exact-Chao)/Exact"
 do i=1,6
 do j=1,6
   xij=0.d0
   do k=1,3
     xij= Ea(i,j,k)*emi(k) + xij                                 ! (16)
   enddo
  if(abs(normal_form%s_ij0(i,j))>1.e-15_dp) then
   write(mf,fmd1) "<x_",i," x_",j,"> = ", real(normal_form%s_ij0(i,j)), &
                 xij ,abs(100*(real(normal_form%s_ij0(i,j))-xij)/        &
   real(normal_form%s_ij0(i,j)))," %"
  endif
 enddo
 enddo
```

The matrix `Ea(i,j,k)` is the moment matrix E^k_{ij} of Eq. (4.26). The sum over the three planes is performed at line (16) in the code. The emittances of Eq. (4.95) are substituted into Eq. (4.26). The reader can check on his own that Eq. (4.93) leads to Eq. (4.26) when the equilibrium moments σ^{∞}_{ij} are such that only the three moments of Eq. (4.95) are non-zero.

The exact result is obtained using Eq. (4.93): the equilibrium phasors are transformed into the usual phase space variables. The results of these two calculations are printed and compared in the output file.

```
                      Exact                 Chao              (Exact-Chao)/Exact
<x_1 x_1> =  0.58786612370D-08  0.58846120372D-08        0.101  %
<x_1 x_2> =  0.11190916503D-07  0.11205380039D-07        0.129  %
<x_1 x_5> =  0.34869597207D-07  0.34939089854D-07        0.199  %
<x_1 x_6> = -0.50998666908D-08 -0.51105996231D-08        0.210  %
<x_2 x_1> =  0.11190916503D-07  0.11205380039D-07        0.129  %
<x_2 x_2> =  0.27378434788D-07  0.27412733220D-07        0.125  %
<x_2 x_5> =  0.84803999222D-07  0.84976915933D-07        0.204  %
<x_2 x_6> = -0.13015660559D-07 -0.13040794262D-07        0.193  %
<x_5 x_1> =  0.34869597207D-07  0.34939089854D-07        0.199  %
<x_5 x_2> =  0.84803999222D-07  0.84976915933D-07        0.204  %
<x_5 x_5> =  0.44053599838D-06  0.44143819880D-06        0.205  %
<x_5 x_6> = -0.67544437716D-07 -0.67682428321D-07        0.204  %
<x_6 x_1> = -0.50998666908D-08 -0.51105996231D-08        0.210  %
<x_6 x_2> = -0.13015660559D-07 -0.13040794262D-07        0.193  %
<x_6 x_5> = -0.67544437716D-07 -0.67682428321D-07        0.204  %
<x_6 x_6> =  0.19987734916D-04  0.19946945340D-04        0.204  %
```

The agreement between the exact results and the Chao calculation, based on the three emittances, is remarkable. The Chao calculation can only brake down if the damping decrements are large compared to the distance from the linear resonances. This is very unlikely since linear resonances are avoided like the plague.

The example code of Appendix H also computes the crab crossing angle using various formulae. Here is the actual code fragment:

```
sij(1)=emi(1)*Ea(1,6,1)+emi(2)*Ea(1,6,2)+emi(3)*Ea(1,6,3)      ! (17a)
sij(2)=emi(1)*Ea(6,6,1)+emi(2)*Ea(6,6,2)+emi(3)*Ea(6,6,3)      ! (17b)
sij(3)=emi(1)*Ea(1,1,1)+emi(2)*Ea(1,1,2)+emi(3)*Ea(1,1,3)      ! (17c)

write(mf,*)
write(mf,'(a56)') " Exact Crab Angle with beam envelope                "
write(mf,'((5x,D18.11))') real(normal_form%s_ij0(1,6)) &
/(real(normal_form%s_ij0(6,6))-real(normal_form%s_ij0(1,1)))            ! (18)
write(mf,'(a25)') " Chao Exact Crab Angle "
write(mf,'((5x,D18.11))') sij(1)/(sij(2)-sij(3))                        ! (19)
write(mf,'(a56)') " Almost Exact Crab Angle ignoring transverse emittances "
write(mf,'((5x,D18.11))') -Ka(5,2,3)/(Ka(5,5,3)-Ka(2,2,3))              ! (20)
write(mf,'(a56)') " Approximate Crab Angle ignoring transverse emittances  "
write(mf,'((5x,D18.11))')  Ha(1,6,3)-Ka(5,6,3)*Ha(1,5,3)/Ka(5,5,3)      ! (21)
```

Lines (18)–(21) are in order of "degrading exactness." First line (18) is the exact linear theory of Eqs. (4.92) and (4.93). The next line, line (19), uses the three Chao emittances of Eq. (4.95) to compute the relevant beam sizes at lines (17a, b, c). It is a very accurate theory which normally requires a "Courant-Snyder" loop within the code as does Sands theory. Here I simply extract the Chao emittances from the 21 equilibrium moments.

The next two lines are the crab angles neglecting the contribution from the transverse emittances: line (20) uses the exact lattice functions. Line (21) is a formula derived by Erlichman et al. [5]. It can be derived easily using de Moivre's formula as shown at Eq. (4.44).

4.4.3.3 The Map at Infinity

Perhaps the simplest property of the map for the stochastic moments is revealed in Eq. (4.89): the map at infinity is the *null* matrix $M = 0$ and the stochastic kick is $E_{ij} = \Sigma_{ij}^{\infty}$.

I can check this easily using the map concatenation as shown in the following code fragment

```
write(mf,*);
write(mf,*) "!!!  Raising the maps with moments to the power 2**100  !!!"

do i=1,100
  one_turn_map=one_turn_map*one_turn_map   ! (22a)
enddo

call print(one_turn_map,mf)                ! (22b)

endif
```

The result is printed at line (22b). The orbital map is the null map and the moment kicks are the equilibrium moments as computed by the normal form. This confirms that any initial distribution will collapse to the equilibrium distribution after an "infinite" number of turns.

References

1. D. Edwards, L. Teng, IEEE Trans. Nucl. Stud. **NS-20**(3) (1973)
2. K. Ohmi, K. Hirata, K. Oide, Phys. Rev. E. **49**, 751 (1994)
3. A.W. Chao, J. Appl. Phys. **50**, 595 (1979)
4. M. Sands, Technical Report No. SLAC-121, Stanford Linear Accelerator (unpublished)
5. M. Ehrlichman, Phys. Rev. ST Accel. Beams **17**, 0044002 (2014)
6. A. Dragt, F. Neri, G. Rangarajan, Phys. Rev. A **45**, 2572 (1992), (N.Y.)
7. E. Forest, Phys. Rev. **58**, 2481 (1998)
8. P. Nishikawa, JINST **7** (2012)
9. E. Forest, *Beam Dynamics: A New Attitude and Framework* (Harwood Academic Publishers, Amsterdam, The Netherlands, 1997)
10. R. Tomàs, Phys. Rev. ST-Accel. Beams **5**, 054001 (2002)
11. H. Wiedemann, *Particle Accelerator Physics* (Springer, Berlin, 2007)

References

Chapter 5
Nonlinear Normal Forms

Abstract I describe in some details how a nonlinear normal form is implemented in a library like my own FPP. I also discuss the first departure from the harmonic normal form: the one-resonance normal form. A very simple magnet modulation is treated theoretically. I leave accelerator physics with the introduction of a map which displays a limit cycle which breaks into what looks like a strange attractor.

Keywords Nonlinearities · Magnet modulations · Isolated resonances · Limit cycles

5.1 What Do I Mean by Nonlinear?

The word nonlinear can apply to two things: nonlinear in the parameters or nonlinear in phase space. The algorithm that is implemented in the code, namely in FPP, does not discriminate between the two "nonlinearities." However the issues of convergence or even existence are very much affected by the nature of the nonlinearity.

For example, Eq. (5.1) displays an approximate Hamiltonian in 2 degrees of freedom that is often used to describe an accelerator in standard textbooks and is responsible for the false belief in the existence of "linear elements":

$$H = \frac{p_x^2 + p_y^2}{2(1+\delta)} + k(s)\frac{x^2 - y^2}{2} + \left(b_1 - \frac{1}{\rho_s}\right)x + \frac{b_1}{2\rho_s}x^2 - \frac{x\delta}{2\rho_s}. \tag{5.1}$$

Usually, this Hamiltonian represents the motion on the design orbit and hence $b_1 = \frac{1}{\rho_s}$. I can rewrite Eq. (5.1) in terms of the dummy variable z that correspounds to the phase space vector in the code PTC and set b_1 to $\frac{1}{\rho_s}$ for simplicity:

$$H = \frac{z_2^2 + z_4^2}{2(1+z_5)} + k(s)\frac{z_1^2 - z_3^2}{2} + \frac{1}{2\rho_s^2}z_1^2 - \frac{z_1 z_5}{2\rho_s}. \tag{5.2}$$

In this Hamiltonian, the fifth variable, $z_5 = \delta$, is a parameter because there is no temporal (longitudinal) focusing and I am not interested in the time of flight.

E. Forest, *From Tracking Code to Analysis*,
DOI 10.1007/978-4-431-55803-3_5

I can imagine performing a normal form on the one-turn map or equivalently on the Hamiltonian. This normal form would be a δ dependent *linear* normal form and, to all orders, can be made to obey the Courant-Snyder condition. The canonical transformation A_s such $R_s = A_s^{-1} \circ M_s \circ A_s$ (dropping the position label s) is given by:

$$A = F \circ A_1 \tag{5.3}$$

where

$$F: \begin{array}{l} \overline{z}_1 = z_1 + \eta_1 (z_5) \\ \overline{z}_2 = z_2 + \eta_2 (z_5) \\ \overline{z}_3 = z_4 \\ \overline{z}_4 = z_4 \end{array} \tag{5.4}$$

and

$$A_1: \begin{array}{l} \overline{z}_1 = \sqrt{\beta_x (z_5)}\, z_1 \\ \overline{z}_2 = -\alpha_x (z_5) / \sqrt{\beta_x (z_5)}\, z_1 + z_2 / \sqrt{\beta_x (z_5)} \\ \overline{z}_3 = \sqrt{\beta_y (z_5)}\, z_3 \\ \overline{z}_4 = -\alpha_y (z_5) / \sqrt{\beta_y (z_5)} z_3 + z_4 / \sqrt{\beta_y (z_5)}. \end{array} \tag{5.5}$$

If Eq. (5.5) is applied to Eq. (5.2), one gets:

$$K = \frac{z_1^2 + z_2^2}{2\beta_x (z_5)} + \frac{z_3^2 + z_4^2}{2\beta_y (z_5)} \,. \tag{5.6}$$

The normalised one-turn map is simply a rotation of angles $\mu_{x,y} = \oint_0^C \frac{1}{\beta_{x,y}(z_5)} ds$ in the horizontal and vertical planes. That is an exact result.

In the code, the normalisation is performed numerically. If a global order N_0 is specified, say $N_0 = 3$, then all the terms depending on the parameter z_5 will be expanded to order $N_0 = 3$. Is this a convergent series? *The answer is yes!* The maps M, F and A_1 are all convergent series in z_5.

The radius of convergence might be less than infinity, but the series will converge within a finite domain. Actually the series will diverge at $z_5 = 1$ even though trajectories and stable maps for values of $\delta > 1$ might exist and are computable by the code. This is one of the reasons I consider the integrator to be the ultimate arbiter of "truth" in an accelerator rather than a code based on Taylor series expansions.

What about nonlinear Hamiltonians? For example, we can consider the map discussed in Sect. 3.3, the cavity map or standard map. It is displayed in the code of Appendix F lines (2a, b, c):

$$
\begin{aligned}
(z_1, z_2) &\longrightarrow (\overline{z}_1, \overline{z}_2) \\
z_1^t &= z_1 + \frac{dt}{2} z_2 \\
\overline{z}_2 &= z_1^t - 2dt(2\pi f)^2 \cos\left(\frac{z_1^t}{2}\right) \sin\left(\frac{z_1^t}{2}\right) \\
\overline{z}_1 &= z_1^t + \frac{dt}{2} \overline{z}_2 .
\end{aligned}
\tag{5.7}
$$

For infinitesimally small dt, the standard map of Eq. (5.7) is just the pendulum. This system is nonlinear in the phase space vector $z = (z_1, z_2)$. What about the normal form? Although the map itself can be represented by a convergent Taylor series, it is not true for the normal form because a normal form precludes the existence of islands and chaos. The standard map produces islands and chaos. Therefore we can say only a few things:

1. In the case of the pendulum, $dt \to 0$, the nonlinear normal form may converge to the separatrix of the pendulum.
2. For finite but small dt, the normal form appears to be convergent but is in reality an asymptotic series. If the order N_0 is pushed too high, the normal form will start to "misbehave." This is very much the case of a mild accelerator. This is a consequence of KAM theory: only the invariants associated to a fractal set of tunes exist. Thus the normal form transformation A, which implies a continuous family of invariants, cannot exist.
3. For large values of dt, low order islands appear and the Taylor based normal form is complete nonsense beyond these islands. This situation is also common in accelerators.
4. For yet larger dt, the standard map is completely chaotic. All hell breaks loose and any normal form is a totally nonsensical representation of the original map.

I have summarize the basic outcomes of normal forms. Linear maps produce normal forms that are convergent power series in the parameters. Truly nonlinear maps *usually or generically* give rise to asymptotic series or worse. Now let us look at our catalogue of examples following more or less the linear classification of Chap. 4.

5.2 The (Damped) Pseudo-Harmonic Oscillator Case

The difference between the 1-d-f case of Sect. 3.3.2 and the most general case of this section is really just a matter of indices. Therefore we encourage the serious reader to master fully Sect. 3.3.2 and, while at it, all of Chap. 3. The present section is mechanically very similar. From Eq. (3.87) to Eq. (3.92), there are no differences besides the fact that the linear part might now be a damped oscillator. As before I write the normal form as in Eq. (3.92)

$$\rho_n = \exp\left(T_1 \cdot \nabla\right) \cdots \exp\left(T_n \cdot \nabla\right) I, \tag{5.8}$$

but we now have:

$$\begin{aligned} T_1 \cdot \nabla &= \begin{pmatrix} \left(-i\mu_{0;1} - \alpha_1\right)\varphi_{1+} \\ \left(i\mu_{0;1} - \alpha_1\right)\varphi_{1-} \\ \cdot \\ \cdot \\ \cdot \\ \left(i\mu_{0;N} - \alpha_N\right)\varphi_{N+} \end{pmatrix} \cdot \nabla \\ &= \sum_{i=1,N} \left(-i\mu_{0;i} - \alpha_i\right)\varphi_{i+}\frac{\partial}{\partial \varphi_{i+}} + \left(i\mu_{0;i} - \alpha_i\right)\varphi_{i-}\frac{\partial}{\partial \varphi_{i-}}. \end{aligned} \tag{5.9}$$

The vector field of Eq. (5.9) reproduces the matrix I discussed in the section on the de Moivre's representation (Sect. 4.1).

Everything proceeds as before until the reader hits Eq. (3.102). This equation now contains N phasors:

$$\begin{aligned} F^a_{n+1} &= \sum_m F^a_{n+1;m}\varphi_{1+}^{m_1}\varphi_{1-}^{m_2}\cdots\varphi_{N+}^{m_{2N-1}}\varphi_{N-}^{m_{2N}} \\ &\text{with } F_{n+1} = \left(F^1_{n+1}, F^2_{n+1}, \cdots, F^{2N}_{n+1}\right). \end{aligned} \tag{5.10}$$

Once more I can simplify the vector field of the map into a rotation:

$$\begin{aligned} &\sum_{k=1,N} \left|m_{2k-1} - m_{2k} + (-1)^a\delta_{ak}\right| \neq 0 \\ &\downarrow \\ &F^a_{n+1;m} = \frac{G^a_{n+1;m}}{1 - \exp\left(\sum_{k=1,N}\left\{i\mu_{0;k}\left(m_{2k-1} - m_{2k} + (-1)^a\delta_{ak}\right) - \left(m_{2k-1} + m_{2k}\right)\alpha_k\right\}\right)}. \end{aligned} \tag{5.11}$$

Equation (5.11) is the multi-dimensional equivalent of Eq. (3.105). I should point out that, in the presence of damping, it is possible to remove all nonlinear terms; that precludes the existence of limit cycles, so it is certainly not done in Sect. 5.5.1.

As I did below Eq. (3.106), in the gray box, I examine the Hamiltonian case where Lie operators based on polynomials, rather than vector fields, are used.

The algorithm on Lie polynomials

In the symplectic case, particularly in analytic calculations, one may elect to work on Hamiltonian operators. In that case, Eq. (3.101) is a polynomial equation in the phasors' basis.

$$\exp\left(\colon\!-t_1\!\colon\right) f_{n+1} - f_{n+1} + f_{n+1} = \text{Tune shifts as in Eq. (3.101).} \quad (5.12)$$

The polynomial t_1 is given by the multivariable extension of Eq. (3.94):

$$\begin{aligned} T_1 \cdot \nabla &= \begin{pmatrix} -i\mu_{0;1}\varphi_{1+} \\ i\mu_{0;1}\varphi_{1-} \\ \cdot \\ \cdot \\ \cdot \\ i\mu_{0;N}\varphi_{N+} \end{pmatrix} \cdot \nabla \\ &= \sum_{i=1,N} -i\mu_{0;i}\varphi_{i+}\frac{\partial}{\partial\varphi_{i+}} + i\mu_{0;i}\varphi_{i-}\frac{\partial}{\partial\varphi_{i-}} \\ &= -\sum_{i=1,N} \frac{\mu_{0;i}}{2}\colon\!\varphi_{i+}\varphi_{i-}\!\colon \end{aligned} \quad (5.13)$$

The polynomial f_{n+1} is also expanded in phasors:

$$f_{n+1} = \sum_k f_{n+1;k}\varphi_{1+}^{k_1}\varphi_{1-}^{k_2}\cdots\varphi_{N-}^{k_{2N}} \text{ as in Eq. (3.102).} \quad (5.14)$$

I then proceed with the action of $-t_1$ on the polynomial f_{n+1}:

$$\begin{aligned} \colon\!-t_1\!\colon f_{n+1;k}\varphi_{+1}^{k_1}\varphi_{2-}^{k_2}\cdots\varphi_{N-}^{k_{2N}} &= -if_{n+1;k}\sum_{i=1,N}\mu_{0;i}\left[\varphi_{i+}\varphi_{i-},\varphi_{1+}^{k_1}\varphi_{1-}^{k_2}\cdots\varphi_{N-}^{k_{2N}}\right] \\ &= if_{n+1;k}\left\{\sum_{i=1,N}\mu_{0;i}\left(k_{2i-1}-k_{2i}\right)\right\}\varphi_{+1}^{k_1}\varphi_{2-}^{k_2}\cdots\varphi_{N-}^{k_{2N}}. \end{aligned} \quad (5.15)$$

The final result for f_{n+1} is:

$$\sum_{i=1,N}\mu_{0;i}\left(k_{2i-1}-k_{2i}\right) \neq 0 \rightarrow f_{n+1;k} = \frac{g_{n+1;k}}{1-\exp\left(i\sum_{i=1,N}\mu_{0;i}\left(k_{2i-1}-k_{2i}\right)\right)} \text{ as in Eq. (5.11).} \quad (5.16)$$

Of course, Eq. (5.16) must reproduce Eq. (5.11). This is done by computing the vector field associated with f_{n+1}. Let us look at the effect on φ_{i+}:

$$\colon\!\varphi_{1+}^{k_1}\varphi_{1-}^{k_2}\cdots\varphi_{N-}^{k_{2N}}\!\colon \varphi_{i+} = 2i\varphi_{+}^{m_1}\varphi_{-}^{m_2}\varphi_{N-}^{m_{2N}} \quad (5.17)$$

where $m_{2i} = k_{2i} - 1$ and $j \neq 2i \Rightarrow m_j = k_j$.

Thus one concludes from Eq. (5.17) that $F^1_{n+1;m}$ is

$$F^1_{n+1;m} = 2if_{n+1;k} \quad \text{with} \quad \begin{array}{l} m_1 = k_1 \\ m_2 = k_2 - 1 \\ m_3 = k_3 \\ \vdots \end{array} . \tag{5.18}$$

Equation (5.18) tells us that the condition $k_{2j} - k_{2j-1} = 0$ for the tune shift is indeed equivalent to the condition for the vector field.

5.3 Derivation of Eq. (4.74): Magnet Modulation Revisited

This section is a small analytic treatment which obviously sits outside the tools of the code; if anything it parallels the treatment by Tomàs in Ref. [1].

To derive Eq. (4.74), I need to rewrite Eq. (4.72) in terms of Lie operators.

$$\mathscr{M} = \mathscr{R}_{\omega L} \underbrace{\exp\left(-\frac{L}{2} : p_x^2 :\right) \exp\left(-\frac{Lk}{2} : x^2 :\right)}_{\mathscr{M}_0} \exp\left(-\frac{L\kappa_0 q}{2} : x^2 :\right) \tag{5.19}$$

$$z = \left(x, p_x, \underbrace{q, p}_{clock}\right) \quad \text{and} \quad \mathscr{R}_{\omega L} = \exp\left(: -\frac{\omega L}{2}\left(q^2 + p^2\right) :\right).$$

In Eq. (5.19), I deliberately moved the variable q outside the Lie operator to remind everyone that it is **not** a symplectic variable and therefore it does not enter in the Poisson bracket. All Poisson brackets involve only the variable (x, p_x). Of course the reader can check that this produces the correct vector fields; however it is often simpler to use Poisson brackets whenever possible.

The first steps consists in normalizing the problem we already know how to solve, namely the map $\mathscr{M}_0$ without modulation. This is done most easily in the matrix representation (which acts in the opposite order):

$$\begin{aligned} M_0 &= \begin{pmatrix} 1 & 0 \\ -kl & 1 \end{pmatrix} \begin{pmatrix} 1 & L \\ 0 & 1 \end{pmatrix} \\ &= \begin{pmatrix} 1 & L \\ -kl & 1 - kL^2 \end{pmatrix} \\ &= \begin{pmatrix} \cos(\mu) + \alpha \sin(\mu) & \beta \sin(\mu) \\ -\gamma \sin(\mu) & \cos(\mu) - \alpha \sin(\mu) \end{pmatrix} \quad \text{(per Eq. (2.28))}. \end{aligned} \tag{5.20}$$

The Courant-Snyder' parameters and the tune can all be computed:

$$\begin{aligned}\beta &= k^{\frac{1}{2}} L \sqrt{1-\frac{1}{4}kL^2}\\ \gamma &= k\beta\\ \alpha &= \frac{1}{2}kl\beta\\ \mu &= \sin^{-1}\left(L\beta^{-1}\right).\end{aligned} \tag{5.21}$$

I can then construct the Courant-Snyder transformation of Eq. (2.51) and let its "function acting" version[1] $\mathscr{A}$ act on the map $\mathscr{M}$ by similarity transformation:

$$\begin{aligned}\mathscr{N}_0 &= \mathscr{A}\mathscr{M}\mathscr{A}^{-1}\\ &= \mathscr{A}\underbrace{\mathscr{R}_{\omega L}\exp\left(-\frac{L}{2}:p_x^2:\right)\exp\left(-\frac{Lk}{2}:x^2:\right)}_{\mathscr{M}_0}\exp\left(-\frac{L\kappa_0 q}{2}:x^2:\right)\mathscr{A}^{-1}\\ &= \mathscr{R}_{\omega L}\underbrace{\mathscr{A}\exp\left(-\frac{L}{2}:p_x^2:\right)\exp\left(-\frac{Lk}{2}:x^2:\right)\mathscr{A}^{-1}}_{\mathscr{R}_\mu}\exp\left(-\frac{L\kappa_0 q}{2}:\mathscr{A}x^2:\right)\\ &= \underbrace{\mathscr{R}_{\omega L}\mathscr{R}_\mu}_{\mathscr{R}}\exp\left(-\frac{L\kappa_0\beta q}{2}:x^2:\right).\end{aligned} \tag{5.22}$$

I now perform a nonlinear transformation whose purpose is to eliminate the operator $q : x^2 :$. Calling this operator $\exp(: f :)$ where f is a function of the clock (q, p) and usual phase space (x, p_x):

$$\begin{aligned}\mathscr{N}_1 &= \exp(: f :)\mathscr{N}_0\exp(- : f :)\\ &= \exp(: f :)\underbrace{\mathscr{R}_{\omega L}\mathscr{R}_\mu}_{\mathscr{R}}\exp\left(-\frac{L\kappa_0\beta q}{2}:x^2:\right)\exp(- : f :)\\ &= \mathscr{R}\exp\left(:\mathscr{R}^{-1}f:\right)\exp\left(-\frac{L\kappa_0\beta q}{2}:x^2:\right)\exp(- : f :).\end{aligned} \tag{5.23}$$

As is always the case in perturbation theory, we first solve for the canonical transformation by neglecting second-order terms, i.e., by assuming temporarily that all Lie operators commute.

$$(\mathscr{R}^{-1}-1)f - \frac{L\kappa_0\beta q}{2}x^2 = 0. \tag{5.24}$$

---

[1] I remind the reader who skipped too many sections of this book that it is precisely the map that acts on *functions* by substitution that can be represented by Lie operators. In the linear case, one is often content with the usual matrix acting on rays. The Lie operators are not that useful. The one-turn map is the critical exception: the Courant-Snyder invariant is the Lie operator!

Equation (5.24) is just a special case of Eq. (3.101) but here the cubic term qx^2 as no tune shifts and thus can be removed completely. To get the answer we go into a phasor basis for both the dynamical and clock variables. To avoid a proliferation of indices, I give special names to the phasors:

$$h = x + ip_x \quad \overline{h} = x - ip_x \quad v = q + ip \quad \overline{v} = q - ip. \tag{5.25}$$

Then Eq. (5.24) becomes:

$$\begin{aligned} f &= -\frac{L\kappa_0\beta}{16}\left(C_{1,2}vh^2 + C_{1,-2}v\overline{h}^2 + 2C_{1,0}vh\overline{h} + \text{ c.c.}\cdots\right) \\ C_{m,n} &= \frac{1}{1-\exp\left(i\left\{m\omega L + n\mu\right\}\right)}. \end{aligned} \tag{5.26}$$

Using the BCH formula, namely $e^A e^B = e^{A+B+1/2[A,B]+\cdots}$, I can approximate Eq. (5.23) to second-order:

$$\mathcal{N}_1 = \mathcal{R}\exp\left(\frac{L\kappa_0\beta}{4}\left[qx^2, f\right] + \cdots\right). \tag{5.27}$$

Since we want second order tune shifts, it suffices to extract from Eq. (5.27) the tune shifts terms. Again this is best performed in phasors basis provided we do not forget the factor of $-2i$ which enters in my definition of the Poisson bracket in phasors:

$$\begin{aligned} \frac{L\kappa_0\beta}{4}\left[qx^2, f\right] &\Rightarrow -2i\left(\frac{L\kappa_0\beta}{4}\right)\left(-\frac{L\kappa_0\beta}{16}\right) \\ &\times \frac{1}{8}\left[vh^2 + v\overline{h}^2 + 2vh\overline{h},\ C_{1,2}vh^2 + C_{1,-2}v\overline{h}^2 + C_{1,0}vh\overline{h} + \text{ c.c.}\right]. \end{aligned} \tag{5.28}$$

The factor of $-2i$ having been taken into account, we take the regular Poisson bracket *not including the clocks*:

$$\begin{aligned} \frac{L\kappa_0\beta}{4}\left[qx^2, f\right] &\Rightarrow \frac{i}{64}(L\kappa_0\beta)^2\left(C_{-1,-2} - C_{1,2} + C_{1,-2} - C_{-1,2}\right)v\overline{v}h\overline{h} \\ &= \frac{1}{32}(L\kappa_0\beta)^2\left(\frac{\sin(2\mu+\omega L)}{1-\cos(2\mu+\omega L)} + \frac{\sin(2\mu-\omega L)}{1-\cos(2\mu-\omega L)}\right)J_x. \end{aligned} \tag{5.29}$$

In Eq. (5.29), I used the fact that the length of the clock arms equals one ($v\overline{v} = 1$) and that $h\overline{h}$ is twice the action J_x. Since minus the derivative of the action is the tune, I just proved Eq. (4.74)!

If one computes the tune as if q was a constant and then averages over q, then I leave it to the reader to show that the leading order correction provided by this sloppy calculation is Eq. (5.29) with $\omega L = 0$. It is the leading order adiabatic result. It is possible to develop a perturbation theory which attempts to remove all time dependence (clock dependence) from the map or the Hamiltonian. This type of perturbation theory is used to compute the so-called ponderomotive force (see for example [2]).

5.4 One Resonance Orbital Normal Form

The following figure was produced using the ring in `one_resonance_map`, a program found in Appendix K.
The phase space picture of Fig. 5.1 displays a situation which cannot be globally represented by a normal form of the type described in Sect. 5.2. The black curves depict the three-turn map for reasons that will become obvious in Sect. 5.4.1.

Below the three islands, inside the triangular curve, we could imagine a deformation of the curves into circles. Analogously, I can deform the curves above the islands into circles. But it is not possible for obvious topological reasons to deform the islands into concentric circles.

Consequently it is not possible to write the map which produced Fig. 5.1 as a single Lie exponent, neither through normal form nor through the logarithm process described in supplemental Chap. 11. This can be seen by using *reductio ad absurdum*. Indeed imagine we could find a single Lie exponent for the map:

$$\mathcal{M} = \exp(: -H :)\,. \tag{5.30}$$

Then the map

$$\mathcal{M}(t) = \exp(: -tH :) \tag{5.31}$$

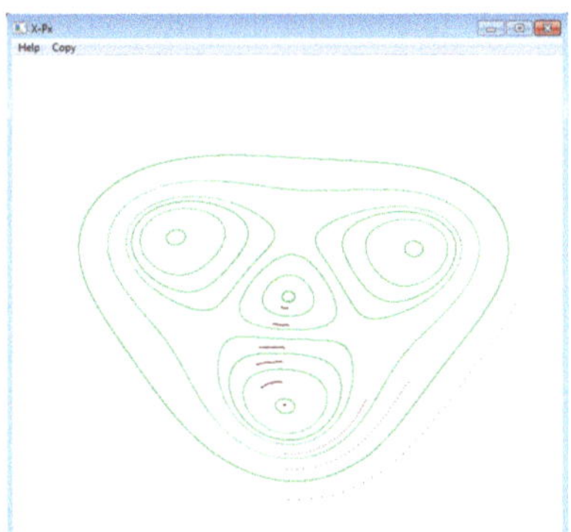

Fig. 5.1 A $3\nu_x$ resonance

has also the same invariants. However t can be made as small as I desire. Thus the map is connected to the identity. This cannot be true in the island region since the particle is "teletransported" from one island to the other by a jump of approximately $2\pi/3$ in the example of Fig. 5.1.

It follows that we cannot compute an invariant by the logarithm or by a normal form technique.

5.4.1 The Naive Dragt Approach: Raising the Map to a Power

In the above example, I can certainly look at the map stroboscopically every 3 turns. The 3-turn map will always be near the identity even in the three islands where particles are trapped. Moreover the map will have the same invariants as the single turn map for obvious reasons. Therefore it is possible to envisage the following equation:

$$\mathscr{M}^3 = \exp\left(: -3H :\right). \tag{5.32}$$

The reader can examine the black curves on Fig. 5.1: whether they are in the inner region, the three islands region or on the outside, all these curves display motion near the identity.

Analytically, the invariant can be computed using a regular normal form. Obviously the 3-turn map has a small tune, say ε. Therefore the normal form will look like this:

$$\mathscr{A}_{1/\varepsilon}\mathscr{M}^3\mathscr{A}_{1/\varepsilon}^{-1} = \exp\left(: -2\pi\varepsilon J \; + \; h(J) \; :\right). \tag{5.33}$$

The normal form of Eq. (5.33) is just as badly behaved as the normal form on the original map $\mathscr{M}$ since it attempts the impossible: turning 3 islands into circles around the origin. Algebraically it is seen by the presence of small denominators in the canonical transformation $\mathscr{A}$. In fact this canonical transformation also normalises the original map $\mathscr{M}$. It is the next trick, due to Dragt [3], which saves the day. I need simply to compute the three-turn map as a single Lie exponent using Eq. (5.33):

$$\begin{aligned}\mathscr{M}^3 &= \mathscr{A}_{1/\varepsilon}^{-1}\exp\left(: -2\pi\varepsilon J \; + \; h(J) \; :\right)\mathscr{A}_{1/\varepsilon} \\ &= \exp\left(: -2\pi\varepsilon\mathscr{A}_{1/\varepsilon}^{-1}J \; + \; h(\mathscr{A}_{1/\varepsilon}^{-1}J) \; :\right).\end{aligned} \tag{5.34}$$

The actual divergence is cancelled out in the exponent of Eq. (5.34). This can be shown explicitly on analytical models: the reader is invited to look at Sect. 5.2 of Ref. [4]. In that book, I treated a rather generic sextupole system and showed how all the divergences disappear. The calculation is carried to second-order in sextupoles to ensure that the important tune shift with amplitude is computed correctly.[2]

Numerically, the calculation is shown in Appendix K. Here is the relevant fragment:

```
three_turn_map = one_turn_map**3          !3-turn map     ! (2)

call  c_normal(three_turn_map,normal_form)               ! (3)
write(mf,*);write(mf,*)"tune",normal_form%tune(1);write(mf,*);

Fh=0
do i=1,normal_form%ker%n    ! Rotation so all exponents commute
 Fh=normal_form%ker%f(i)+Fh                              ! (4)
enddo

h3=(cgetpb(Fh)*to_phasor())*normal_form%a_t**(-1)        ! (5)

write(mf,*);write(mf,*)"Invariant of the 3-turn map";
write(mf,*);
call print(h3,mf,prec)

Write(mf,*); Write(mf,*) "Checking that it is indeed invariant"
Write(mf,*) "h3*one_turn_map - h3 ";write(mf,*);
h3t=h3*one_turn_map
h3t=h3t-h3
```

This entire code fragment emphasises the generality of the procedure. Here the map `one_turn_map` was produced by the example code PTC. However this map could have been produced by any code including a code which produces primarily Taylor maps such as Cosy-Infinity [5].

Because line (3) is a total normal form into circles, the Lie exponents stored in `one_turn_map%ker` all commute with each other (see Sect. 3.4.1, p. 71, for the definition of the Fortran 90 types). Thus they can be added into one big Lie exponent `Fh` at line (4).

Since I am dealing with a symplectic map in canonical variables, I might as well look at the Lie polynomial also known as the pseudo-Hamiltonian of the map. Thus on line (5), the routine call `cgetpb(Fh)` turns the phasors-based vector field into a Lie polynomial. This requires the special routine `cgetpb` in which a factor of $-2i$ is added to the definition of the Poisson bracket. This pseudo-Hamiltonian is converted into the original variables at the end of line (5). I invite the reader to read supplemental Chap. 10 where I explain why I chose "non-symplectic" phasors incurring this annoying factor of $-2i$!

---

[2] Any sextupolar system requires *at least* a second-order calculation of the tune shifts. They determine the existence of islands.

Finally, I check that it is indeed an invariant. The reader is encouraged to "break the code" by changing line (2) into `three_turn_map = one_turn_map`. The resulting pseudo-Hamiltonian will still be an invariant, however with larger coefficients. Indeed the power series of this Hamiltonian must diverge at the separatrix just below the islands.

The code in Appendix K also shows the same calculation using the logarithm of the three-turn map as explained in supplemental Chap. 11. Here it works well because the map is near the identity. In general, the other degrees of freedom undergoing harmonic motion, might not be near the identity. One must be aware of this fact before taking a logarithm. This problem and that of coupled resonances I will now address in the next sections.

5.4.2 General Approach: Leaving One Resonance

The general approach consists of three steps. It was first applied on Taylor maps by Turchetti. In his approach to the one-resonance normalization, Turchetti [6, 7] imposed symmetries directly on the Taylor representation of the map. I developed independently an equivalent approach [8] using Lie operators. It is simpler in the Hamiltonian case since it parallels Hamiltonian theory. So here are the steps:

1. Perform a normal form where a "single" resonance is left in the map. I will define later what this means.
2. Multiply the resulting normal form by a rotation that commutes with this resonance and brings the map close to the identity. Because all the maps commute, they share common invariants. I call this map "the co-moving map" because it is always close to the identity and is analogous to the co-moving Hamiltonian of standard accelerator theory. In fact the co-moving Hamiltonian generates a co-moving map; the two theories are equivalent.
3. I compute the invariant associated to the resonance by a complete normalisation followed with the action of $\mathscr{A}^{-1}$ on the normalised invariant as in Eq. (5.34) or by taking the logarithm of the co-moving map. The resulting Lie operator should be valid through the island region.
   The first method is to be avoided in numerical calculations since it potentially divides vanishingly small quantities. However it is well-suited for analytical calculations such as the one presented in Sect. 5.2 of my other book [4]. The logarithm method is useful in a numerical computation but impossible to use in analytical calculation beyond vague guesses valid if the linear part of the map is very near the resonance.

First I must explain the meaning of a "single" resonance. For simplicity, as in the previous section, I will use Lie polynomials rather than general vector field operators.

Consider a system with two degrees of freedom and let me concentrate on the $3\nu_x$ resonance. The following monomial is left invariant by a rotation of $2\pi/3$:

$$|mkn\rangle = u^m \overline{u}^{m+3k} v^n \overline{v}^n \quad \text{where} \quad u = x + ip_x \text{ and } v = y + ip_y. \tag{5.35}$$

The "bar" above the letter denotes the complex conjugate. Following Eq. (2.13) for example, one sees that

$$\mathscr{R}_x\left(\frac{\lambda 2\pi}{3}\right)|mkn\rangle = e^{\lambda k 2\pi}\;|mkn\rangle \tag{5.36}$$

and therefore, if λ is an integer, the monomial $|mkn\rangle$ is invariant under this *discrete* rotation.

Moreover the Poisson bracket of two monomials such as $|mkn\rangle$ is also a similar monomial:

$$\left[u^m \overline{u}^{m+3k} v^n \overline{v}^n,\, u^a \overline{u}^{a+3c} v^b \overline{v}^b\right] = -6i\,\{c-k\}\,u^{m+a-1}\overline{u}^{m+a-1+3(k+c)} v^{n+b}\overline{v}^{n+b}. \tag{5.37}$$

I invite the reader to confirm these results with the more common action-angle variables, namely using Eq. (2.21):

$$|mkn\rangle = u^m \overline{u}^{m+3k} v^n \overline{v}^n \;=\; (2J_x)^{\frac{2m+3k}{2}}\left(2J_y\right)^n\;\exp\left(i3k\Phi_x\right). \tag{5.38}$$

Therefore the set of monomials such as those of Eq. (5.35) is an invariant set under the Poisson bracket and so is its associated set of Lie operators under commutation.

This invariance implies that if Hans in Germany decides to leave one family of resonances in the map of some machine in the USA, and Mieko in Japan does the same calculation using a different algorithm, they will both agree that *irrespective* of the algorithm, their normalised maps contain only a single family.

N.B. This result is not as strong as that of a total normal form. If the map is turned into circles, then the normal form of the one-turn map will be identical if canonical transformations are used. This amounts to saying that the tunes are universal functions of the actions (J's) in Hamiltonian theory. In the one-resonance case, the normalised map will be invariant under a discrete group of rotations in the plane of the resonance. For example, the $3\nu_x$ resonance will produce normalised phase plots which are invariant under 120° and 240° rotations in the plane of the resonance. So Hans and Mieko's graphs could certainly be rotated with respect to one another. They might also be stretched, but the areas of the islands will be the same. This is because they can differ not only by a rotation but by a generator of the resonance itself.

In practice, that is in the code, the resonance is left during the computation of the canonical transformation. In the general vector field case, *henceforth removing damping for tidiness,* this was shown at Eq. (5.11) which I reproduce here with the small modification needed for a single resonance normal form:

$$m_{2k-1} - m_{2k} + (-1)^a \delta_{ak} \neq m_k^r$$
$$\downarrow$$
$$F_{n+1;m}^a = \frac{G_{n+1;m}^a}{1 - \exp\left(i \sum_{k=1,N} \mu_{0;k} \left(m_{2k-1} - m_{2k} + (-1)^a \delta_{ak} \right) \right)}. \tag{5.39}$$

The vector of indices $\mathbf{m}^r$ describes our family of resonances. In our example, it is just given by:

$$\mathbf{m}^r = \lambda\,(3, 0) \quad \text{where} \quad \lambda = \cdots - 2, -1, 0, 1, 2 \cdots . \tag{5.40}$$

The inclusion of $\lambda = 0$ in Eq. (5.40) insures that the tunes (shifts) are also in the set. Once more I included the appropriate code fragment from Appendix K:

```
normal_form%nres=0;normal_form%m=0;
do i=1,map_order+1                                         ! (7a)
  if(mod(i,3)==0) then
   normal_form%nres=normal_form%nres+1
   normal_form%m(1,normal_form%nres)=i
  endif
enddo

call  c_normal(one_turn_map,normal_form)                   ! (7b)

h2=(pi/3.d0)*((1.d0.cmono.1)**2+(1.d0.cmono.2)**2)        ! (8a)
F2=getvectorfield(h2)                                      ! (8b)

id=normal_form%a_t**(-1)*one_turn_map*normal_form%a_t ! (8c)
co_moving_map=exp(F2)*id                                   ! (8d)
Fh=log(co_moving_map)                                      ! (8e)
h3c=getpb(Fh)*normal_form%a_t**(-1)                        ! (8f)
```

At line (7a), all the multiples of $3\nu_x$ are included as resonant terms to be left in the map. The normalisation is done at (7b).

On lines (8a, b, d) I compute the rotation of Eq. (5.36) with $\lambda = -1$. This rotation achieves two things. First it commutes with the normalised map and thus does not change its invariants. Secondly it brings the map near the identity and permits the computation of a global invariant by "diverging" normal form or by a logarithm. Here in lines (8e) I used the logarithm which is more suited to a numerical calculation. Line (8f) is the final result in the original variables.

5.4.3 *The Co-moving Map*

I will illustrate the co-moving map approach on a simple coupling resonance, $\nu_x + 2\nu_y = p$, where actually $p = 1$. The general theory can be found in Sect. 5.2 of Ref. [4] in more abstract details. Here I hope that the reader will be able to extrapolate[3] inductively from $\nu_x + 2\nu_y = p$ to the general coupled resonance.

I write the monomials of the $\nu_x + 2\nu_y$ resonance in the form of Eqs. (5.35) and (5.38):

$$|mkn\rangle = u^m \overline{u}^{m+k} v^n \overline{v}^{n+2k} = (2J_x)^{m+\frac{k}{2}} \left(2J_y\right)^{n+k} \exp\left(i\, k\left\{\Phi_x + 2\Phi_y\right\}\right). \quad (5.41)$$

Following the exact same procedure as before, I normalise the map and leave only the family represented by Eq. (5.41). Using the polynomial symplectic operators, I can write the normalised map as

$$\mathscr{N} = \exp\left(: -\left\{\mu_x J_x + \mu_x J_x\right\} :\right) \exp\left(: \sum_{m,n,k} A_{mkn}\, |mkn\rangle : \right). \quad (5.42)$$

The idea is to re-express the linear part of $\mathscr{N}$ in Eq. (5.42) in terms of the resonance vector $\mathbf{m}^r = (1, 2)$ and its perpendicular plane(s). This can always be done. In our simple example, I can immediately write two orthogonal vectors:

$$\mathbf{m}^r = (1, 2) \quad \text{and} \quad \mathbf{a} = (2, -1). \quad (5.43)$$

Using these two vectors, I can rewrite the Lie operator of the linear part of $\mathscr{N}$:

$$\mu_x J_x + \mu_y J_y = \frac{\mathbf{m}^r \cdot \boldsymbol{\mu}}{5} \mathbf{m}^r \cdot \mathbf{J} + \frac{\mathbf{a} \cdot \boldsymbol{\mu}}{5} \mathbf{a} \cdot \mathbf{J}. \quad (5.44)$$

The crucial step is to recognise that

$$\mathbf{m}^r \cdot \boldsymbol{\mu} = 2\pi p + 2\pi \varepsilon \quad (5.45)$$

and to rewrite Eq. (5.44). Using

$$\frac{\mathbf{m}^r \cdot \boldsymbol{\mu}}{5} \mathbf{m}^r \cdot \mathbf{J} = \frac{2\pi}{5} \{p + \varepsilon\}\, \mathbf{m}^r \cdot \mathbf{J}, \quad (5.46)$$

Equation (5.44) is rewritten as

---

[3]The reader will find a description of this resonance at the lowest order in a paper by Schmidt and Franchetti in Ref. [9].

$$\mathcal{N} = \exp\left(: -\frac{2\pi}{5} p\mathbf{m}^r \cdot \mathbf{J} : \right) \exp\left(: -\frac{\mathbf{a} \cdot \boldsymbol{\mu}}{5} \mathbf{a} \cdot \mathbf{J} : \right) \mathcal{N}_c$$

$$\text{where } \mathcal{N}_c = \exp\left(: -\frac{2\pi}{5} \varepsilon \mathbf{m}^r \cdot \mathbf{J} : \right) \exp\left(: \sum_{m,n,k} A_{mkn} \left| mkn \right\rangle : \right). \tag{5.47}$$

The reader should verify that the two rotations multiplying $\mathcal{N}_c$ commute with $\mathcal{N}_c$. They do so for different reasons:

- $\exp\left(: -\frac{2\pi}{5} p\mathbf{m}^r \cdot \mathbf{J} : \right)$ commutes because it is part of a discrete subgroup like the rotation of 120 degrees in the case of the $3\nu_x$ resonance. The driving terms are rotated by a multiple of 2π.
- $\exp\left(: -\frac{\mathbf{a} \cdot \boldsymbol{\mu}}{5} \mathbf{a} \cdot \mathbf{J} : \right)$ commutes with $\mathcal{N}_c$ for the same reason that a rotation around the y-axis commutes with the generators of $3\nu_x$ resonances: the planes are orthogonal. If the reader works it out, he will find that all hinges on the fact that $\mathbf{m}^r \cdot \mathbf{a} = 0$. The Lie exponent on the right hand side of Eq. (5.47) contains the resonant terms and produces a pseudo-pendulum motion in the plane of that resonance. It also depends parametrically on the value of $\mathbf{a} \cdot \mathbf{J}$. The pseudo-pendulum can be stable or unstable depending on the value of $\mathbf{a} \cdot \mathbf{J}$, a fact emphasized in Ref. [9].

Because the map $\mathcal{N}_c$ is near the identity, its Lie operator can be found by a simple logarithm as described in supplemental Chap. 11.

I conclude with the code fragment concerning the resonance $\nu_x + 2\nu_y = p$.

```
call   c_normal(one_turn_map,normal_form)                      ! (C)

mdotmu=2*pi/5.d0                                                ! (D1)
adotmu=2*pi*(2*normal_form%tune(1)-normal_form%tune(2))/
5.d0 ! (D2)

jx=(0.5d0.cmono.'2')+(0.5d0.cmono.'02')                       ! (E1)
jy=(0.5d0.cmono.'002')+(0.5d0.cmono.'0002')                   ! (E2)
jm=(jx+2*jy)                                                   ! (E3)
ja=(2*jx-jy)                                                   ! (E4)

h2=mdotmu*jm+adotmu*ja                                          ! (F1)
F2=getvectorfield(h2)                                           ! (F2)

id=normal_form%a_t**(-1)*one_turn_map*normal_form%a_t ! (F3)
co_moving_map=exp(F2,id)                                        ! (F4)

Fh=log(co_moving_map)                                           ! (F5)
h3c=getpb(Fh)*normal_form%a_t**(-1)                             ! (F6)

h3t=h3c*one_turn_map
h3t=h3t-h3c                                                     ! (G)
```

From line (D1) to line (F2), I construct the commuting rotation which brings the normalised map close to the identity. At line (F3), the map is actually put into a one-resonance normal form. The co-moving map is constructed at line (F4). At (F5) I compute the logarithm and at line (F6), I put this logarithm into the basis of the original map. Line (G) checks that I have indeed an invariant.

5.4.4 The Instructive Resonant Case $\mathcal{M} = \mathcal{R}_{\nu=\frac{1}{4}+\delta} \exp\left(: \frac{k_s}{3}x^3 + \frac{k_o}{4}x^4 :\right)$

In the case of orbital resonances, it is very easy to give a geometrical definition: a resonance destroys the invariant tori by creating islands where the tunes are frozen into a linear relation with integer coefficients. For example the right-hand sides of Figs. 5.2 and 5.4 depict such a case. I will examine both cases because they are quite revealing and will somewhat clarify what is meant by an "isolated resonance." It will also dispel some false beliefs. For example the divergence of perturbation theory for the computation of an invariant is not necessarily connected to a "physical" effect. Also we will see that it is far from obvious that an isolated resonance can be properly computed by a naive retention of the leading order terms as often suggested in elementary discussions. All my assertions concerning formal manipulations can be verified by a Taylor series based package such as my own FPP. They can also be checked, more painfully, by pure analytical computations on the map.

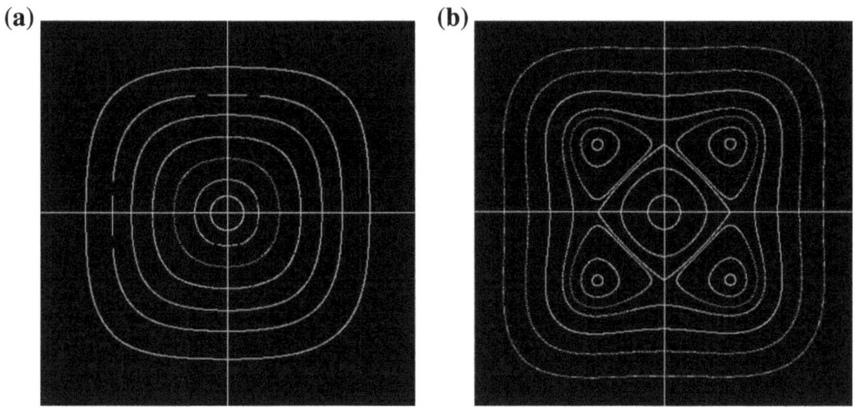

Fig. 5.2 Octupole driven resonance. **a** Below the resonance. **b** Above the resonance

5.4.4.1 The Octupole Map $\mathscr{M} = \mathscr{R}_{\nu=\frac{1}{4}\mp 0.001} \exp\left(: \frac{1}{4}x^4 :\right)$: Exact Results on Convergence

In Fig. 5.2, the map used is just

$$\mathscr{M} = \mathscr{R}_{\nu=\frac{1}{4}\mp 0.001} \exp\left(: \frac{1}{4}x^4 :\right). \tag{5.48}$$

Looking at Fig. 5.2b, anyone would say: "Ha! There is a fourth-order resonance!" Some would even say: "it is an octupole resonance." Moreover the four-fold symmetry seems perfect, and therefore in variables where the linear motion is a circle, we have two reasons to call this resonance isolated: only one set of islands and the absence of distorting effects from other "potential" resonances. So I summarize:

1. There are islands, therefore there is a topological destruction of the tori.
2. They are alone and thus, this is an isolated resonance.
3. There is a near perfect four-fold symmetry. Potential driving terms of other resonances have a very tiny effect. In that sense, the resonance is even more "isolated." The perpendicular planes, not present in this example, move happily on circles[4] unaffected by the resonance.

If item 2 is true and a normal form is successfully computed, then item 3 is always true in the normalised space. It is true in the un-normalised[5] variables only if the resonance is unperturbed by the other nonlinearities and thus isolated in the sense of 3.

Because the map is so close to the resonance, i.e., $\delta = 0.001$, it can be shown that the fourth power of the map, to leading order, is almost, but not exactly, given by:

$$\begin{aligned} \mathscr{M}^4 &= \exp(: -H :) \\ \text{where } H &= 8\pi\delta J - \frac{J^2}{2}\left(3 + \cos(4\Phi)\right) + O(\delta J^2)\cdots. \end{aligned} \tag{5.49}$$

The fixed points of H can be easily computed and are given by:

$$J_{\mp} = \frac{8\pi\delta}{3 \pm 1} \quad \text{with} \quad \Phi_- = k\frac{\pi}{2} \quad \text{and} \quad \Phi_+ = \Phi_- + \frac{\pi}{4}. \tag{5.50}$$

In the case of Fig. 5.2b which corresponds to $\delta = 0.001$, the predictions given by Eq. (5.50) are remarkably accurate. Of course it is essential to compute the fourth power of the map or to use the co-moving map methods. The islands preclude the

[4] In the case of a couple resonance, $\mathbf{a} \cdot \mathbf{J}$ of Eq. (5.44) is an invariant by virtue of being orthogonal to the plane of the resonance.

[5] Here by "un-normalised" I mean "normalised by a linear map" only.

existence of global circular-like invariants. For larger values of δ, for example $\delta = 1/10$, I do expect distortions coming from the "other resonances" and indeed the four fold symmetry is slightly violated.

Because of the islands, I do not expect a standard normal form to converge beyond the inner separatrix. Any attempt to turn Fig. 5.2b into circles will diverge badly at the inner separatrix. This can actually be checked numerically on such a simple map. However what about the map for $\delta = -0.001$ that is depicted on Fig. 5.2a?

This map does not display any islands. All the curves are closed and I can imagine deforming them with a symplectic map into circles. So will our normal form algorithm, which is a Taylor series in the normalised phase space variables $x + ip_x = \sqrt{2J}e^{-i\Phi}$, converge in a greater region of phase space? The answer to this is a *categorical no*. This is not too surprising looking at the pseudo-Hamiltonian H of Eq. (5.49): this Hamiltonian generates the fourth power of the map; it was derived independently of the sign of δ. Therefore it is hard to imagine that the radius of convergence of any canonical transformation that normalises H would depend on the sign of δ. But there is a stronger argument which applies rigorously the map of Eq. (5.48). Consider the application of a complex symplectic transformation on the map with a linear tune of $\nu = \frac{1}{4} + \delta$:

$$\begin{aligned}
&\text{Suppose } \mathscr{A}x = ix \;\text{ and }\; \mathscr{A}p = -ip \\
&\quad\Rightarrow \mathscr{N} = \mathscr{A}\exp\left(: -\frac{\mu}{2}\left(x^2+p^2\right) :\right)\exp\left(: \frac{1}{4}x^4 :\right)\mathscr{A}^{-1} \\
&\quad\quad\;\; = \exp\left(: \frac{\mu}{2}\left(x^2+p^2\right) :\right)\exp\left(: \frac{1}{4}x^4 :\right).
\end{aligned} \tag{5.51}$$

The angle μ changed into $-\mu$. I can multiply this map by minus the identity and the invariant will not change because this map as well as the original one commutes with $-I$.

$$\begin{aligned}
\mathscr{N}_c &= (-\mathscr{I})\exp\left(: \frac{\mu}{2}\left(x^2+p^2\right) :\right)\exp\left(: \frac{1}{4}x^4 :\right) \\
&= \exp\left(: \frac{\mu+\pi}{2}\left(x^2+p^2\right) :\right)\exp\left(: \frac{1}{4}x^4 :\right) \\
&= \exp\left(: -\frac{\pi/2 - 2\pi\delta}{2}\left(x^2+p^2\right) :\right)\exp\left(: \frac{1}{4}x^4 :\right) = \mathscr{M}_{\nu=\frac{1}{4}-\delta}.
\end{aligned} \tag{5.52}$$

By rotating the map $\mathscr{M}_{\nu=\frac{1}{4}+\delta}$ in the complex plane, I produced a map which has *exactly* the same real invariants as $\mathscr{M}_{\nu=\frac{1}{4}-\delta}$. Therefore, I can claim rigorously that the map with no real islands has islands located at exactly the same value of J in the complex plane. Therefore the Taylor series of a canonical transformation on the no-island map will diverge exactly at the same J location in phase space. This is illustrated on Fig. 5.3.

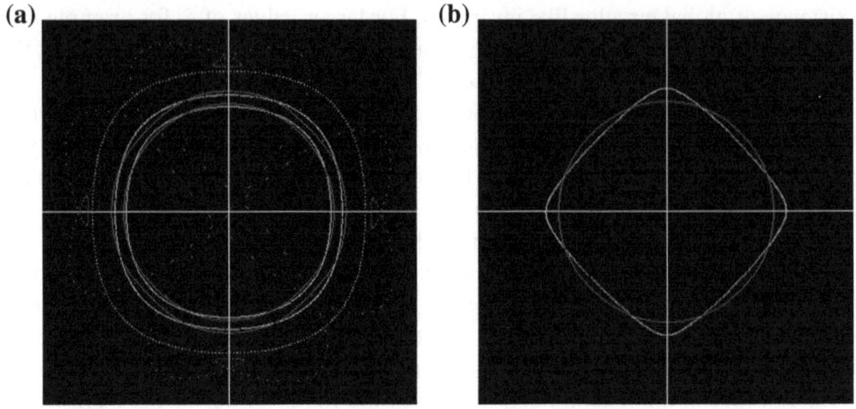

Fig. 5.3 Octupole driven resonance with 20th-order normalisation in *red*. **a** Below the resonance. **b** Above the resonance

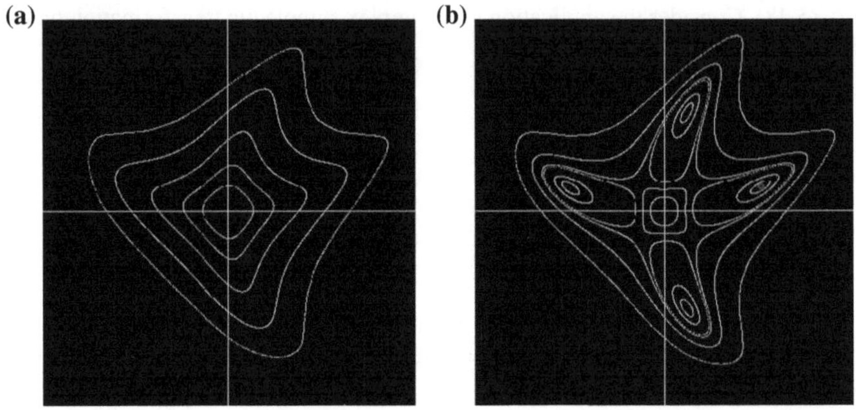

Fig. 5.4 Sextupole driven resonance. **a** Below the resonance. **b** Above the resonance

Using the FPP package, I computed a full normalisation of the map to 20th-order below and above the resonance. Both pictures have the same scale. On Fig. 5.3b, I display in blue a trajectory just below the separatrix; the corresponding normalised curve is red. As one can see, it is a circle. On Fig. 5.3a I display 3 curves. The inner one is normalised into a circle with radius just slightly smaller than the near separatrix curve on Fig. 5.3b. On the second curve, the normalisation fails. On the outer curve we witness a grotesque failure.

Thus we have checked numerically that Taylor series normalisation can failed for physically sound reasons (real islands) or mathematical reasons (islands in the complexified map). Therefore, in the case of Fig. 5.3a, only a "one-resonance" normalisation has a chance of working if the canonical transformation is represented by a Taylor series.

The situation of Fig. 5.3a is analogous to a spin-orbit resonance affecting the spin as we will see in Sect. 6.5.1.

The spin map is a linear spectator map, a matrix in $SO(3)$ in the case of PTC, which depends nonlinearly on the orbital variables. We will see that nothing qualitatively new happens near a resonance because there are no "spin islands" in a linear map. However attempts to perform a full spin normalisation using an expanded Taylor series will also fail. For this reason, spin will also require a one-resonance normal form and the computation of a co-moving map. In the case of spin, due to its greater simplicity, the co-moving map—or the Abell-Barber map as I call it—will turn out to be exactly solvable. Its exact solution will expose the futility of a Taylor expansion.

The next map I will now present is somewhat pathological but it displays very strikingly the presence of islands greatly affected by neighbouring resonances.

5.4.4.2 The Sextupole Map $\mathscr{M} = \mathscr{R}_{\nu=\frac{1}{4}\mp 0.001} \exp\left(: x^3 :\right)$: not as Trivial...

I will now study the following map:

$$\mathscr{M} = \mathscr{R}_{\nu=\frac{1}{4}\mp 0.001} \exp\left(: x^3 :\right). \tag{5.53}$$

Figure 5.4b clearly shows the map of Eq. (5.53) with four huge islands. These four islands appear isolated and therefore we hope that Taylor based perturbation theory will work well. Moreover, with a bit of luck, the leading order Lie operator for the $\mathscr{M}^4$ (or the pseudo-Hamiltonian) should provide us, at least, with a qualitative estimate of the island size and positions. Well, it will not; our luck runs out! Nevertheless we will understand exactly the causes of this failure for the simple case of Eq. (5.53).

It is easy to compute the leading order pseudo-Hamiltonian with the help of a package such as my FPP package. Here is the result to fourth-order:

$$\begin{aligned}\mathscr{C}\mathscr{A}\mathscr{M}^4\mathscr{A}^{-1}\mathscr{C}^{-1} &= \exp\left(\overset{\star}{\underset{\star}{:}} -\frac{1}{4\pi} H_4 \overset{\star}{\underset{\star}{:}}\right) \\ \text{where } H_4 &= \nu_0 u\overline{u} + A\left(u\overline{u}\right)^2 + Bu^4 + \overline{B}\overline{u}^4.\end{aligned} \tag{5.54}$$

The map $\mathscr{A}$ transforms away the terms not belonging[6] to the family of $4\nu_x$, the map $\mathscr{C}$ puts us in phasors' basis (see Eq. (2.10)) and the operator $\overset{\star}{\underset{\star}{:}} H_4 \overset{\star}{\underset{\star}{:}}$ defined by Eq. (3.74) takes Poisson brackets with phasors (adding an extra factor of $-2i$). For our example, the coefficients ν_0, A, B and C are just (to 4 digits of accuracy):

[6] ν_x, $2\nu_x$ and $3\nu_x$ driving terms are cleansed to second-order in sextupole strength.

$$\nu_0 = \frac{1}{100}$$
$$A = -8.783\ 10^{-2}$$
$$B = 4.505\ 10^{-2} + 5.661\ 10^{-4} i. \tag{5.55}$$

The fixed points can be easily computed using action-angle as in Eq. (5.49). I leave it to the reader to show that the formula for the actions of the fixed points is:

$$u\bar{u}|_{\pm} = 2J_{\pm} = -\frac{1}{2}\frac{\nu_0}{A \pm 2\sqrt{B\bar{B}}}. \tag{5.56}$$

The reader will notice that one fixed point, J_+ is located at $J_+ = 1.41\ 10^{-3}$ according to Eq. (5.56). If I average the position of the four fixed points on the non-normalised graph (which should nullify some of the effect of the other resonances), then I come up with a value of $J_+ = 1.37\,10^{-3}$. This is not too bad. Looking at the plot of Fig. 5.4b, I can see that the inner separatrix is "fairly" square and should be predicted more or less by a fourth-order naive calculation.

As for the other fixed point, the leading order calculation fails completely. It predicts a negative value of $J_- = -0.11$. Therefore this fixed point does not exist according to second-order *analytic* perturbation theory. Incidentally, on p. 210 of [10], it is claimed that the system we are now looking at, can be described by a quartic Hamiltonian (Eq. 2.394 of Ref. [10]). If the author of that book is performing analytical canonical transformations, that is impossible. Most likely he is fitting tracking or experimental data. If the resulting canonical transformation is *not analytic*, it may indeed produce different results at second-order; see Ref. [1] where our little map is treated differently and theoretical results consistent with non-analytic transformations are presented. Unfortunately, the author of Ref. [10] only says:

> ...the resonance strength $G_{4,0,\ell}$ can be obtained from the Fourier transformation of the effective particle Hamiltonian in the synchrotron.

Well may be so, but this is all Greek to me[7]—or shall I say in this case more appropriately—*c'est du chinois*!

In any event, we should not be too surprised by this annoying result: the island is a very long banana shaped like object. Near the unstable fixed point, it does look indeed like an unstable system. It is only when considering a higher order calculation and the proper treatment of neighbouring resonances (beyond trivial averaging), that one can get a result that is approximately correct.

[7] I need to be very explicit here because Lee's book [10] is famous, especially in Asia where I work, and the results presented here seem to contradict his pronouncements. Since his pronouncements are couched in the usual East Asian vagueness, I felt the need to clarify the origin of the apparent discrepancy.

While this is not easy to do purely analytically, it is easy to try with a package like FPP. For example, I can compute the 6th-order extension to the pseudo-Hamiltonian of Eq. (5.54): It is given by

$$H_6 = H_4 + u\overline{u}\ \left(Cu^4 + \overline{C}\overline{u}^4\right)\ + D(u\overline{u})^3. \tag{5.57}$$

The additional coefficients are

$$\begin{aligned} C &= 0.1005 - 1.263\ 10^{-3} i \\ D &= 0.3362. \end{aligned} \tag{5.58}$$

Here are the results of a calculation using the FPP package on this Hamiltonian.

```
               Order              6
 exact fixed point
0.164426706049449          4.055421249340808E-002
 approximate fixed point in normalised units
0.162343786856694          5.100197260436692E-004
 approximate fixed point  in original units
0.164457491808164          4.052997676408270E-002
```

The first line contains the fixed point represented by a red circle on Fig. 5.4b. I computed it with a Newton search on the fourth power of the map. It is exact to machine precision. The second line contains the fixed point obtained by a Newton search on the one-resonance vector field of H_6. Finally that point is mapped into the original phase space by $\mathscr{A}$. The result is close to the exact fixed point. Accidentally, it is possible to compute the fixed point of H_6 exactly because it can be rotated into a Hamiltonian which does not have any imaginary terms. Thus the modulus of the fixed point is given by:

$$\begin{aligned} u\overline{u}|_{\pm} &= \frac{-(A \mp 2B) + \sqrt{(A \mp 2B)^2 - 3\nu_0\,(D \mp 2C)}}{3\,(D \mp 2C)} \\ \longrightarrow\ u\overline{u}|_{+} &= 0.162344587994465. \end{aligned} \tag{5.59}$$

For this map, it is possible to go to a higher order with improved results.

```
              Order              12
 exact fixed point
 0.164426706049450         4.055421249340815E-002
 approximate fixed point in normalised units
 0.162264710980367         5.097713010342348E-004
 approximate fixed point  in original units
 0.164426502511187         4.055370460794808E-002
```

```
                Order              18
exact fixed point
0.164426706049449        4.055421249340808E-002
approximate fixed point in normalised units
0.162264366343349        5.097606276932037E-004
approximate fixed point  in original units
0.164426683146872        4.055418432358079E-002
```

The moral of this story is that it is not always true that a lowest order calculation will provide a qualitatively correct picture. In this case, if analytic transformations are used, one is forced to go beyond the lowest reasonable order. The presence of neighbouring resonances, more precisely sextupolar distortions, is sometimes impossible to neglect despite tacit assertions by some to the contrary.

5.5 A Map with a Limit Cycle: Akin to a Resonance

In Sect. 9.2 I mention topics that are germane to this book which are excluded because I do not have the ideal tools to illustrate the theory. Here it is the opposite. The map of this section will not be found[8] in an accelerator. It is a very non-symplectic map which displays limit cycles and what appears to be a strange attractor. I included this map to show that it is possible to compute limit cycles with normal forms.

$$\begin{aligned} x^{new} &= \left(1+\alpha-x^2\right)\left(\cos\left(2\pi\nu\right)\ x+\sin\left(2\pi\nu\right)\ p\right) \\ p^{new} &= \cos\left(2\pi\nu\right)\ p-\sin\left(2\pi\nu\right)\ x. \end{aligned} \tag{5.60}$$

The map of Eq. (5.60) in both examples of Fig. 5.5 contains linear anti-damping represented by a positive value of α. This damping is mitigated by the term $-x^2$ in Eq. (5.60).

5.5.1 The Computation of Limit Cycles

For very small α, I expect the anti-damping effect to be balanced on average by $-x^2$. By naive averaging techniques, it is possible to derive an equilibrium radius. The formula is given by

[8] I am not a specialist in electronics, but limit cycles could be perhaps found in the electronic circuitry of an accelerator. But that is pushing things a bit far...Look at Ref. [11] for example.

(a) (b)

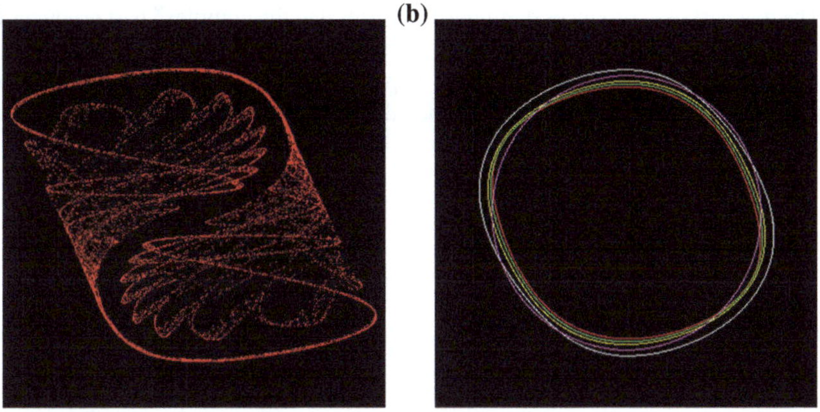

Fig. 5.5 Strange attractor and limit cycle: $\nu = 0.4433$. **a** $\alpha = 0.9$: scale of the graph from -1.54 to 1.54. **b** $\alpha = 0.2$: scale of the graph from -0.78 to 0.78

$$r_\infty = \sqrt{\frac{4\alpha}{1 + 2\cos^2(2\pi\nu)}}. \tag{5.61}$$

On Fig. 5.5b, the red curve is the limit cycle: it was obtained by tracking a ray from the interior and plotting only after the particle reaches the red curve. The purple curve is the circle obtained using the formula of Eq. (5.61). The reader should check that for a tiny α, Eq. (5.61) hits the target with a precision of order α!

The other curves are the result of a normal form calculation which bears some resemblance with the one-resonance normal form of Sect. 5.4. The white curve is the result of a third-order calculation (third in the degree of the polynomial). The yellow curve is a fifth-order result and finally the green curve is a seventh-order calculation. Higher order calculations are practically on top of the limit cycle. So how does one do this calculation with normal forms? Here is the theory.

I first perform a standard normal form on this nonlinear map denoted as $\mathscr{M}$, where I leave all tune shifts and all amplitude dependent damping terms. The result has the form:

$$\mathscr{R} = \mathscr{A}\mathscr{M}\mathscr{A}^{-1} = \exp(T_1 \cdot \nabla)\exp(T_3 \cdot \nabla)\exp(T_5 \cdot \nabla)\cdots. \tag{5.62}$$

The vector field T_n, in the phasors' basis, is given by:

$$\begin{aligned} T_{n;1} &= \bar{t}_n(\varphi_+\varphi_-)^{\frac{n-1}{2}}\varphi_+ \\ T_{n;2} &= t_n(\varphi_+\varphi_-)^{\frac{n-1}{2}}\varphi_- \\ \text{where } t_n &= \alpha_n + i2\pi\nu_n. \end{aligned} \tag{5.63}$$

The real part of t_n produces damping-like terms and the complex[9] part produces tune shifts.

The analogy to the one resonance map of Sect. 5.4 lies in the non-commutative properties of the Lie exponents T_n. They do not commute but they form a closed set under commutation. Therefore we know that $\mathscr{R}$ can be re-expressed as:

$$\mathscr{R} = \exp\left(\left\{T_1' + T_3' + T_5' + \cdots\right\} \cdot \nabla\right) \tag{5.64}$$

where all the T_n''s have the same form as the T_n's and actually $T_1' = T_1$.

$$\begin{aligned} T_{n;1}' &= \bar{t}_n' (\varphi_+ \varphi_-)^{\frac{n-1}{2}} \varphi_+ \\ T_{n;2}' &= t_n' (\varphi_+ \varphi_-)^{\frac{n-1}{2}} \varphi_- \\ \text{where } t_n' &= \alpha_n' + i 2\pi \nu_n' \end{aligned} \tag{5.65}$$

I leave it to the reader to check that the Lie operators T_n''s form a closed set under the Lie bracket given by Eq. (3.38). Assuming that we have obtained Eq. (5.64) from Eq. (5.62), the equation for the limit cycle is now simple. I first rewrite T' as the sum of a damping and a tune shift term

$$T_\pm' = \left\{ \underbrace{\sum_{n=1} \alpha_n' (\varphi_+ \varphi_-)^{\frac{n-1}{2}}}_{\text{Damping}=0} \mp i 2\pi \sum_{n=1} \nu_n' (\varphi_+ \varphi_-)^{\frac{n-1}{2}} \right\} \varphi_\pm \tag{5.66}$$

and I set the damping to zero:

$$\sum_{n=1} \alpha_n' r_\infty^{n-1} = 0 \text{ where } r_\infty = \varphi_+ \varphi_- . \tag{5.67}$$

Once we have found r_∞, then the limit cycle in the original variables is gotten by letting the map $a = \mathscr{A} I$ act on all the points of the circle of radius r_∞. Mathematically, using the jargon of sets, I can write:

$$\text{Limit Cycle} = \{ z \mid z = a\left(r_\infty \cos(\theta), r_\infty \sin(\theta)\right) \text{ such } \theta = 0 \to 2\pi \} . \tag{5.68}$$

[9]Consult supplemental Chap. 10 to see that one must choose the phasors judiciously to insure this simple interpretation beyond linear matrices.

5.5.2 Computation of $\mathscr{R}$ of Eq. (5.62): Co-sinking Map

On p. 132 at item 3, I mentioned that there are two ways to get an invariant in the one-resonance case. One can take the logarithm of a map (as described in supplemental Chap. 11) or one can perform a diverging normal form and then get the invariant nevertheless. The same is true here.

Here the concept of a co-moving map is simple: any linear rotation commutes with the map of Eq. (5.62). One can check this directly with Lie brackets or just by using a little logic: the normalised map is a sink near the origin which is rotationally invariant and therefore it should commute with a rotation.

Therefore I consider the map

$$\mathscr{R}_c = \exp\left(\begin{pmatrix} i2\pi\nu_1\varphi_+ \\ -i2\pi\nu_1\varphi_- \end{pmatrix}\cdot\nabla\right)\mathscr{R}. \tag{5.69}$$

The first factor in Eq. (5.69) is a rotation of angle $-2\pi\nu_1$ and consequently $\mathscr{R}_c$ has no linear tune and thus is near the identity. If you think of this map as a sink (or toilet bowl), in this co-moving frame the water flows, linearly, straight into the sink hole. In a numerical calculation it is best to take the logarithm of this vortex free sink following supplemental Chap. 11.

$$T'\cdot\nabla = \begin{pmatrix} i2\pi\nu_1\varphi_+ \\ -i2\pi\nu_1\varphi_- \end{pmatrix}\cdot\nabla + \log\left(\mathscr{R}_c\right). \tag{5.70}$$

I can also do a complete normalisation of $\mathscr{R}$ removing the tune shifts and nonlinear damping and then use the canonical transformation to get T'. This technique is useful in analytical calculations. The resulting normal form diverges at the limit cycle; however, as in the case of a single resonance, the divergence is removed in the calculation of the single Lie exponent. The reader is invited to look at p. 130–131 where I outline the logic of this procedure.

The next step consists in computing r_∞ which requires the solution of an algebraic polynomial equation with real coefficients.

5.5.3 Example Program for the Limit Cycle of Fig. 5.5b

The program which analyses Eq. (5.60) and computes its limit cycle can be found in Appendix L. I will only show here the normalisation part and the computation of the limit cycle. I first start with the normalisation:

```
call  c_normal(m,n)            ! (4)

if(normalise) then             ! (5a)
```

```
    remove_tune_shift=.true.
    m_n=n%a_t**(-1)*m*n%a_t
    call  c_normal(m_n,n_n)
    m_n=to_phasor()*m_n*from_phasor()
    a1= to_phasor()*n_n%a_t**(-1)*from_phasor()
    call  flatten_c_factored_lie(n_n%ker,vf)
    T_prime=a1*vf
  else                               ! (5b)
    m_n=to_phasor()*n%a_t**(-1)*m*n%a_t*from_phasor()
    vf%v(1)=+i_*(1.d0.cmono.1)*(twopi*n%tune(1))
    vf%v(2)=-i_*(1.d0.cmono.2)*(twopi*n%tune(1))
    R_c=exp(vf,m_n)
    T_prime=log(R_c)
  endif

  radius=sqrt(4*alpha/(2*cos(twopi*nu)**2+1))  ! (6)
  rad0=radius; write(6,*)"naive average radius =",radius;
```

The map m is normalised at line (4). The logical (boolean) parameter `normalise` is defaulted to false which forces the logarithm to be computed rather than performing a secondary "diverging" normalisation. So following the "else" branch, we can see that the map m is put into the normalised phasors' basis and the result is plugged into `m_n`. On the next three lines, the vector field of the linear part is created and used to produce the map $\mathscr{R}_c$. This implements Eq. (5.69). As for the logarithm, i.e., Eq. (5.70), it is implemented on the next line where `T_prime` is extracted. Notice that the addition of the linear vector field to produce the full T' is not necessary if only the computation of the limit cycle is desired. Therefore it is left out. Line (6) evaluates the naive average formula of Eq. (5.61).

In the next part, the code solves Eq. (5.67) using 10 steps of a Newton search at line (8). I did not use any special tools of the library FPP but constructed "by hand" the polynomial of Eq. (5.67) at line (7) and its derivative which is needed in a Newton search.

The final step is the construction of the limit cycle set at lines (9a) and (9b). First a circle of radius r_∞ is created with 1000 rays and then the canonical transformation `n%a_t` is applied on each ray. This implements the set of Eq. (5.68). This set of 1000 rays is printed in the file `plot.dat`.

```
  call kanalnummer(mf,"plot.dat")
  call kanalnummer(mf1,"naive.dat")
  ns=1000
  do i=1,ns
  m%v(1)=radius*cos(twopi*i/ns)    ! results of perturbation
   theory               ! (9a)
  m%v(2)=radius*sin(twopi*i/ns)
  m=n%a_t.o.m !A_t(z) where z=(radius*cos(twopi*i/ns),radius*sin
```

```
(twopi*i/ns)) ! (9b)
z(1)=m%v(1).sub.'0'
z(2)=m%v(2).sub.'0'
write(mf,*) z
 z(1)=rad0*cos(twopi*i/ns)    ! naive result
 z(2)=rad0*sin(twopi*i/ns)
write(mf1,*) z
enddo
close(mf)
close(mf1)
```

It is the file `plot.dat` which was plotted on Fig. 5.5b and produced the white, yellow and green curves.

References

1. F. Willeke, F. Schmidt, Proceedings of the First European Particle Accelerator Conference (Joint Accelerator Conferences Website, CERN, 1988), p. 908
2. Y. Wu, E. Forest, KEK Preprint 2005–108 A (2006)
3. A.J. Dragt, *Lectures on Nonlinear Orbit Dynamics*, vol. 87, p. 147 (American Institute of Physics, New York, 1981)
4. E. Forest, *Beam Dynamics: A New Attitude and Framework* (Harwood Academic Publishers, Amsterdam, 1997)
5. K. Makino, M. Berz, Nucl. Instr. Meth. **A558**, 346 (2005)
6. A. Bazzanni, P. Mazzanti, G. Servizi, G. Turchetti, Nuovo Cimento B **51** (1988)
7. A. Bazzani, E. Todesco, G. Turchetti, G. Servizi, Physica D **64** (1993)
8. E. Forest, J. Irwin, in *Nonlinear Problems in Future Particle Accelerators, Capri, Italy*, ed. by W. Scandale, G. Turchetti (World Scientific, Singapore, 1990), p. 46
9. F. Schmidt, G. Franchetti, Phys. Rev. Lett. **114** (2015)
10. S.Y. Lee, *Accelerator Physics* (World Scientific Publishing, Singapore, 2004)
11. E. Tamaševičiūtė et al., Nonlinear Analysis: Modell. Control **13**(2), 241 (2008)

Chapter 6
Spin Normal Form

Abstract In this chapter I extend the concept of normal form to the spin-orbit map when spin is a spectator. This leads to the introduction of the invariant spin field or ISF. Every concept of the harmonic system extends to spin: Courant-Snyder, phase advance, one resonance, etc. I do not discuss the addition of modulated magnets: they are trivially added on top as additional dimensions.

Keywords Spin · Linear · Nonlinear · Invariant spin field (ISF) · Isolated resonance · Abell-Barber factorisation

6.1 Introductory Verbiage on Spin in the Code PTC

In accelerator physics it is sufficient to consider spin as a spectator: spin is affected by the orbital motion but the orbital trajectory is not affected by spin. In the example code PTC, three spin directions are carried by the type probe_8 (or probe for the usual real numbers):

```
type probe_8
   type(real_8) x(6)      ! Polymorphic orbital ray
   type(rf_phasor_8) AC   ! Modulation of magnet
   real(dp) E_ij(6,6)    !  Envelope for stochastic radiation
   type(spinor_8) s(3)    ! Polymorphic spin s(1:3)
    ...
```

Type spinor_8 (as well as type spinor) contains the 3 spin directions (s_x, s_y, s_z):

```
type spinor
       real(dp) x(3)  ! x(3) = (s_x, s_y, s_z)   with  |s|=1
end type spinor

type spinor_8
   type(real_8) x(3)  ! x(3) = (s_x, s_y, s_z)   with  |s|=1
end type spinor_8
```

To get an accurate one-turn Taylor map, probe_8 is initialised with the orbital closed orbit (with or without radiation). Because spin is a spectator (in PTC), the

E. Forest, *From Tracking Code to Analysis*,
DOI 10.1007/978-4-431-55803-3_6

closed orbit does not depend[1] on spin. Spin is initialised with the three independent directions $\mathbf{e}_x = s(1)\%x(1:3) = (1,0,0)$, $\mathbf{e}_y = s(2)\%x(1:3) = (0,1,0)$ and $\mathbf{e}_z = s(3)\%x(1:3) = (0,0,1)$. If the spin was not a spectator, PTC would have to be structured differently. In that case, in the SO(3) configuration, spin would be represented by three polymorphs whose constant part would be the spin closed orbit.

Notwithstanding this spectator *caveat*, everything proceeds as usual in the tracking code. To get the one-turn map, I add the closed orbit to the identity map in the usual fashion:

```
!!!! Polymorphic probe is created in the usual manner
   id_s=1

  xs=xs0+id_s ! xs is probe_8, xs0 is a probe with the closed orbit.

!!!! get spin polymorphic probe after one turn
call propagate(als,xs,state,fibre1=1)
!!!!!!!!!!!!!!!!!!!!!!!!!!!!!!!!!!!!!!!!!!!!!!!!!

! Copy probe_8 into a complex damap
! because we use the complex package for normal forms
c_map=XS
!!!!!!!!!!!!!!!!!!!!!!!!!!!!!!!!!!!!!!!!!!!!!!!!!

!!!! one resonance is left in the map
c_n%nres=1
c_n%m(2,1)=1    !  the
c_n%ms(1)=1     ! spin tune
!     so the resonance is    n_y + n_spin = integer
call c_normal(c_map,c_n,dospin=my_true)
```

In the above code fragment, the c_damap `ID_S` is set to unity. I recall for the reader that a c_damap is defined as:

```
type c_damap
 type (c_taylor) v(lnv)
 integer :: n=0
 type(c_spinmatrix) s
 complex(dp) e_ij(6,6)
end type c_damap
```

The c_spinmatrix `id_s%s` contains the nine entries of the spin matrix `id_s%s%s (3,3)`. The map is added to the orbit and I proceed as usual. The object is tracked for a single turn. I then put back the probe_8 `XS` into the map `c_map` which is then normalised.

In this particular example, one resonance is left in the map: $\nu_y + \nu_{spin} = n$. This is done, for example, if the purpose of the calculation is to estimate the loss of polarisation due resonance crossing via the Froissart-Stora formula. I will first concentrate on the full normalisation: no resonances.

---

[1]This is a bit analogous to the case of the cavity free Jordan normal form described in Sect. 4.2 because time (or z_6) is also a spectator. But the analogy stops here because the normal form for the spin will be akin to an oscillator as we will see.

6.2 The Normal Form for Spin on the Closed Orbit: n_0

Let me introduce some notation for the map which is reflected in the library FPP. I start with a map of type c_damap. Mathematically it can be written as follows:

$$T = (m, S) \tag{6.1}$$

where m is an orbital map and S is a spin matrix that depends on the orbit. This map acts on a ray z and a spin vector $\mathbf{s}$ as follows:

$$T(z, \mathbf{s}) = (m(z), S(z)\mathbf{s}) . \tag{6.2}$$

The matrix for the spin is evaluated at z and multiplies onto the vector $\mathbf{s}$.

Next I need to tell you how one concatenates two such maps. If a beam line #1 is followed by beam line #2, then the spin map for the full beam line is given by:

$$\begin{aligned} T_2 \circ T_1 &= (m_2, S_2) \circ (m_1, S_1) \\ &= (m_2 \circ m_1, S_2 \circ m_1 S_1) . \end{aligned} \tag{6.3}$$

The matrix $S_2 \circ m_1 \ S_1$ is simply the product of $S_2 S_1$ where $S_2(\overline{z})$ is evaluated at $\overline{z} = m_1(z)$ with $z = (x, p_x, y, p_y, z_5, z_6)$. The phase space can be, depending on the situation, the transverse phase space or the full 6-dimensional one. Modulation clocks can also be added. For example, if we have 3 modulation clocks, the phase space dimension[2] can be potentially 12.

From now on, if a single index is used (or if none are used), the map is the one-turn map as some position s. For example the one-turn map around the closed orbit (or fixed point) at s, I will denote as follows:

$$T_s = (m_s, S_s) . \tag{6.4}$$

The matrix $S_s(z)$ is a function of the orbital+clock variables. However on the fixed point whose coordinates will be $\mathbf{0} \equiv (0, 0, \ldots, 0)$ without loss of generality, it is really easy to raise T_s to a power

$$T_s^k (\mathbf{0}, \mathbf{s}) = (\mathbf{0}, S_s^k(\mathbf{0})\mathbf{s}) . \tag{6.5}$$

Equation (6.5) simply reflects the fact that on the closed orbit, the matrix S for the spin is a constant matrix turn after turn. This matrix is a rotation and thus contains an invariant direction denoted as $\mathbf{n}_0$. We have

$$S_s^k(\mathbf{0})\mathbf{n}_0 = \mathbf{n}_0. \tag{6.6}$$

---

[2]FPP allows as many dimensions as the computer memory permits; PTC, the example code, has only one clock at the moment, so the maximum dimension is 8.

I drop, for the time being, the index s keeping in mind that I am dealing with the one-turn map at some arbitrary s. The matrix $S(\mathbf{0})$ can be expressed in terms of $\mathbf{n}_0$ and its rotation angle θ_0 around $\mathbf{n}_0$:

$$S_s^k(\mathbf{0}) = \exp\left(k\,\theta_0\,\mathbf{n}_0 \cdot \mathbf{L}\right). \tag{6.7}$$

The matrices L_i are the usual generator of rotations obeying the commutation relations of the rotation group:

$$\left[L_i, L_j\right] = \varepsilon_{ijk} L_k. \tag{6.8}$$

They are

$$L_1 = \begin{pmatrix} 0 & 0 & 0 \\ 0 & 0 & -1 \\ 0 & 1 & 0 \end{pmatrix} \quad L_2 = \begin{pmatrix} 0 & 0 & 1 \\ 0 & 0 & 0 \\ -1 & 0 & 0 \end{pmatrix} \quad L_3 = \begin{pmatrix} 0 & -1 & 0 \\ 1 & 0 & 0 \\ 0 & 0 & 0 \end{pmatrix}. \tag{6.9}$$

I will refer to $L_{1,2,3}$ as $L_{x,y,z}$ most of the time.[3]

As long as the particle is on the closed orbit, $\mathbf{n}_0$ is an invariant direction; the spin of a particle on this processes around $\mathbf{n}_0$. At this point it is so simple that it is not necessary to introduce the concept of a normal form. But since things will get a little obscure later, I will introduce immediately a normal form for the spin matrix around the closed orbit. Indeed, let us pick a direction, say the vertical direction, it is possible to rotate our frame of reference so as to convert $\mathbf{n}_0$ into the vertical direction $\mathbf{e}_y$, in other words:

$$D_0^{-1} S(\mathbf{0}) D_0 = \exp\left(\theta_0 L_y\right). \tag{6.10}$$

Two very important things should be noted:

1. The normal form is a commutative group, i.e., a rotation around a single axis. It is completely analogous to the orbital normal form.
2. The matrix D_0 is **not** unique because we can add a fixed rotation around the y-axis. This is analogous to the orbital case.

Therefore it is not surprising that the concept of phase advance will be **identical** to that of the orbital case. In fact, once we define the full normal form around any orbit, everything we know about the orbital motion will carry over to the spin.

Additionally, as first noticed by Dan Abell,[4] everything we know about the one-resonance normal formin the orbital case will also carry over to spin! It is really

[3] My choice for θ_0 produces, unfortunately, a negative spin tune. It is not wrong, but too late to change. The reader is invited to put a minus sign if he compares with other treatments.

[4] Chinese whispers via Desmond Barber.

conceptually powerful that all the results of Sect. 5.4 carry over. As I admit later, I only realised this fact while writing this book!

6.3 The Nonlinear Normal Form for the Invariant Spin Field $\mathbf{n}(z)$

6.3.1 *Short History and Comments on the Mysterious ISF $n(z)$*

I must point out that the initial elucidation of this topic, based on Soviet work, was done by Yokoya of KEK in Ref. [1]. Yokoya published a Hamiltonian treatment using generating functions and later a treatment with Lie methods. I do not recommend a first reading of these Hamiltonian treatments, since they are in my view less faithful to the "code." This is the message of this book. But I feel compelled to give credit to Yokoya since he was the first person to explain to the "West" the normal form for the spin and the mysterious vector **n** first introduced by Soviet physicists. This vector is called the invariant spin field (ISF) by Barber and collaborators.

My own understanding of the ISF was born when I was privileged to listen to a private lecture on the topic by Dr. Desmond Barber of DESY in 2009. It started with the obvious definition of $\mathbf{n}_0$, the invariant spin direction on the closed orbit, which Barber generalised for my own benefit.

Barber's explanations were not based on normal forms. Here I will start with normal form and state Barber's result, namely that the ISF **n** is a generalisation of $\mathbf{n}_0$ and that it obeys a very special equation. The ISF obeys an equation which we will deduce from the normal form later:

$$S(z)\mathbf{n}(z) = \mathbf{n}(m(z)) \quad \text{or} \quad S\mathbf{n} = \mathbf{n} \circ m. \tag{6.11}$$

Equation (6.11) states that there exists (perhaps) a vector $\mathbf{n}(z)$ whose transformation under the spin matrix $S(z)$ is the same as its transformation under the map m. This equation trivially applies to $\mathbf{n}_0$ since it is a constant under the application of $S(\mathbf{0})$ and the closed orbit is by definition a constant, i.e., $m(\mathbf{0}) = \mathbf{0}$.

For an arbitrary z, Eq. (6.11) implies that if we follow **n** after k turns, the answer is simply $\mathbf{n} \circ m^k$. Thus the Fourier spectrum of $\mathbf{n} \circ m^k$ will not contain the spin frequency. This object behaves as if spin motion did not exist. If viewed as a vector field, the entire three dimensional field $\mathbf{n}(z)$ is left invariant under the action of the full spin-orbital map T.

Obviously, if a particle at coordinate z starts with a spin slightly different from $\mathbf{n}(z)$, the actual spin will move around the axis $\mathbf{n}(z)$ and its spectrum will contain the spin tune as well as the orbital tunes.

6.3.2 The Normal Form and Why We Get n(z) from It

Allow me to define the normal form for the full nonlinear spin map and deduce the ISF from the normal form. First, since the spin is a spectator, I can initially normalise the orbital part using a transformation U_{orb}:

$$U_{orb} = (a, I)\,. \tag{6.12}$$

The result will be

$$N_{orb} = U_{orb}^{-1} \circ (m, S) \circ U_{orb} = (r, S_{orb}) \tag{6.13}$$

where r is the usual amplitude dependent rotation of the normalised orbital phase space. S_{orb} is a matrix which depends on the normalised orbital phase space:

$$S_{orb}(z) = \exp\left(\boldsymbol{\omega}(z) \cdot \mathbf{L}\right). \tag{6.14}$$

The first delusion one might entertain is to assume that $\omega(z)$ is proportional to $\mathbf{n}(z)$ because it is certainly true that $\omega(0)$ is just $\theta_0\, \mathbf{n}_0$. To check that $\omega(z)$ is not invariant, I apply the square of the map on $\omega(z)$ using Eq. (6.3):

$$\begin{aligned} T \circ T(z, \boldsymbol{\omega}(z)) &= (r, S_{orb}) \circ (r, S_{orb})\,(z, \boldsymbol{\omega}(z)) \\ &= \left(r^2, S_{orb} \circ r\; S_{orb}\right)(z, \boldsymbol{\omega}(z)) \\ \Rightarrow\; \boldsymbol{\omega}^f(z) &= \exp\left(\boldsymbol{\omega}(r(z)) \cdot \mathbf{L}\right) \boldsymbol{\omega}(z). \end{aligned} \tag{6.15}$$

Because $\boldsymbol{\omega}(r(z))$ is not necessarily the same as $\boldsymbol{\omega}(z)$, I generally expect that $\boldsymbol{\omega}^f(z) \neq \boldsymbol{\omega}(z)$. The idea of the normal form is to extend Eq. (6.10) to a position dependent $D(z)$ and write it as a product

$$D(z) = D_0 d(z) \tag{6.16}$$

and apply it to the map N_{orb} of Eq. (6.13):

$$\begin{aligned} \left(I, d^{-1} D_0^{-1}\right) \circ N_{orb} \circ (I, D_0 d) &= \left(I, d^{-1}\right) \circ \left(I, D_0^{-1}\right) \circ (r, S_{orb}) \circ (I, D_0) \circ (I, d) \\ &= \left(I, d^{-1}\right) \circ \left(r, D_0^{-1} S_{orb} D_0\right) \circ (I, d)\,. \end{aligned} \tag{6.17}$$

The spin part of Eq. (6.17) can be factored as

$$D_0^{-1} S_{orb} D_0 = \exp\left(\mathbf{a}(z) \cdot \mathbf{L}\right) \exp\left(\theta_0\, L_y\right), \tag{6.18}$$

where $\mathbf{a}(z)$ does not contain any constant part.

Now I state what the normal form ought to be: one must select $d(z)$ so as to align the spin in the vertical direction and reduce the orbital dependence on invariants alone. Mathematically, it can be written as

$$
\begin{aligned}
\left(I, d^{-1}\right) \circ \left(r, D_0^{-1} S_{orb} D_0\right) \circ (I, d) = \left(r,\ d^{-1} \circ r\ \ D_0^{-1} S_{orb} D_0 d\right) \\
\Downarrow \\
\left(r,\ d^{-1} \circ r\ \ D_0^{-1} S_{orb} D_0 d\right) = \left(r,\ \exp\left(\theta(\mathbf{J})\, L_y\right)\right) \\
\Downarrow \text{ or equivalently} \\
(r, S_{orb}) = (I, D_0 d) \circ \left(r,\ \exp\left(\theta(\mathbf{J})\, L_y\right)\right) \circ \left(I, d^{-1} D_0^{-1}\right).
\end{aligned}
\tag{6.19}
$$

The symbol $\mathbf{J}$ represents all the invariants of the orbital normal form r, namely the action variables, the energy in a cavity-less ring, and the length of the clock hands if present. As usual, in the normalised space it is easier to extract invariant quantities. It should be obvious from Eq. (6.19) that the y-direction is the invariant direction or the ISF in normalised coordinates. Moreover spin precesses around the y-direction with a *constant* amplitude dependent spin angle $\theta(\mathbf{J})$. As in the case of the orbital motion, the existence of $\theta(\mathbf{J})$, different from linear spin tune θ_0, is a generic property of the orbital dependent spin map. It can be seen at least 3 different ways:

1. First, one can Fourier transform the spin motion on an arbitrary trajectory and measure the spin precession numerically. The result is not θ_0.
2. Secondly, we can simulate resonance crossing and clearly see that it does not happen at the position in tune space predicted by the zeroth-order tunes if the orbital amplitude is "large."
3. Mathematically, as will be obvious at Eq. (6.45), it is impossible to select a canonical transformation which removes these "tune shift" terms from the map. They are a fundamental property of the map. This is why, as it is the case of the orbital motion, these terms are measurable by Fourier transforms. Unlike the phase advances, they are not an artefact of the normal form algorithm, but an intrinsic property of the map.

Assuming that this normal form is indeed computable using the spin map (I, d), I can construct the vector $\mathbf{n}$. First let me rewrite the original map in terms of the various transformations in Eqs. (6.13), (6.12) and (6.19):

$$
\begin{aligned}
T = (m, S) &= U \circ \left(r,\ \exp\left(\theta(\mathbf{J})\, L_y\right)\right) \circ U^{-1} \\
&= \underbrace{(a, I) \circ (I, D)}_{U} \circ \left(r,\ \exp\left(\theta(\mathbf{J})\, L_y\right)\right) \circ (I, D)^{-1} \circ (a, I)^{-1} \\
&= \underbrace{(a, I) \circ (I, D_0 d)}_{U=(a,D)} \circ \left(r,\ \exp\left(\theta(\mathbf{J})\, L_y\right)\right) \circ (I, D_0 d)^{-1} \circ (a, I)^{-1} \\
&= \left(I, D_0 \widetilde{d}\right) \circ (a, I) \circ \left(r,\ \exp\left(\theta(\mathbf{J})\, L_y\right)\right) \circ (a, I)^{-1} \circ \left(I, D_0 \widetilde{d}\right)^{-1}
\end{aligned}
\tag{6.20}
$$

where $\widetilde{d} = d \circ a^{-1}$ and $\widetilde{D} = D_0 \widetilde{d}$

Now I consider the action of the map T on the spin-orbit[5] vector $\mathbf{n}$ defined as follows

$$\mathbf{n} = \underbrace{D_0 \tilde{d}}_{\tilde{D}}\, e_y = \tilde{D} e_y, \tag{6.21}$$

and let me act on it using the one-turn map T in its trivial form:

$$T \circ (z, \mathbf{n}(z)) = (m\ ,\ S) \circ (z, \mathbf{n}(z)) = (m(z), S\mathbf{n}(z)) \,. \tag{6.22}$$

Next, I use the normal form description of the map T, that is to say, Eq. (6.20), and act on $\mathbf{n}$ again:

$$\begin{aligned}
T \circ \left(I, \tilde{D}\right) \circ \left(z, e_y\right) &= \left(I, \tilde{D}\right) \circ (a, I) \circ \left(r, e^{\theta(\mathbf{J})\, L_y}\right) \circ (a, I)^{-1} \circ \left(I, \tilde{D}\right)^{-1} \circ \left(I, \tilde{D}\right) \circ \left(z, e_y\right) \\
&= \left(I, \tilde{D}\right) \circ (a, I) \circ \left(r,\ \exp\left(\theta(\mathbf{J})\, L_y\right)\right) \circ (a, I)^{-1} \circ \left(z, e_y\right) \\
&= \left(I, \tilde{D}\right) \circ \left(m,\ \exp\left(\theta(\mathbf{J} \circ a^{-1})\, L_y\right)\right) \circ \left(z, e_y\right) \\
&= \left(I, \tilde{D}\right) \circ \left(m(z),\ e_y\right) \\
&= \left(m(z), \left\{\tilde{D} \circ m\right\} e_y\right) \\
&\downarrow \ \text{because } e_y \text{ is a constant} \\
&= \left(m(z),\ \tilde{D} e_y \circ m\right) .
\end{aligned} \tag{6.23}$$

Comparing the last line in (6.23) and Eq. (6.22), I conclude the famous equation obeyed by the ISF $\mathbf{n} = \tilde{D} e_y$:

$$S\mathbf{n} = \mathbf{n} \circ m \,. \tag{6.24}$$

The two perpendicular directions to $\mathbf{n}$, $\mathbf{l} = \tilde{D} e_x$ and $\mathbf{m} = \tilde{D} e_z$ define the axes around which spin phase advance is measured.

6.3.3 *The Algorithm for the Spin Normal Form*

The first step is the computation of D_0. I will not say much about it except that it must be done exactly to machine precision. It is really an exercise in linear algebra and Gram-Schmidt orthogonalisation. Of course, as in the orbital case, we have an infinite number of choices. This is only relevant if a phase advance is needed and in analytic calculations: this situation mirrors the Courant-Snyder choice. At present all we need to know is that we have some D_0. Therefore I am at Eq. (6.18) and need to go to the next step.

[5] "Orbit-spin" vector might be a better name given the order I chose!

First it is necessary to factorise the matrix $d(z)$ of Eq. (6.13):

$$d = d_1 d_2 \cdots d_{i-1} d_i \cdots . \tag{6.25}$$

Each factor d_i removes the ith-order in the perturbative parameter. The order can be defined in many[6] ways. Here it will be the order of the homogeneous TPSA polynomial.

I will start with the first step of the iteration and, in a second phase, write the ith step. At the first step of the iteration the map is given by

$$\begin{aligned} N_0 &= \left(r, \exp\left(\mathbf{a}_0(z)\cdot \mathbf{L}\right) \exp\left(\theta_0\, L_y\right)\right) \\ \mathbf{a}_0(z) &= \sum_{k=1,\infty} \mathbf{a}_0^k(z). \end{aligned} \tag{6.26}$$

In practice the ∞ in the summation of Eq. (6.26) is the order of truncation of the calculation predetermined by a call to a subroutine called `init` in the FPP[7] package.

The vector $\mathbf{a}_0^k(z)$ has kth degree homogeneous monomials in the variable z, the parameters and the clock variables for magnet modulation. The role of d_1 is to simplify $\mathbf{a}_0^1$ into terms proportional to invariants. Generally, at order i, the minimal and only needed role of d_i is to simplify $\mathbf{a}_{i-1}^i$. So let me start with d_1:

$$N_1 = \left(I, d_1^{-1}\right) \circ N_0 \circ \left(I, d_1\right) = \left(I, d_1^{-1}\right) \circ \left(r, \exp\left(\mathbf{a}_0(z)\cdot \mathbf{L}\right) \exp\left(\theta_0\, L_y\right)\right) \circ \left(I, d_1\right). \tag{6.27}$$

Writing d_1 using an exponent:

$$\begin{aligned} d_1 &= \exp\left(\delta^1 \cdot \mathbf{L}\right) = \exp\left(\delta_x^1 L_x + \delta_y^1 L_y + \delta_z^1 L_z\right) \\ &= \exp\left(\delta_y^1 L_y + \lambda^1 \Lambda + \overline{\lambda^1 \Lambda}\right) \quad \text{where } \Lambda = \frac{L_x + i L_z}{2}. \end{aligned} \tag{6.28}$$

Now I multiply all the factors in Eq. (6.27) to get S_1:

$$\begin{aligned} N_1 &= (r, S_1) \\ \text{where } S_1 &= d_1^{-1} \circ r\ \ e^{\mathbf{a}_0(z)\cdot \mathbf{L}} e^{\theta_0\, L_y}\ d_1. \end{aligned} \tag{6.29}$$

---

[6] In the normalisation library FPP, it is the degree of the monomial that defines the order of the perturbation. There is nothing sacred about this: the map D can be "canonised," that is to say, put into whatever form the user desires. This is analogous to forcing a matrix into the Courant-Snyder definition after it has been computed by a nondescript algorithm.

[7] Again I must emphasize that this is a natural choice in FPP given the TPSA tools upon which it is built. A better choice would be to discriminate between phase space and parameters.

I evaluate Eq. (6.29) to leading order in the exponents:

$$\begin{aligned} S_1 &= e^{-\delta^1 or\cdot \mathbf{L}}\, e^{\mathbf{a}_0(z)\cdot \mathbf{L}} e^{\theta_0\, L_y} e^{\delta^1\cdot \mathbf{L}} \\ &= e^{-\delta^1 or\cdot \mathbf{L}}\, e^{\mathbf{a}_0(z)\cdot \mathbf{L}} e^{\theta_0\, L_y} e^{\delta^1\cdot \mathbf{L}} e^{-\theta_0\, L_y} e^{\theta_0\, L_y} \\ &= e^{-\delta^1 or\cdot \mathbf{L}}\, e^{\mathbf{a}_0(z)\cdot \mathbf{L}} e^{e^{\theta_0\, adL_y}\delta^1\cdot \mathbf{L}} e^{\theta_0\, L_y} \\ &= \exp\left(-\delta^1 \circ r \cdot \mathbf{L} + \mathbf{a}_0(z)\cdot \mathbf{L} + e^{\theta_0\, adL_y}\delta^1 \cdot \mathbf{L} + \cdots\right) e^{\theta_0\, L_y}. \end{aligned} \tag{6.30}$$

In Eq. (6.30), I use

$$\exp(ad A)\, B = \exp(A)\, B \exp(-A) \quad \text{where} \quad ad A\; B = AB - BA. \tag{6.31}$$

I need to select the vector function δ^1 in Eq. (6.30) so that to leading order the entire exponent reduces to a single function $\theta_0^1(\mathbf{J})$ multiplying L_y. In other words, I must have

$$\begin{aligned} S_1 &= \exp\left(-\delta^1 \circ r \cdot \mathbf{L} + \mathbf{a}_0(z)\cdot \mathbf{L} + e^{\theta_0\, adL_y}\delta^1 \cdot \mathbf{L} + \cdots\right) e^{\theta_0\, L_y} \\ &\approx \exp\left(\theta_0^1(\mathbf{J})\, L_y\right) e^{\theta_0\, L_y} = \exp\left(\left\{\theta_0^1(\mathbf{J}) + \theta_0\right\} L_y\right). \end{aligned} \tag{6.32}$$

The matrix S_1 is then to leading order a rotation around the y-axis with terms that depend on the orbital amplitude and parameters. This process must be repeated order by order until the order of truncation.

I now explain this recursive procedure in more details. The operator adL_y operates by taking commutators. For example,

$$adL_y\, L_z = \left[L_y, L_z\right] = L_x. \tag{6.33}$$

Consequently, the operator Λ is an eigen-operator of adL_y:

$$adL_y\, \Lambda = \left[L_y, \frac{L_x + iL_z}{2}\right] = \frac{-L_z + iL_x}{2} = i\Lambda. \tag{6.34}$$

It follows from Eq. (6.34) that

$$\exp\left(\theta_0\, adL_y\right)\, \Lambda = \exp(i\theta_0)\, \Lambda. \tag{6.35}$$

The last line of Eq. (6.30) contains the Lie operator representation of the non-constant part of the rotation to leading order. It is permissible to replace the nonlinear rotation r by its linear part R without affecting[8] the final result.

[8] An irrelevant phase will be introduced in the map d; this phase will be modified anyway when the transformation is "canonised", i.e., put into a special form such as the Courant-Snyder form in the case of the linear orbital motion.

$$-\delta^1 \circ R \cdot \mathbf{L} + \mathbf{a}_0^1(z) \cdot \mathbf{L} + e^{\theta_0\, adL_y} \delta^1 \cdot \mathbf{L} = \theta_0^1(\mathbf{J})\, L_y\,.$$

$$-\delta^1 \cdot \mathbf{L} + \left(\mathbf{a}_0^1(z) \circ R^{-1}\right) \cdot \mathbf{L} + e^{\theta_0\, adL_y} \delta^1 \circ R^{-1} \cdot \mathbf{L} = \theta_0^1(\mathbf{J})\, L_y. \tag{6.36}$$

$\theta_0^1(\mathbf{J})$ is the part of $a_{0;y}^1$ which depends on the orbital invariants only, it is the only term that remains after normalisation.

Equation (6.36) is solved by going into the phasors' basis and into the spin eigen-operators of Eq. (6.28).

$$a_{0;x}^1 L_x + a_{0;z}^1 L_z + a_{0;y}^1 L_y = \alpha_0^1 \Lambda + \overline{\alpha}_0^1\, \overline{\Lambda} + a_{0;y}^1 L_y$$
$$\text{where } \alpha_0^1 = a_{0;x}^1 - i a_{0;z}^1. \tag{6.37}$$

A similar expansion applies to δ^1:

$$\delta_{0;x}^1 L_x + \delta_{0;z}^1 L_z + \delta_{0;y}^1 L_y = \lambda_0^1 \Lambda + \overline{\lambda}_0^1\, \overline{\Lambda} + \delta_{0;y}^1 L_y. \tag{6.38}$$

I can now solve. First I look at the term proportional to L_y.

$$\delta_{0;y}^1 = \frac{1}{1 - \mathscr{R}^{-1}} \left\{ a_{0;y}^1 - \theta_0^1(\mathbf{J}) \right\}. \tag{6.39}$$

A similar calculation applies to the term proportional to Λ (and $\overline{\Lambda}$):

$$\lambda_0^1 = \frac{1}{1 - e^{i\theta_0} \mathscr{R}^{-1}} \alpha_0^1 \tag{6.40}$$

where θ_0 is the constant part of the spin angle (see Eq. (6.35)).

The entire process is repeated order by order. Now that I have δ^1, I can compute d_1 and feed it in Eq. (6.27).

$$\begin{aligned} N_1 &= \left(1, d_1^{-1}\right) \circ N_0 \circ (1, d_1) \\ &= \left(1, d_1^{-1}\right) \circ \left(r, \exp\left(\mathbf{a}_0(z) \cdot \mathbf{L}\right) \exp\left(\theta_0\, L_y\right)\right) \circ (1, d_1) \\ &= \left(r, \exp\left(\mathbf{a}_1(z) \cdot \mathbf{L}\right) \exp\left(\{\theta_0 + \theta_1\}\, L_y\right)\right). \end{aligned} \tag{6.41}$$

The next iteration can proceed with N_2:

$$N_2 = \left(1, d_2^{-1}\right) \circ N_1 \circ (1, d_2) \tag{6.42}$$

and so on and so forth!

At this stage it is more revealing to write Eqs. (6.39) and (6.40) at the ith-order and go deeper into the details:

$$\delta_{i-1;y}^i = \frac{1}{1 - \mathscr{R}^{-1}} \left\{ a_{i-1;y}^i - \theta_{i-1}^i(\mathbf{J}) \right\} \quad \text{and} \quad \lambda_{i-1}^i = \frac{1}{1 - e^{i\theta_0} \mathscr{R}^{-1}} \alpha_{i-1}^i. \tag{6.43}$$

To see more explicitly what Eq. (6.43) entails, I write it for phase space in two degrees of freedom and no parameters. Thus assuming that I have

$$\{a^i_{i-1;y} - \theta^i_{i-1}(\mathbf{J})\} = \sum_{|k|=i} A^i_{y;\,k}(x+ip_x)^{k_1}(x-ip_x)^{k_2}(y+ip_y)^{k_3}(y-ip_y)^{k_4}, \tag{6.44}$$

then it follows that

$$\delta^i_{i-1;y} = \sum_{|k|=i} \frac{A^i_{y;\,k}}{1-\exp\left(i\left\{(k_1-k_2)\mu_x+(k_3-k_4)\mu_y\right\}\right)} u^{k_1}\overline{u}^{k_2}v^{k_3}\overline{v}^{k_4} \tag{6.45}$$

where u is the horizontal phasor $x + ip_x$ and v is the vertical phasor $y + ip_y$. A similar equation applies to λ^i_{i-1}:

$$\text{if} \quad \alpha^i_{i-1} = \sum_{|k|=i} B^i_k(x+ip_x)^{k_1}(x-ip_x)^{k_2}(y+ip_y)^{k_3}(y-ip_y)^{k_4}, \tag{6.46}$$

then it follows that

$$\lambda^i_{i-1} = \sum_{|k|=i} \frac{B^i_k}{1-\exp\left(i\left\{\theta_0+(k_1-k_2)\mu_x+(k_3-k_4)\mu_y\right\}\right)} u^{k_1}\overline{u}^{k_2}v^{k_3}\overline{v}^{k_4}. \tag{6.47}$$

The reader will notice that the denominator of Eq. (6.47) contains the spin angle. Sometimes, when computing the effects of a spin resonance, one deliberately leaves certain terms out of Eq. (6.47). But, in the case of a complete normalisation, everything goes away besides the amplitude/parameter dependent angle shifts represented by $\theta^i_{i-1}(\mathbf{J})$.

6.3.4 A Code Implementation for the Spin Normal Form

In this section I flesh out the implementation using a code fragment of the example library FPP[9] and correlate it with the theory outlined in Sect. 6.3.3. The incoming map `m1` is already normalised in the orbital part and is in phasors' basis.

```
call c_normal_spin_linear(m1,m1,n%AS,n0)   ! (1)

ri=1 ; ri%s=m1%s.sub.0 ; ! exp(theta_0 L_y)   (2)
egspin(3)=ri%s%s(1,1)-i_*ri%s%s(1,3)
egspin(2)=1.0_dp
```

---

[9]All my code fragments, so far, were from programs in the appendices. You will need to find FPP and open the appropriate file called Ci_tpsa.f90 if you want to see beyond this book. Of course, if you are running the examples, you already have FPP.

```
        egspin(1)=ri%s%s(1,1)+i_*ri%s%s(1,3)
!!! tune is taken from egspin(1) or egspin(3)   spin_tune_def= +/- 1
        n%spin_tune=aimag(log(egspin(2-spin_tune_def))/twopi)
! because  exp(a L_y) x = x- a z + O(a**2)
        ri=ri**(-1) ! exp(-alpha_0 L_y)   (3)
```

The first step at line (1) is the computation of the zeroth-order transformation D_0, Eq. (6.10), which is stored in `n%AS`. This storage space will eventually contain all the spin transformation $D(z)$ of Eq. (6.16). `m1` is transformed by D_0 inside `c_normal_spin_linear`. The map `ri` of line (2) and (3) will eventually contain at line (3) the map $\exp(-\theta_0 L_y)$.

```
         nonl=m1.sub.1 ; nonl%s=1 ;nonl=nonl**(-1)  ! R_0^-1       (4)

        do i=1,no+2

         mt=m1*ri !  S*exp(-theta_0 L_y)     (5)
           call c_find_om_da(mt%s,n0)  ! exp(n0.L)     (6)
           call c_n0_to_nr(n0,n0)    ! n0 = > eigen-operator of spin    (7)
           n0=n0*nonl                !  no * R^-1      (8)

           nr=0
        do k=1,3
          j=1
         do while(.true.)
           call  c_cycle(n0%v(k),j,v ,je); if(j==0) exit;
           call check_kernel_spin(k,xy%n,je,removeit)
                 if(n%nres>0.and.removeit) then
                  do kr=1,n%nres
                  call check_resonance_spin(k,xy%n,je,kr,n%ms,n%m,removeit)
                  if(.not.removeit) then
                    exit
                  endif
                 enddo
                enddo
                endif
              if(removeit) then
                 lam=egspin(k)
                  do l=1,xy%n
                   if(coast(l)) cycle
                     lam=lam*eg(l)**je(l)
                  enddo
                nr%v(k)=nr%v(k) +(v.cmono.je)/(1.0_dp-lam)    ! (9)
               endif
         enddo ! cycle
        enddo ! k
           call c_nr_to_n0(nr,nr)  !   (10)

         AS=1 ; AS%s=exp(nr)*AS%s           ! (11)

         n%AS=n%AS*AS              ! (12)

         m1=c_simil(AS,m1,-1)

        enddo

        n%AS=from_phasor()*n%AS*from_phasor(-1)
        n%AS=n%A_t*n%AS*n%a_t**(-1)
```

At line (4), I store the inverse of the linear rotation of the orbital map. Then the "do loop" over the order begins. Line (5) extracts the non-constant part of the spin matrix: namely the first factor of Eq. (6.26), $\exp(\mathbf{a}_{i-1}(z)\cdot\mathbf{L})$ where i is the loop index. Equation (6.26) described the first iteration $i = 1$.

Line (6) performs the actual computation of $\mathbf{a}_{i-1}(z)$ which it stores in `n0`. Line (7) puts `n0` in the eigen-operator basis as in Eq. (6.37).

At line (8), the linear orbital rotation R^{-1} is substituted into $\mathbf{a}_{i-1}(z)$ as in the second line of Eq. (6.36).

The routine `c_cycle` cycles over all the monomial of the polynomial being analysed. The routine `check_kernel_spin` returns "removeit=false" if the term belongs to a tune shift. Finally, the routine `check_resonance_spin` scans for spin resonances the user may want to keep in the map.

Most important is line (9): it reproduces Eqs. (6.45) and (6.47) for $k = 2$ and $k = 1, 3$ respectively. Line (10) returns the result into the L_x, L_y and L_z basis. Finally line (12) increments $d(z)$ as in Eq. (6.25).

6.4 Leaving One Resonance in the Spin Map

In this section I will present two types of resonances which might affect the spin. The first one is a pure orbital resonance. I will not discuss it in detail. The second one is the classical spin resonance which involves the spin tune. I will apply the theory of the co-moving map on the spin resonance. I call the resulting factorisation the "Abell-Barber factorisation" for reasons that will become clear.

6.4.1 *Two Cases for the Spectator Spin: Orbital and Spin-Orbit*

As in the case of the orbital map, it is possible to leave a single resonance in the map. There are two distinct possibilities:

1. A single orbital resonance is left in the map. This was discussed in Sect. 5.4. For example the beam can be on the orbital resonance $3\nu_x = 1$ but away from any spin resonance.
2. The beam is near a spin resonance or slowly crossing a spin resonance, for example $-\nu_x + \nu_y + \nu_{spin} = 0$. Because the spin is a spectator, I expect such a resonance to affect only the spin.

I will now dispatch case 1. When on an orbital resonance, the ISF $\mathbf{n}$ can be computed; the normalised direction exists and it can be set to e_y.

Can the spin angle rotation around e_y be expressed in terms of the invariants of the orbital map? The answer is a semi-learned: "I am not sure." As I have shown in Sect. 5.4, the orbital invariants are $N - 1$ actions and an invariant with the topology of that single resonance. The spin angle, just like the orbital map, will depend on the $N - 1$ circular invariants and the generators of that single resonance. For example, in the case of the $3\nu_x$ in two degrees of freedom, the spin angle will depend on objects of the form

$$u^{k_1}\overline{u}^{k_2}v^{k_3}\overline{v}^{k_4} \qquad k_1 - k_2 = 3m \quad \text{and} \quad k_3 - k_4 = 0, \tag{6.48}$$

where u, $\overline{u}$, v and $\overline{v}$ are the phasors defined at and below Eq. (6.45).

The problem here is that the invariant function of the orbital map and the functional form of the angle spin rotation, while being made of objects belonging to the same resonance family, will in general be different and non-commuting. So the angle of rotation will change along the orbital trajectory.

So the angle of rotation will depend on the position in phase space but only through functions that have the symmetry[10] of the resonance.

Case 2 is what interests us here. In this case, the problem of small denominators will arise at Eq. (6.47). The algorithm must be modified to avoid the resonance.

$$\lambda_{i-1}^{i} = \sum_{\substack{|k| = i \\ k \neq k^{res}}} \frac{B_k^i}{1 - \exp\left(i\left\{\theta_0 + (k_1 - k_2)\mu_x + (k_3 - k_4)\mu_y\right\}\right)} u^{k_1}\overline{u}^{k_2}v^{k_3}\overline{v}^{k_4}. \tag{6.49}$$

For example, in the case of $-\nu_x + \nu_y + \nu_{spin} = 0$, the terms to avoid in Eq. (6.49) are of the form

$$k^{res} \quad \text{such that} \quad k_1 - k_2 = -1 \quad \text{and} \quad k_3 - k_4 = 1. \tag{6.50}$$

I complete the recursive algorithm as done before, the final result, equivalent to Eq. (6.20), is:

$$\begin{aligned} T = (m, S) &= \left(I, D_0\widetilde{d}\right) \circ (a, I) \\ &\circ \left(r\ ,\ \exp\left(\theta(\mathbf{J})\, L_y\right) \exp\left(\alpha \Lambda + \overline{\alpha}\overline{\Lambda}\right)\right) \circ (a, I)^{-1} \circ \left(I, D_0\widetilde{d}\right)^{-1} \end{aligned} \tag{6.51}$$

where $\widetilde{d} = d \circ a^{-1}$.

The Lie exponent $\alpha\Lambda + \overline{\alpha}\overline{\Lambda}$ drives a single resonance. I explore the implication of this in the next section.

---

[10]The one-resonance normal form is known up to a transformation which depends on the family of the resonance. It would be an interesting problem to determine if one can use this freedom to "align" the spin tune with the orbital invariant. In that case they would commute and the spin tune would stay constant on the trajectory as in a full normalisation of the spin. I think that the answer is yes but I do not know for sure.

6.5 The Abell-Barber Co-moving Map for the Spin Orbit Case

In Sect. 5.4, I discussed the one-resonance normal form. In particular, I introduced the concept of a co-moving map. This co-moving map can be computed easily in the normalised form of the map.

In the case of a single spin resonance, the theory of Sect. 5.4 applies. In fact it is even simpler because the spin is a spectator: orbital planes are automatically orthogonal in the normalised frame; moreover the one-resonance spectator map is exactly solvable.

Unknown knowns

U.S. Secretary of Defence Donald Rumsfeld in a famous verbal "sortie" once said:

> Reports that say that something hasn't happened are always interesting to me, because as we know, there are known knowns; there are things we know we know. We also know there are known unknowns; that is to say we know there are some things we do not know. But there are also unknown unknowns – the ones we don't know we don't know. And if one looks throughout the history of our country and other free countries, it is the latter category that tend to be the difficult ones.

The Marxist philosopher Slavoj Žižek also stated that there are also "unknown knowns" where we do not discern what we already know. In my case, plain stupidity is what caused the following "unknown known."

The factorisation I will present here is just the factorisation of the one-resonance orbital map of Sect. 5.4 applied to the spin-orbit map. If I had believed my own propaganda, I would have found this out by myself and this section would have started with the ever arrogant: "*it obviously follows after a few trivial steps that …blah blah blah* ".

However because I lack expertise in spin physics, my brain has a tendency to focus too much on the differences which I must learn than on the similarities. This allowed something that was known to become literally unknown!

So, I must admit that it is through the unpublished work of Abell and Barber that this section emerged. So if the reader has trouble in following, he can be reassured that the author could not follow his own theory to its logical end!

For all these reasons, I shall refrain from taking any credit and call the factorisation of the spectator spin single resonance map the "Abell-Barber" factorisation. I also elevated this entire discussion to a section of its own rather than a subsection of Sect. 6.4 so as to emphasize my own lack of foresight.

6.5.1 N_0 and the Co-moving Spin Map

The starting point must be the map with a single spin resonance which I can write as:

$$N = (r, S) = (r, I) \circ (I, S)$$
$$\text{where S} = \exp\left(\Delta L_y + \alpha \Lambda + \overline{\alpha \Lambda}\right). \tag{6.52}$$

The map r is an orbital amplitude dependent rotation. The function α contains a single resonance. For example, for the $-\nu_x + \nu_y + \nu_{spin} = p$ resonance, the function α has the form:

$$\begin{aligned}
\alpha &= \sum_{m,n\geq 0} \alpha_{mn} u^m \overline{u}^{m+1} v^{n+1} \overline{v}^n \\
&= \sum_{m,n\geq 0} \alpha_{mn} (2J_x)^{m+\frac{1}{2}} \left(2J_y\right)^{n+\frac{1}{2}} \exp\left(-i\left\{-\Phi_x + \Phi_y\right\}\right) \\
&= \left(4J_x J_y\right)^{\frac{1}{2}} \sum_{m,n\geq 0} \alpha_{mn} (2J_x)^m \left(2J_y\right)^n \exp\left(-i\left\{-\Phi_x + \Phi_y\right\}\right).
\end{aligned} \tag{6.53}$$

The form of Eq. (6.53) reveals the difference with the orbital motion. First one notices that it has a single phase dependence; multiples of that phase are not allowed. This follows from the linearity of the motion with respect to the spin directions. Secondly, all the orbital actions are invariant and decoupled from the spin. This follows from the spectator aspect of the spin motion. I will now take full advantage of these features following Abell-Barber.

Since all the actions J are invariant, I can construct a map made of invariants which sit exactly on the resonance by choosing the proper spin tune. For example, in the case of the $-\nu_x + \nu_y + \nu_{spin} = p$ resonance, let me assume that the orbital map r is generated by the Lie operator $-H(J_x, J_y)$, i.e.,

$$r = \exp\left(: -H\left(J_x, J_y\right) :\right) I. \tag{6.54}$$

Thus for $-\nu_x + \nu_y + \nu_{spin} = p$, θ_r is selected at a given J_x and J_y to satisfy

$$-\frac{\partial H}{\partial J_x} + \frac{\partial H}{\partial J_y} + \theta_r = 2\pi p. \tag{6.55}$$

I then define a spin-orbit map N_0 which has exactly a spin tune of θ_r and the same orbital tune as N:

$$N_0 = (r, I) \circ (I, S_0) = \underbrace{(I, S_0) \circ (r, I)}_{\substack{\text{Commutes because} \\ \theta_r \text{ depends only on actions}}} \tag{6.56}$$

where $S_0 = \exp\left(\theta_r L_y\right)$.

Quite obviously the spin angle θ_r is generally a function of the orbital actions and other invariants. That completely defines N_0. I can re-expressed the map N using N_0 to isolate the resonance.

$$\begin{aligned} N &= N \circ N_0^{-1} \circ N_0 = (r,S) \circ N_0^{-1} \circ N_0 \\ &= N_0^{-1} \circ \underbrace{N_0 \circ (r,S) \circ N_0^{-1}}_{=(r,S)} \circ N_0 \\ &= \underbrace{\left(r^{-1}, S_0^{-1}\right) \circ (r,S)}_{(I, S_0^{-1} S)} \circ N_0 = (I, S_c) \circ N_0 \end{aligned} \tag{6.57}$$

$$\text{where } S_c = \exp\left(2\pi \left\{\delta L_y + \varepsilon \Lambda + \overline{\varepsilon \Lambda}\right\}\right). \tag{6.58}$$

In Eq. (6.57), I used the fact that N_0 commutes with the one-resonance operator left in the map. Moreover since the commutators of such resonances form a closed set, we are indeed assured that the form of ε in Eq. (6.57) is identical to that of α in Eq. (6.52), that is to say:

$$\begin{aligned} \varepsilon &= \sum_{m,n\geq 0} \varepsilon_{mn} u^m \overline{u}^{m+1} v^{n+1} \overline{v}^n \\ &= \underbrace{\left\{ (4J_x J_y)^{\frac{1}{2}} \sum_{m,n\geq 0} \varepsilon_{mn} (2J_x)^m (2J_y)^n \right\}}_{|\varepsilon|} \exp\left(-i \underbrace{\left\{-\Phi_x + \Phi_y\right\}}_{\psi} \right) = |\varepsilon| \exp(-i\psi). \end{aligned} \tag{6.59}$$

Equation (6.57) is totally equivalent to Eq. (5.47) of the orbital resonance. This factorisation was first revealed to me by Abell[11] and Barber. So I call it the Abell-Barber factorisation: the reader appreciates that it is an application of the one-resonance normal form approach of Sect. 5.4.3 shamefully overlooked by yours truly…

Now I will proceed to solve this map exactly: this is due to Barber.

6.5.2 *Solution for the ISF n of the One-Resonance Map of Eq. (6.57)*

Let me rewrite the co-moving map S_c of Eq. (6.57):

[11]Chinese whisper via Barber.

$$S_c = \exp\left(2\pi\left\{\delta L_y + \varepsilon\Lambda + \overline{\varepsilon}\overline{\Lambda}\right\}\right) = \exp\left(2\pi\nu\mathbf{n}\cdot\mathbf{L}\right)$$
$$\nu = \sqrt{\delta^2 + |\varepsilon|^2} \text{ and } \mathbf{n} = \frac{1}{\sqrt{\delta^2 + |\varepsilon|^2}}\left(|\varepsilon|\cos(\psi), \delta, |\varepsilon|\sin(\psi)\right). \tag{6.60}$$

On a trajectory, δ, $|\varepsilon|$ and thus ν are constants since they depend only on the actions. The phase ψ is the angle of the resonance and contains the basic orbital phase space dependence on a given orbit. In our example $-\nu_x + \nu_y + \nu_{spin} = p$, the angle ψ is given by

$$\psi = -\Phi_x + \Phi_y. \tag{6.61}$$

The claim of this section, which I will now prove, is that the vector $\mathbf{n}$ defined in Eq. (6.60) is the ISF of the full one-resonance map $N = S_c \circ N_0$ of Eq. (6.57). This fact is a direct consequence of the commutator relation $[N_0, S_c] = 0$ which is how N_0 was constructed in the first place.

There are two ways to prove that $\mathbf{n}$ is the ISF. One is the normal form route and the other approach consists in showing directly that Eq. (6.11), i.e., "$S\mathbf{n} = \mathbf{n}\circ m$" is satisfied.

Simple Lemma

I will quote a property of rotations which I leave to the reader to prove. It is very useful in purely analytical calculations. Consider two spin rotations S and R defined in terms of their Lie exponents, then the following is true:

$$\begin{aligned} RSR^{-1} &= \exp(\boldsymbol{\rho}\cdot\mathbf{L})\exp(\boldsymbol{\sigma}\cdot\mathbf{L})\exp(-\boldsymbol{\rho}\cdot\mathbf{L}) \\ &= \exp(\boldsymbol{\sigma}\cdot\exp(ad\boldsymbol{\rho}\cdot\mathbf{L})\mathbf{L}) = \exp((R\boldsymbol{\sigma})\cdot\mathbf{L}). \end{aligned} \tag{6.62}$$

A similarity transformation defined by R that acts on the Lie operators of a matrix S, i.e. on the angular momentum matrices $L_{x,y,z}$, is equivalent to its matrix R acting on the invariant direction $\boldsymbol{\sigma}$ of the matrix S.

6.5.2.1 Proof by Assuming a Normal Form

I will start with the normal form approach. First, ignoring N_0, I will imagine a rotation which normalises S_c alone:

$$S_c = \exp(2\pi\nu\mathbf{n}\cdot\mathbf{L}) = \exp(\boldsymbol{\omega}\cdot\mathbf{L})\circ\exp\left(2\pi\nu L_y\right)\circ\exp(-\boldsymbol{\omega}\cdot\mathbf{L}). \tag{6.63}$$

It follows from the lemma that

$$\mathbf{n} = \exp(\boldsymbol{\omega}\cdot\mathbf{L})\,e_y. \tag{6.64}$$

However I still need to prove that it is the ISF.

I now use a very important fact: the vector ω is also made of the same resonance family because this family forms a closed set under commutation. Therefore $\exp(\boldsymbol{\omega}\cdot\mathbf{L})$ commutes with N_0. I can use this to rewrite the total map N:

$$\begin{aligned}
N &= S_c \circ N_0 \\
&= \exp(\boldsymbol{\omega}\cdot\mathbf{L}) \circ \exp\left(2\pi\nu L_y\right) \circ \exp(-\boldsymbol{\omega}\cdot\mathbf{L}) \circ N_0 \\
&= \exp(\boldsymbol{\omega}\cdot\mathbf{L}) \circ N_0 \circ \exp\left(2\pi\nu L_y\right) \circ \exp(-\boldsymbol{\omega}\cdot\mathbf{L}) \\
&= \exp(\boldsymbol{\omega}\cdot\mathbf{L}) \circ r \circ S_0 \circ \exp\left(2\pi\nu L_y\right) \circ \exp(-\boldsymbol{\omega}\cdot\mathbf{L}) \\
&= \exp(\boldsymbol{\omega}\cdot\mathbf{L}) \circ r \circ \exp\left(2\pi\left\{\nu+\nu_r\right\} L_y\right) \circ \exp(-\boldsymbol{\omega}\cdot\mathbf{L}) \\
&\equiv (I, \exp(\boldsymbol{\omega}\cdot\mathbf{L})) \circ \left(r, \exp\left(2\pi\left\{\nu+\nu_r\right\} L_y\right)\right) \circ (I, \exp(-\boldsymbol{\omega}\cdot\mathbf{L}))\,.
\end{aligned} \tag{6.65}$$

In Eq. (6.65), for a pure spin map, I dropped the cumbersome notation (I, S) and use S. The same is done for the pure orbital map (r, I) where I use r. In Eq. (6.65), I have made extensive use of the commutative property of N_0 with a single resonance type operator. Equation (6.65) has exactly the form of Eq. (6.20)—the normalised form—and it is thus completely normalised. It follows from previously obtained results that $\mathbf{n}$ is the ISF of this single resonance map.

The magic of the Abell-Barber representation of the Lie map with one spin resonance operates at two levels:

1. First it provides a map "near the identity" which could have an approximate invariant even if Stern-Gerlach forces were present. This is totally equivalent to the orbital case and does not merit further elaboration.
2. In the spectator case, as pointed by Barber, the resulting map is trivially solved as shown above.

Before examining item 2 in Sect. 6.5.3, I will provide a direct proof that $\mathbf{n}$ is the ISF using Eq. (6.11), i.e., "$S\mathbf{n} = \mathbf{n} \circ m$."

6.5.2.2 Direct Proof that "$\mathbf{Sn} = \mathbf{n} \circ m$"

I start with $N = S_c \circ N_0 = N_0 \circ S_c$ (Eq. (6.57)) and apply it on vector $\mathbf{n}$ at position (Φ_x, Φ_y) in orbital space:

$$\begin{aligned}
N\begin{pmatrix}\Phi_{x,y} \\ \mathbf{n}_{\Phi_{x,y}}\end{pmatrix} &= N_0 \circ S_c\left(\begin{pmatrix}\Phi_{x,y} \\ \mathbf{n}_{\Phi_{x,y}}\end{pmatrix}\right) \\
&= r \circ S_0 \circ S_c\begin{pmatrix}\Phi_{x,y} \\ \mathbf{n}_{\Phi_{x,y}}\end{pmatrix} \\
&= r\begin{pmatrix}\Phi_{x,y} \\ S_0 S_{c;\Phi_{x,y}}\mathbf{n}_{\Phi_{x,y}}\end{pmatrix} = \begin{pmatrix}\Phi_{x,y}+\mu_{x,y} \\ S_0\mathbf{n}_{\Phi_{x,y}}\end{pmatrix}.
\end{aligned} \tag{6.66}$$

The map S_c evaluated at the initial phases leaves the ISF evaluated at the same phases invariant: in fact this is how I defined the vector $\mathbf{n}$. Thus I concluded that

$$\mathbf{n}^{final} = S_0 S_c \mathbf{n}^{initial} = S_0 \mathbf{n}^{initial}. \tag{6.67}$$

Eq. (6.67) is the plain action of the one-turn spin matrix on the ISF. I must now show that it is the same substituting the orbital map in the ISF.

To get the result I impose the resonance condition on the Lie operator of S_c as written in Eq. (6.60):

$$\begin{gathered}
S_0 \nu \mathbf{n} \cdot \mathbf{L} S_0^{-1} \circ r^{-1} = \nu \mathbf{n} \cdot \mathbf{L} \\
\Downarrow \text{ because } \nu \text{ is invariant under } r^{-1} \\
S_0 \mathbf{n} \cdot \mathbf{L} S_0^{-1} \circ r^{-1} = \mathbf{n} \cdot \mathbf{L} \\
\Downarrow \text{ because of the lemma} \\
(S_0 \mathbf{n}) \cdot \mathbf{L} \circ r^{-1} = \mathbf{n} \cdot \mathbf{L} \Rightarrow (S_0 \mathbf{n}) \cdot \mathbf{L} = \mathbf{n} \cdot \mathbf{L} \circ r \\
\Longrightarrow \qquad S_0 \mathbf{n} = \mathbf{n} \circ r.
\end{gathered} \tag{6.68}$$

Equation (6.68) in conjunction with Eq. (6.67) proves the desired result: the transformation of $\mathbf{n}$ by the spin matrix is the same as the transformation by the orbital map.

6.5.3 Numerical Behaviour of the One-Resonance ISF

The results derived in the previous section permit a complete solution of the motion of the ISF in phase space. Combining Eqs. (6.60) and (6.68), I conclude that the ISF after k turns is

$$\begin{aligned}
\mathrm{N}^{\mathrm{k}} \mathbf{n} &= \frac{1}{\sqrt{\delta^2 + |\varepsilon|^2}} \left(|\varepsilon| \cos(\psi + k \mathbf{m} \cdot \boldsymbol{\mu}), \delta, |\varepsilon| \sin(\psi + k \mathbf{m} \cdot \boldsymbol{\mu}) \right) \\
&= \frac{1}{\sqrt{\delta^2 + |\varepsilon|^2}} \left\{ \delta e_y + |\varepsilon| \cos(\psi + k \mathbf{m} \cdot \boldsymbol{\mu}) e_x + |\varepsilon| \sin(\psi + k \mathbf{m} \cdot \boldsymbol{\mu}) e_z \right\}.
\end{aligned} \tag{6.69}$$

The above formula can be found in Ref. [2]. It is a result valid strictly speaking only in normalised space. Let us explore this fact a bit further.

Although formula Eq. (6.69) applies only to the normalised space, it also tells us what to expect if the effects of the neighbouring distortions (or resonance) are weak. This is pointed out by Barber in Ref. [2]. In normalised space, we expect the vertical component of $\mathbf{n}$ to stay constant and the other components to oscillate sinusoidally.

In the absence of nonlinearities or with very weak nonlinearities, I can go back to the "real" directions of the spin using D_0 and the lemma of Eq. (6.62):

$$\begin{aligned} W_0 &= D_0 \circ N \circ D_0^{-1} \\ &= D_0 \circ \exp\left(2\pi \nu \mathbf{n} \cdot \mathbf{L}\right) \circ r \circ \exp\left(\theta_r L_y\right) \circ D_0^{-1} \\ &= \exp\left(2\pi \nu \left(D_0 \mathbf{n}\right) \cdot \mathbf{L}\right) \circ r \circ \exp\left(\theta_r \mathbf{n}_0 \cdot L\right). \end{aligned} \tag{6.70}$$

Then, Eq. (6.69) can be rewritten:

$$\begin{aligned} \mathrm{W}_0^{\mathrm{k}} \mathbf{n}^{new} &= \mathrm{W}_0^{\mathrm{k}} \mathrm{D}_0 \mathbf{n} \\ &= \frac{1}{\sqrt{\delta^2 + |\varepsilon|^2}} \left\{ \delta \mathbf{n}_0 + |\varepsilon| \cos\left(\psi + k\mathbf{m} \cdot \boldsymbol{\mu}\right) \mathbf{l}_0 + |\varepsilon| \sin\left(\psi + k\mathbf{m} \cdot \boldsymbol{\mu}\right) \mathbf{m}_0 \right\}. \end{aligned} \tag{6.71}$$

In Eq. (6.71) the vectors $\mathbf{l}_0$ and $\mathbf{m}_0$ are defined by the *linear* normalisation process and they define implicitly the spin phase advance. The reader can use a "canonisation" routine to force them into a form that pleases the Holy See. Again this is strictly analogous to the orbital phase advance where one must impose the Courant-Snyder condition. If for some reasons you do not like my "canonisation" routine, you are free to write your own without fear of excommunication.

What follows now is due to Barber [2]: assuming we are close enough to the closed orbit, Eq. (6.71) and the formula for the tune, i.e. Eq. (6.60), teach us that we ought to be able to extract the resonance strength $|\varepsilon|$ by some clever manipulation of tracking data. In particular, if we displace the spin tune until the component of $\mathbf{n}$ (say obtained by averaging techniques) along $\mathbf{n}_0$ is $\frac{1}{\sqrt{2}}$, then the difference in spin tune from the resonance condition will be the resonance strength $|\varepsilon|$.

Additionally, if you follow the spin exactly on the resonance, the vector $\mathbf{n}$ will rotate in the plane perpendicular to $\mathbf{n}_0$. In his work, Barber virtually defines these experimental facts as the condition for a resonance to be isolated. This is very analogous to the item 3, p. 138 where I stated that an orbital resonance is certainly isolated if its phase plot has the symmetry of the resonance once it is put in *linear* Floquet variables. In the spin case, the linear transformation D_0 puts the spin map in "linear Floquet variable."

I would like here to go back to the map of Sect. 5.4.4.2 and produce some plots again of this sextupole map near a quarter integer resonance. This will complete the analogy.

The maps on Fig. 6.1 represent the same system below and above the resonance. In Fig. 6.1b four bubbles are present. I can say geometrically that the resonance is isolated and that phase space has been grotesquely deformed. The equivalent of our $\mathbf{n}$ vector, namely a family of closed curves around the origin, has been destroyed and replaced by something new. There is no doubt looking at Fig. 6.1b that any attempt to normalise the map into circles will fail beyond the inner separatrix. I do not need fancy mathematics, just my eyes, to call this case an isolated $4\nu_x$ resonance.

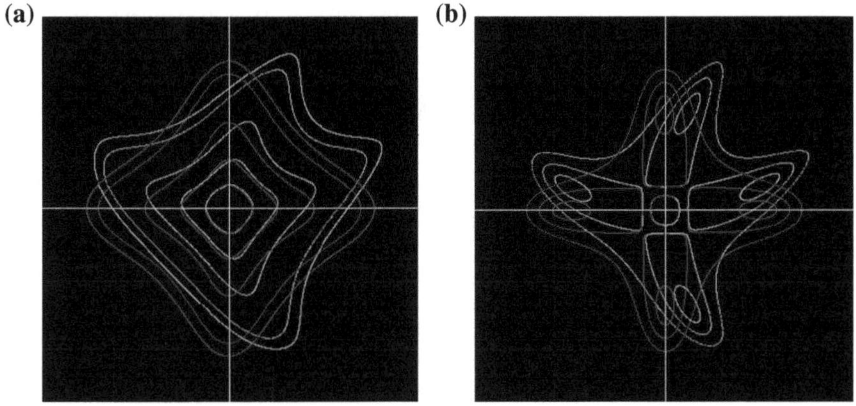

Fig. 6.1 Sextupole driven resonance in *blue* with normalised curves in *red*. **a** Below the resonance. **b** Above the resonance

Despite the obvious presence of four islands, the original system depicted by the blue curves, does not have the symmetry of the resonance, namely invariance under 90° rotations. In fact there are nonlinear polluting terms which distort the simple "one-resonance" symmetry. To remove these I perform a nonlinear canonical transformation[12] to 16th-order. The results are the red curves which are obtained by transforming the blue curves.

The red curves have a perfect four-fold symmetry and therefore the co-moving map can be computed, i.e., the one-resonance map. The canonical transformation that permits this is the equivalent of the spin transformation $d = d_1 \circ d_2 \circ \ldots$.

The case of the spin map is analogous to the case depicted on Fig. 6.1a. On that figure, there are no islands. Therefore the concentric invariant curves exist. But if you attempt to compute them using a Taylor map, this map will diverge at a location very near to the location of the unstable fixed points in Fig. 6.1b. This was explained in detail in Sect. 5.4.4.1. The best you can do with a Taylor map is to remove the polluting influence of "other" resonance. You then obtained the red curves which have perfect four fold symmetries.

The situation is analogous with the spin. If we are near a spin resonance which is greatly affected by the orbital nonlinearities and the "other spin resonances," the best we can hope with a Taylor series is to remove these resonances with the map $d = d_1 \circ d_2 \circ \ldots$. This is necessary even though the vector $\mathbf{n}$, just like the invariants of Fig. 6.1a, still exist. In the spin case, I have an exact solution given by Eq. (6.69) or Eq. (6.71), therefore I can see immediately why a Taylor solution will fail very near the resonance. Attempts to include the removal of the resonant term in Eq. (6.69), is tantamount to an expansion of Eq. (6.69) in powers of ε with an obvious divergence if $|\varepsilon| > \delta$.

[12]This is excessive but since it works why not!

In summary, if the nonlinearities are large, the full map $D_0 \circ d(z)$ must be included. In that case, we have:

$$W = D_0 \circ d \circ N \circ d^{-1} \circ D_0^{-1}. \tag{6.72}$$

Just as a reminder, the original total map T prior to normalisation can be written easily in terms of W. Going back to Eq. (6.51), I can write:

$$\begin{aligned} T &= (a, I) \circ (I, D_0 d) \circ \underbrace{\left(r, \exp\left(\theta(\mathbf{J})\, L_y\right) \exp\left(\alpha \Lambda + \overline{\alpha} \overline{\Lambda}\right)\right)}_{\text{See Eqs. (6.53) and (6.58)}} \circ (I, D_0 d)^{-1} \circ (a, I)^{-1} \\ &= (a, I) \circ \underbrace{(I, D_0 d) \circ N \circ (I, D_0 d)^{-1}}_{W} \circ (a, I)^{-1}. \end{aligned} \tag{6.73}$$

W is the full spin-orbit map expressed in the full nonlinear Floquet variables of the orbital map m where $m = a \circ r \circ a^{-1}$.

Now I will define three vectors. First I imagine that by some miracle the entire resonance, the function ε, vanishes like a fart in the air! If that was possible, the ISF would be given by:

$$\mathbf{n}_{0r}(z) = D_0 d(z)\, \mathbf{e}_y. \tag{6.74}$$

Secondly, I define two perpendicular vectors to $\mathbf{n}_{0r}$, $\mathbf{l}_{0r}$ and $\mathbf{m}_{0r}$:

$$\mathbf{m}_{0r}(z) = D_0 d(z)\, \mathbf{e}_x \quad \text{and} \quad \mathbf{l}_{0r}(z) = D_0 d(z)\, \mathbf{e}_z. \tag{6.75}$$

In Eqs. (6.75) and (6.74), I explicitly wrote the orbital dependence to emphasize that these vectors are functions of phase space.

Then, in that basis, I can track the new ISF vector $\mathbf{n}^{new}$:

$$\begin{aligned} \mathrm{W}_{0\mathrm{r}}^{\mathrm{k}} \mathbf{n}^{new} &= \mathrm{W}_{0\mathrm{r}}^{\mathrm{k}} D_0 d \mathbf{n} \\ &= \frac{1}{\sqrt{\delta^2 + |\varepsilon|^2}} \left\{ \delta \mathbf{n}_{0r}^k + |\varepsilon| \cos\left(\psi + k \mathbf{m} \cdot \boldsymbol{\mu}\right) \mathbf{l}_{0r}^k + |\varepsilon| \sin\left(\psi + k \mathbf{m} \cdot \boldsymbol{\mu}\right) \mathbf{m}_{0r}^k \right\}. \end{aligned} \tag{6.76}$$

The vectors $\mathbf{n}_{0r}$, $\mathbf{l}_{0r}$ and $\mathbf{m}_{0r}$ are not constants any more; they depend on the position in phase space.

The evolution described by Eq. (6.76) is analogous to that of the map depicted in Fig. 6.1a. The evolution is controlled by a dominant term in the map. But it is not easy to separate that term from the rest of the map without a normal form algorithm. Certainly if I examine the map away from resonance numerically, I will obtain an ISF that will approach $\mathbf{n}_{0r}$, but it is not exactly $\mathbf{n}_{0r}$. Thus Eq. (6.76) is a generalisation of Eq. (6.71) that is correct, but it does not lead itself to purely numerical methods.

At least, I do not know how to proceed purely numerically following the prescription outlined by Barber; therefore it is not without merit to say, as Barber does, that the resonance is not really "isolated" in this case.

References

1. K. Yokoya, Technical Report No. DESY 86–057, DESY (unpublished)
2. D.P. Barber, Pre-existing betatron motion and spin flipping with RF fields in storage rings, talk at Juelich, SPIN2010, 30 Sept 2010

Chapter 7
The Nonlinear Spin-Orbital Phase Advance: The Mother of All Algorithms

Abstract I discuss the phase advance in its grandiose generality. I also include the concept of "time slip advance." The phase advance results from the selection of a particular canonical transformation. I call this selection and factorisation by the religious term "canonisation." I show how this is a completely general concept which extends to nonlinearities and spin. I also contrast the map based theory of the code with the Hamiltonian case where the concept of phase advance is somewhat superfluous. Of course the general canonisation routine can be simplified in codes where speed is important. Here generality is emphasised.

Keywords Generalised Courant-Snyder loop · Phase advance · Factorisation and canonisation · Lattice functions · Formula evaluation

7.1 Introductory Verbiage on Phase Advance and "Canonisation"

The reader should focus his attention on the program in Appendix M and attempt to run it. What follows is a theoretical explanation of that code based on Fig. 7.1: the universal Courant-Snyder (or Twiss) loop in accelerator jargon. The phase advance depends on the process for choosing the canonical transformation. It is described in Sect. 7.7 using the usual step-by-step approach of this book.

This chapter encapsulates all of Courant-Snyder theory as it pertains to the symplectic nonlinear spin-orbital map. It is one theory. It can also include a linearly damped machine (radiation) if the normal form is made of three linear affine dilations as in Sect. 4.1.1. But I do not discuss this trivial albeit rather unknown case in this book. I will simply say that in that case there is a phase advance which includes stretches and rotations and that the only periodic lattice functions are defined in Sect. 4.1.2: the matrices of the de Moivre representation.

The generality of the Courant-Snyder (or Twiss) loop of this chapter is to be contrasted with standard treatments where linear, nonlinear, coupled, uncoupled, spin and no spin, etc. are all scattered over different chapters and if not different books. While there is nothing wrong in providing special treatments for the case

E. Forest, *From Tracking Code to Analysis*,
DOI 10.1007/978-4-431-55803-3_7

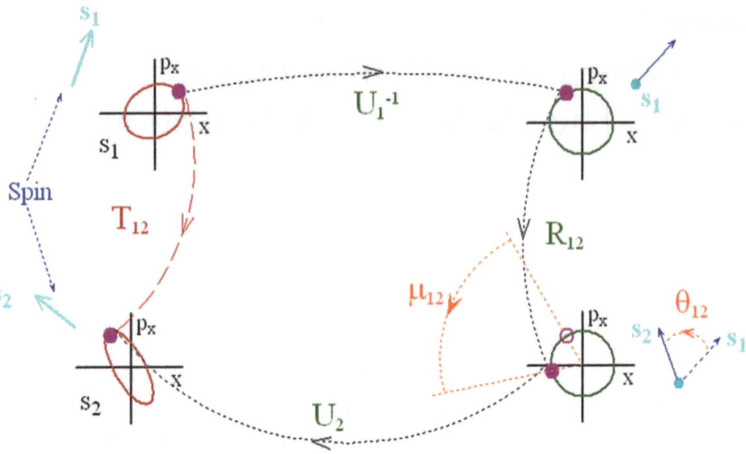

Fig. 7.1 Pictorial view of the phase advance

of analytical perturbation theory and for speeding up certain ubiquitous calculations (say linear lattice functions), it is quite criminal in my view that the general structure is never presented in its full glory, fully consistent with, and subservient to the "Code." And, as it turns out, compatible with analytical calculations and even Guignard style perturbation theory: see Chap. 8. So here it is.

As it was for the rest of this book, it is expected that you have, or you imagine having, an analysis package such as my own FPP and that you can program with it or even modify it. There is nothing revolutionary about this concept: it is the concept of the numerical recipe library on matrices applied to a large class of nonlinear maps. So let us start!

Figure 7.1 is the analog of Fig. 2.1 except that I added spin. The entire theory of "phase advance" follows from the fact that the normalised map is a product of commuting rotations.[1] The orbital motion plus spin is made of the stable part of $Sp(2N)$ and $SO(3)$: a subgroup of $Sp(2N) \times SO(3)$. This is symbolically depicted on the left-hand side of Fig. 7.1 which can be quite complicated to grasp even in the linear case. However the normal form is simple: it is the group $\underbrace{SO(2) \times \cdots \times SO(2)}_{N+1 \text{ times}}$.
This is depicted on the right-hand side of Fig. 7.1.

The topology of the nonlinear normal form is derived from the same trivial structure except that the angles of rotations depend on all the amplitudes, i.e., on the radii in the plane of these rotations. In the spectator framework for the spin, nothing depends on the spin vector itself.

This mathematical jargon appears difficult but it is actually very simple provided someone has gone to the trouble of producing a normal form algorithm leading to a factorisation such as that of Eq. (6.20). The next step is to provide what I call in

[1]Commuting dilations if you have radiation: I do not discuss this case in this code.

my usual politico-religious jargon a "canonisation" routine that uniquely defines the canonical transformation. In the case of my package FPP, the normal form algorithm or the process of tracking the canonical transformation will spew out a nondescript canonical transformation which I denote here as:

$$U = (a, D_0 d) \quad \leftarrow \quad \text{See Eq. (6.20).} \tag{7.1}$$

U fully normalises the one-turn map $T = (m, S)$. The canonisation procedure rewrites U into a precise form chosen by the scientist, myself in the case of FPP. To be exact, the map U is factorised as

$$U = \; U_c \circ R \tag{7.2}$$

where U_c has a unique form, a "canonical" form in the original ecclesiastical sense. The linear orbital part of U_c is grounded in physics and mathematical ease. It is the Courant-Snyder-Teng-Edwards[2] choice and I will justify it in Sect. 7.4.

So without further ado here is the actual factorisation in my library FPP for the symplectic spin-orbit map:

$$U_c = D \circ F \circ A \circ b. \tag{7.3}$$

The map D is a pure spin matrix that depends on all the other variables. The map F carries us to the fixed point as a function of all the parameters of the systems including the energy z_5 if it is a constant. The map A is linear in the oscillatory variables and makes the linear map around the fixed point into circles. The map b is purely nonlinear in the orbital variables.

This factorisation is quite generic and still does not tell you what U_c is. Therefore I will describe my "canonical" choice for these maps one by one starting with the fixed point map F. But first I need to set the notation straight.

7.2 A Little Notation Hurdle Due to the Jordan Normal Form

When all the orbital phase space undergoes harmonic motion, I can write the "state space" vector w in my own "PTC" order:

$$\begin{aligned} w &= (z, k) \\ \text{where } z &= \left(x, p_x, y, p_y, \delta, t\right) \\ \text{and } k &= \left(k_1, \cdots, k_{n_p}\right). \end{aligned} \tag{7.4}$$

[2]However it is the Teng-Edwards choice *without* the useless and noxious factorisation that characterises its exposition in papers and textbooks.

The vector z contains the oscillatory variables and it can have dimension two, four or six in the code PTC. The vector k contains parameters such as quadrupole strengths or any other magnet parameters. In the example code of Appendix M, I declared one single quadrupole strength to be a TPSA parameter, thus $k = (k_1)$.

When the dimension of the oscillatory space is two or four, it is sometimes useful to ignore the time t but retain $z_5 = \delta$ as a parameter. For the sake of the discussion in this chapter, I advise the reader to imagine the following redefinition of w:

$$\begin{aligned} w &= (z, k) \\ \text{where } z &= \left(x, p_x, y, p_y\right) \\ \text{and } k &= \left(\delta, k_2, \cdots, k_{n_p}\right). \end{aligned} \tag{7.5}$$

Of course the tracking code PTC, like most tracking code, always considers δ to be part of phase space. However the analysis tools of FPP, which have no connection with tracking,[3] simply view δ as a parameter if the time z_6 is ignored.

Finally there is the case of the Jordan normal form of Sect. 4.2. For the purpose of the analysis tools, I will mentally break the state vector w in the following unconventional way:

$$\begin{aligned} w &= \left(x, p_x, y, p_y, \delta, t, k_1, \cdots k_{n_p}\right) \\ \text{where we have an extended } k \rightarrow k_e &= \left(\delta, k_1, \cdots, k_{n_p}\right). \end{aligned} \tag{7.6}$$

In Eq. (7.6) I defined an extended vector of parameter k_e that also contains δ which is both a parameter that foliates phase space as well as a canonical variable. The time-like $w_6 = z_6$ depends on everything else but nothing depends on it.

In the Jordan case, the normal form is *not* $\underbrace{SO(2) \times \cdots \times SO(2)}_{N+1 \text{ times}}$ but it is $SO(2) \times SO(2) \times T \cdots \times SO(2)$ where one $SO(2)$ is replaced by the translation group in one dimension. As I showed in Sect. 4.2, the time variable "drifts" or translates. This is the most complex case and it affects the definition of the fixed point map F of Eq. (7.3).

7.3 My Choice for the Fixed Point Map

I will first discuss in Sect. 7.3.1 the cases of Eqs. (7.4) and (7.5) because all the parameters are unrelated to phase space. In the case of (7.5), time is ignored, the analysis package is unaware that $z_5 = \delta$ is actually a canonical variable.

Thus, in this case, all orbital planes are harmonic oscillators. The parameter dependent fixed point can be found by techniques that totally ignore the symplectic nature of the map.

---

[3] FPP is akin to a library for diagonalising matrices as I said several times.

7.3.1 The Trivial Case

I denote the fixed point or "dispersion" as z_c.

$$m(z_c, k) = z_c. \tag{7.7}$$

Here the map is computed around the closed orbit of the machine

$$m(0, 0) = 0. \tag{7.8}$$

Therefore the procedure can be done iteratively. First the map m is expanded and the linearised version of Eq. (7.7) is solved exactly by matrix inversion:

$$Mz_c^1 + Vk = z_c^1 \Rightarrow z_c^1 = (I - M)^{-1} Vk. \tag{7.9}$$

In Eq. (7.9), M is a rectangular matrix that gives the dependence of phase space on itself: a $2N \times 2N$ matrix. The matrix V is $2N \times n_p$ matrix connecting phase space with parameter space. This equation is analogous to Eq. (4.60).

Once Eq. (7.9) is solved by exact inversion, the rest of the z_c is found iteratively to order n_o, the order of truncation. Thus F is defined as:

$$F(z; k) = z + z_c^1 (k) + z_c^2(k) + \cdots z_c^{n_o}(k). \tag{7.10}$$

This map is unique and only depends on external parameters.

Now I will explain the hard case: when z_5 (or δ) is a not only a constant but also the canonical variable conjugate to time (z_6).

7.3.2 The Case of a Jordan Normal Form

In the case of a Jordan normal form, the transverse orbital part of F follows Eq. (7.10) except that the collection of parameter k must be replaced by k_e, a vector that contains k as well as δ as explained at Eq. (7.6).

Hence I can immediately write the transverse part of F restricted to the harmonic planes $\left(z_\perp = \left(x, p_x, y, p_y\right)\right)$:

$$\begin{aligned} F_\perp(z; k) &= z_\perp + z_c^1 (k_e) + z_c^2(k_e) + \cdots z_c^{n_o}(k_e) \\ &= z_\perp + z_c^1 (\delta, k) + z_c^2(\delta, k) + \cdots z_c^{n_o}(\delta, k) = \; z_\perp \; + \; z_c \left(\delta, k\right). \end{aligned} \tag{7.11}$$

I can easily create a Lie operator which will generate Eq. (7.11). The answer was actually given in Eq. (4.62):

$$\mathscr{F}^{pot} = \exp\left(: - \sum_{i,j=1,4} z_{c;i} S_{ij} z_j : \right). \tag{7.12}$$

This is a potential choice. As we track the canonical transformation around the ring for the purpose of computing the phase advance, we need to force this map into a unique "holy" form. As I said the fixed point is uniquely defined, however the time of flight is not uniquely defined. Let us see what is the time of flight shift induced by Eq. (7.12):

$$\begin{aligned} \mathscr{F}^{pot} z_6 &= \exp\left(: - \sum_{i,j=1,4} z_{c;i} S_{ij} z_j : \right) z_6 \\ &= z_6 - \sum_{i,j=1,4} \frac{dz_{c;i}}{d\delta} S_{ij} z_j + \frac{1}{2} \sum_{i,j=1,4} z_{c;i} S_{ij} \frac{dz_{c;j}}{d\delta}. \end{aligned} \tag{7.13}$$

The second term is purely energy dependent and could be removed or modified to suit one's purpose. I decided to leave it as is. Therefore my choice for the "canonical" F is simply F^{pot} which is the function F one gets by creating a Lie map[4] to insure symplecticity.

There is another trivial choice to implement: I remove the terms dependent on energy alone[5] from the time of flight. One can switch back and forth in my example code with the flag `time_lie_choice`. The default value is "false." The only remarkable thing to see is that the total time slip after one turn does not depend on the "canonical" choice for F.

7.4 The Linear Transformation $A(s)$

This is by far the most interesting and important "canonisation" namely the choice of the linear part. There is a good physical reason to choose the Courant-Snyder transformation and its coupled extension, defined by Eq. (7.19). Indeed they facilitate the correlation between BPM measurements and the theory. This merits a section of its own. This parametrization also facilitates analytical calculations.

---

[4]The simplest of mixed generating function, which looks like the exponent of Eq. (7.12), will generate a different pure δ dependence. So it is a bit pretentious to call my choice "the natural symplectic choice." It is only natural if your analysis code uses Lie operators.

[5]This choice may actually be the real "natural choice." Choices induced by the use of a Lie operator or a generating function can add an arbitrary function of energy like the second term of Eq. (7.13). In theory, that could send a particle to the moon and back for no physical or mathematical reasons.

The most famous choice for the transformation $A(s)$ is the Courant-Snyder choice. I can simply look at its matrix.

$$A = \begin{pmatrix} \sqrt{\beta_s} & 0 \\ -\frac{\alpha_s}{\sqrt{\beta_s}} & \frac{1}{\sqrt{\beta_s}} \end{pmatrix}. \tag{7.14}$$

The β function in Eq. (7.14) corresponds to that of Sect. 2.2.2. The first obvious advantage of this choice is that it changes the position minimally; a normalised x is simply the non-normalised x divided $\sqrt{\beta_s}$. If one uses Eq. (7.14), it is a simple result of analytical perturbation theory to show that two thin sextupoles separated by a phase advance of π, can be made to cancel exactly. This is not the case if one chooses the following transformation for $A(s)$:

$$A = \begin{pmatrix} \frac{1}{\sqrt{\gamma_s}} & -\frac{\alpha_s}{\sqrt{\gamma_s}} \\ 0 & \sqrt{\gamma_s} \end{pmatrix}. \tag{7.15}$$

The γ function in Eq. (7.15) corresponds to that of Sect. 2.2.2. If the phase advance is defined on the basis of Eq. (7.15) then it is no longer true that "sextupoles π apart" cancel; physics is not changed, but analytical results involving potentials which depend solely on the position x cannot be expressed so concisely. Again I must emphasize that this is mostly an annoyance for analytical theory and for "textbook" descriptions.

There is another virtue to the Courant-Snyder definition of the phase advance: in the linear regime, it corresponds to the phase difference between beam position monitors and it is directly measurable with BPMs. Consider the displacement of a particle at $s = 1$ after n turns expressed in terms of the initial canonical variables at $s = 0$:

$$\begin{aligned} z^1(n) &= M_{01} M_0^n z^0(0) \\ &= A_1 R_{01} A_0^{-1} A_0 R_0^n A_0^{-1} z^0(0) \\ &= A_1 R_{01} R_0^n A_0^{-1} z^0(0) \\ &= A_1 A_0^{-1} \underbrace{A_0 R_{01} R_0^n A_0^{-1}}_{\text{De Moivre applicable}} z^0(0). \end{aligned} \tag{7.16}$$

I can apply the result of Sect. 2.2.2 to re-express Eq. (7.16) in terms of the lattice functions at $s = 0$:

$$\begin{aligned} z^1(n) &= A_1 A_0^{-1} \begin{pmatrix} \cos(\Phi_n^c) + \alpha_0 \sin(\Phi_n^c) & \beta_0 \sin(\Phi_n^c) \\ -\gamma_0 \sin(\Phi_n^c) & \cos(\Phi_n^c) - \alpha_0 \sin(\Phi_n^c) \end{pmatrix} z^0 \\ &= \begin{pmatrix} \sqrt{\frac{\beta_1}{\beta_0}} & 0 \\ -\frac{(\alpha_1 - \alpha_2)}{\sqrt{\beta_1 \beta_0}} & \sqrt{\frac{\beta_0}{\beta_1}} \end{pmatrix} \begin{pmatrix} \cos(\Phi_n^c) + \alpha_0 \sin(\Phi_n^c) & \beta_0 \sin(\Phi_n^c) \\ -\gamma_0 \sin(\Phi_n^c) & \cos(\Phi_n^c) - \alpha_0 \sin(\Phi_n^c) \end{pmatrix} z^0 \end{aligned} \tag{7.17}$$

where $\Phi_n^c = n\mu + \mu_{01}^c$ and here μ_{01}^c result from Eq. 7.14.

The final step is simply to write the position x at $s = 1$:

$$x^1(n) = \sqrt{\frac{\beta_1}{\beta_0}} \left(\left(\cos\left(\Phi_n^c\right) + \alpha_0 \sin\left(\Phi_n^c\right) \right) x_0 + \beta_0 \sin\left(\Phi_n^c\right) p_0 \right). \quad (7.18)$$

According to Eq. (7.18), the only difference between $s = 0$ and $s = 1$ is the scaling factor $\sqrt{\frac{\beta_1}{\beta_0}}$ and the "Courant-Snyder" phase μ_{01}^c. Therefore if one Fourier analyses the positions at $s = 0$ and $s = 1$, the phase μ_{01}^c will be directly measurable.

This result extends to the coupled $x - y$ matrix provided that A is chosen so as to satisfy

$$A_{12} = A_{34} = 0 \quad \text{with} \quad A_{11} > 0 \text{ and } A_{33} > 0. \quad (7.19)$$

In an electron machine, one can excite the beam with a small kick at a frequency corresponding to a given tune, say tune ν_1. The motion in the other eigen-plane will damp out. If one then performs a measurement of the beam position at $s = 0$ and $s = 1$, the motion will be phase-shifted by an amount corresponding to the phase advance defined by Eq. (2.54).

Some readers may wonder if that corresponds to the so-called Teng-Edwards phase advance. The answer is yes. However I want to state emphatically that the factorization of A into a pseudo-rotation and a Courant-Snyder transformation, as done by Teng-Edwards [1], is unnecessary. Actually this factorization is not globally smooth and must be changed discontinuously across infinitesimal changes of the underlying equations of motion for large coupling. Some people, using the angle-like parameter of this pseudo-rotation, have talked about the angle ϕ between the eigenplanes: this is again a misconception. For arbitrarily small coupling, this number ϕ can be imaginary. Worse, in the real case, this "rotation" is not an element of $SO(4)$! Thus the only necessary and useful condition to impose on the map A is Eq. (7.19).

In conclusion, it is important to state that the special choice of a phase is mainly dictated by analytical ease and the human brain: on the computer it does not matter what A we choose.

The reader can also try, in the example code of Appendix M, the anti-Courant-Snyder phase advance defined in Eq. (2.52). This is activated by the flag `courant_snyder_teng_edwards` which needs to be set to false.

In all the harmonic planes, including longitudinal, I impose the conditions of Eq. (7.19). The matrix A is computed as a function of the parameters k (or $k_e = (\delta, k)$ in the Jordan normal form case).

7.5 My Choice for the Nonlinear Transformation *b*

The nonlinear part of the canonical transformation is selected by going into the phasors basis. Since the transformation is nonlinear, I can write it as a single Lie exponent. I choose b such that it does not have any generators of rotation in its Lie exponent. Therefore the Lie version of b, $\mathscr{B}$, has the form

$$\mathscr{C}\mathscr{B}\mathscr{C}^{-1} = \exp\left(: \sum_{m} B_m\,(\delta, k)\, u^{m_1}\overline{u}^{m_2} v^{m_3}\overline{v}^{m_4} : \right)$$

$$\text{with } |m_1 - m_2| + |m_3 - m_4| \neq 0 \text{ and } \sum_{i=1,4} m_i \geq 3 \tag{7.20}$$

where u and v are the horizontal and vertical phasors respectively as in Eq. (6.46). In Eq. (7.20) I chose to depict the Jordan normal form because it is the most complex. If the three orbital planes are oscillating, please simply add one phasor and remove δ from the parameter list.

Of course, $\mathscr{B}$ is obtained from a tracked canonical transformation by multiplying it by the appropriate *amplitude* dependent rotation. The sum of all the angles of this rotation around the ring will be the total amplitude dependent phase advance.

7.6 My Choice for the Spin Transformation *D*

In the traditional language of spin normal form (see Ref. [2]), the spin phase advance is defined in terms of two vectors perpendicular to the ISF **n**, the vectors **l** and **m**. These vectors are actually properties of the normal form. In the normalised space, the ISF is simply $\mathbf{e}_y = (0, 1, 0)$. I choose quite normally the normalised basis to be $\mathbf{e}_x = (1, 0, 0)$ and $\mathbf{e}_z = (0, 0, 1)$. In the space of the original variables, the three basis vectors are gotten with the help of the full matrix $\widetilde{D}$ of Eq. (6.20) which depends nonlinearly on the state space:

$$\text{state space basis} \rightarrow (\mathbf{l}, \mathbf{n}, \mathbf{m}) = \left(\widetilde{D}\mathbf{e}_x, \widetilde{D}\mathbf{e}_y, \widetilde{D}\mathbf{e}_z\right). \tag{7.21}$$

In the language of normal form, as in the case of orbital motion, the phase advance emerges from a special choice for $\widetilde{D}$. The matrix $\widetilde{D}$ is in the original variables. Thus we should look at the matrix D which is in normalised orbital variables. I can express my choice for the "canonisation" using D in phasors. If we go in phasors and find the rotation operators, i.e.,

$$D \circ c = \exp\left(\mathbf{o} \cdot \mathbf{L}\right), \tag{7.22}$$

then I can impose the following condition:

$$o_y = \sum_m O_m(\delta, k)\, u^{m_1} \overline{u}^{m_2} v^{m_3} \overline{v}^{m_4} \tag{7.23}$$

$$\text{where } |m_1 - m_2| + |m_3 - m_4| \neq 0.$$

7.7 The Canonisation in the Example Code of Appendix M

This program performs a lattice function loop and computes the most general phase advance: nonlinear spin-orbital phase advance. Moreover I compute the lattice function associated to $\langle x^2 \rangle$ as well as the ISF **n** around the ring.

At the end of the lattice loop, I check that **n** obtained by normal form obeys Eq. (6.11). Finally I compare a numerical computation of the ISF by stroboscopic average with an evaluation of the normal form ISF. The stroboscopic average, due to Heinemann and Hoffstätter, is justified theoretically in supplemental Chap. 12.

The example program uses the case of the Jordan normal form because it is the most complex: two harmonic planes and a temporal plane "slipping" away.

I invite the reader to run this program and modify it. It is really at the root of a huge part of accelerator theory.

7.7.1 *Setting Up the Example*

This example uses a misaligned ring. Here is a code fragment:

```
courant_snyder_teng_edwards=.true.
time_lie_choice=.true.
p=>als%start
do i=1,als%n
 if(p%mag%p%nmul>=3) then
 if(first) then
   call add(p,2,0,0.1d0)
   call make_it_knob(p%magp%bn(2),1) ! (0)
  first=.false.
 endif
 endif

 p=>p%next
enddo

x(1:3)=1.d-5; x(5:6)=1.d-6; x(5)=1.d-3; !
```

```
cut=4.d0;
call mess_up_alignment(als,x,cut) ! (1)
x=0.d0;dx_average_dcs=0.d0;
closed=0.d0

state=nocavity0 ! (2)
```

The misalignment is done at line (1). All magnets are displaced randomly using the array `x(1:3)` and rotated using the three angles in `x(4:6)`. Most importantly, I strongly broke the 12-fold symmetry of the ring by adding to the first sextupole a quadrupole strength and I make this strength a parameter of the TPSA using line (0).

Thus my extended parameter vector k_e is simply $(\delta, \Delta K_q)$ where ΔK_q is a deviation from the "design" quadrupole strength of 0.1. This vector k_e is made of the fifth and seventh variable of the TPSA package in use, i.e., $k_e = (w_5, w_7) = (z_5, w_7)$ as defined in Eq. (7.6).

The state is set to "`nocavity0`" which turns off the RF cavities but leaves the full longitudinal phase space in the calculation. Thus the Taylor maps will take into account the time and energy variables. A full Jordan normal form will be performed and the time slip will be computed by normal form. In fact I will compute the full phase advances (transverse+spin) and the temporal slip using canonisation within the phase advance loop.

7.7.2 *Finding the One-Turn Map and Normalising It*

The following code fragment contains the usual initialisation for the computation of the one-turn map except that in line (1) a seemingly useless "+" is placed in front of the `state`. This simply tells the code PTC to include the system parameters in the Taylor calculations if they exist. I can ignore them by omitting the "+."

```
!!!! polymorphic probe is created in the usual manner
xs0=closed
id_s=1
xs=xs0+id_s

!!!! get spin polymorphic probe after one turn
call propagate(als,xs,+state,fibre1=1) ! (4)
!!!!!!!!!!!!!!!!!!!!!!!!!!!!!!!!!!!!!!!!!!!!!!!!

! copy probe_8 into a complex damap
c_map=xs ! (5)
!!!!!!!!!!!!!!!!!!!!!!!!!!!!!!!!!!!!!!!!!!!!!!!!
call c_normal(c_map,c_n,dospin=my_true) ! (6)
```

```
write(6,'(4(1x,g21.14))') c_n%tune(1:3), c_n%spin_tune

u=c_n%as*c_n%a_t ! att=c_n%a_t*c_n%as
! id_s is a rotation
id_s=u**(-1)*c_map*u ! (7a)

! a trick to compute the fractional tunes and time slip
call c_full_canonise(id_s,u_c,d,f,a,b,r,phase,nu_spin) ! (7b)
```

This code fragment has only one bizarre calculation. At line (7a) I compute the normalised map which must be a rotation in the oscillatory planes (1, 2, and spin) and a drift of the time variable. At line (7b) I shove this normalised map, `id_s`, into my canonisation routines whose purpose is to rewrite a general canonical transformation as a product of a "canonised" transformation and a rotation as explained at Eq. (7.2). Line (7b) exploits a convenient feature: if a totally normalised map is fed into the canonisation routine, its canonical form should be the identity map because with my definition a rotation does not need any transformation to become a rotation! Thus the angles of the rotation are fed conveniently into the three Taylor series `phase(1:3)` and the spin tune into `nu_spin`.

Therefore line (7b) returns the fractional tunes and time slip.

7.7.3 *Example of Canonisations in the "Courant-Snyder Loop"*

I will start with the code that merely checks the assertions of the canonisation algorithm. I do not trust Etienne Forest: fair enough, who would? So I check my assertion to make sure I am not pulling a fast one past the reader! All the results are printed in the file `check_canonisation.txt` if you run the program for $n_0 \geq 4$.

First I remind the reader of the concatenation in the code; I rewrite Eq. (7.3) here for convenience:

$$U_c = D \circ F \circ A \circ b \text{ becomes in the code } \rightarrow \texttt{U\_c} = \texttt{D} * \texttt{F} * \texttt{A} * \texttt{b}. \quad (7.24)$$

All my assertions about canonisation are checked in the coming Sects. 7.7.3.1–7.7.3.4.

7.7.3.1 The Dispersive Map

The calculation of F is only "funny" in the case of a Jordan normal form.

```
if(no>=4) then ! (A0) special interlude
call kanalnummer(mf,"check_canonisation.txt")
```

```
write(mf,*);Write(mf,*)" Time slip factor "; write(mf,*);
h_vector_field=log(F)
h_poisson_bracket=getpb(h_vector_field) ! (A1)
write(mf,*);Write(mf,*)"Fixed point map";  write(mf,*);
call print(F,mf,prec)
write(mf,*);Write(mf,*)"Lie exponent of the fixed point map";
write(mf,*);
call print(h_poisson_bracket,mf,prec)


do k=1,4
 id_s%v(k)=F%v(k)-(1.d0.cmono.k) ! (A2)
 enddo
h_poisson_bracket=h_poisson_bracket & ! (A3)
+(id_s%v(1)*(1.d0.cmono.2)-id_s%v(2)*(1.d0.cmono.1)) &
+(id_s%v(3)*(1.d0.cmono.4)-id_s%v(4)*(1.d0.cmono.3))
write(mf,*);Write(mf,*)" Comparing with the canonical form ";
write(mf,*);
call print(h_poisson_bracket,mf,prec)
```

Above line (A1) I compute the vector field of the map by a logarithm as explained in supplemental Chap. 11. At (A1) itself, I extract the function associated to the Poisson bracket operator: its pseudo-Hamiltonian if you prefer. The full map F called `a0` is printed as well as its pseudo-Hamiltonian.

This needs to be compared with the Lie operator in Eq. (7.12). It is computed at lines (A2, 3). At line (A2) I subtract the identity from the transformation F de facto extracting the dispersion z_c. Then I construct the Lie operator per Eq. (7.12) and immediately subtract it from the at line (A3). The result, hopefully zero, is printed on the file `check_canonisation.txt`.

The reader can flip the flag `time_lie_choice` to true and see what happens!

7.7.3.2 The Courant-Snyder-Teng-Edwards Condition

This is the most important canonisation because it is used by all accelerator physicists and it is based on some physical insight. Please look at Sect. 7.4.

```
integer expo(4)
         .
         .
         .
 expo=0;expo(2)=1;
 a12=A%v(1).par.expo    ! (B1)
 expo=0;expo(4)=1;
 a34=A%v(3).par.expo    ! (B2)
```

```
write(mf,*); Write(mf,*)" Checking Courant-Snyder-Teng-Edwards"
write(mf,*);Write(mf,*)" A_12 should be zero "; write(mf,*);
call print(a12,mf,prec)
write(mf,*);Write(mf,*)" A_34 should be zero "; write(mf,*);
call print(a34,mf,prec)
```

The operator `.par.` extract as a Taylor series the coefficient $C(k_e) = C(\delta, w_7)$ of the monomial

$$C(k_e) z_1^{expo(1)} z_2^{expo(2)} z_3^{expo(3)} z_4^{expo(4)} . \tag{7.25}$$

The dimension of the integer array expo is critical: it distinguishes the phase space variables from the parameters. At lines (B1) and (B2) I extract the matrix coefficient A_{12} and A_{34} respectively.

There is nothing much to say except that I invite the reader to mess up the code a little. At the beginning of the main program, the reader will find the line

```
courant_snyder_teng_edwards=.true.
```

I advise him to set it up to "false." This flag toggles the canonisation routine to the "anti-Courant-Snyder" definition of Eq. (2.52). The polynomials for A_{12} and A_{34} are no longer vanishing. I invite the reader to modify the code to produce the polynomials for A_{21} and A_{43}: these coefficients should then be zero!

7.7.3.3 The Nonlinear Condition

Once more I need to find a unique form for the nonlinear map. I decided to choose a form that does not depend on the number of system parameters in the Taylor map. Given the tools I have at my disposal, namely a complex version of the old Berz package—which I have been using since a little before 1987 [3]—the simplest unambiguous nonlinear definition[6] involves a single Lie operator. Thus suppose I write the nonlinear map as a single Lie operator

$$\mathscr{B} = \exp\left(: h :\right) . \tag{7.26}$$

```
h_vector_field=log(b)    ! (C1)
h_poisson_bracket=getpb(h_vector_field) ! (C2)
h_poisson_bracket=h_poisson_bracket*from_phasor() ! (C3)
```

[6]If you have a TPSA package that distinguishes parameters from phase space, you could have other self-consistent definitions easily implemented; for example your canonisation could deal directly with the more natural factored representation of the nonlinear canonical transformation.

```
write(mf,*);Write(mf,*)" Lie exponent of the nonlinear part";
write(mf,*);
call print(h_poisson_bracket,mf,prec)
```

then I can use my rotation freedom to require that the Lie operator : h : does not contain any generators of rotation as explained in Eq. (7.20). The reader will discover that indeed all the generators of rotations have vanished. Additionally terms belonging to the parameter dependent fixed point map or to the Courant-Snyder transformation are not part of the nonlinear transformation. These have been properly assigned to the maps F and A.

7.7.3.4 The Spin Map Condition

My choice for the spin transformation is explained at Eqs. (7.22) and (7.23). My choice for the canonical $SO(3)$ matrix D induces a spin phase advance and implicitly defines the two little basis vectors **l** and **m** discussed in Barber's work [2].

Here is the code that checks my canonisation condition:

```
U_c=f*A*b
D_tilde=to_phasor()*U_c**(-1)*D*U_c*from_phasor()   ! (D1)
O=log(D_tilde%s) ! (D2)
```

```
write(mf,*);Write(mf,*)"Vertical spinor O_y of the canonised D;
write(mf,*);
call print(O%v(2),mf,prec)
```

This condition implies a specific choice for the periodic vectors **l** and **m**.

7.8 The Phase Advance, Its Freedom and the Code of Appendix M

7.8.1 The Case of Hamiltonian Perturbation Theory

The phase advance depends very strongly on the definition of the canonical transformation U_c. In Hamiltonian perturbation theory, the transformed Hamiltonian is the "phase advance." For example, consider the simple quadratic Hamiltonian

$$
\begin{aligned}
H &= \nu \frac{p^2 + x^2}{2} + \frac{\varepsilon}{2} \cos(s)\, x^2 \;=\; \nu J \;+\; \varepsilon J \cos(s) \cos^2(\Phi) \\
&= \nu J \;+\; \frac{\varepsilon}{8} J \left(e^{is} + e^{-is}\right)\left(e^{is2\Phi} + e^{-is2\Phi} + 2\right)
\end{aligned}
$$

$$
\begin{aligned}
&= \nu J + \frac{\varepsilon}{8} J \left(e^{i(s+2\Phi)} + e^{i(s-2\Phi)} + 2e^{is} + e^{-i(s+2\Phi)} + e^{-i(s-2\Phi)} + 2e^{-is} \right) \\
&= \nu J + \varepsilon \underbrace{\sum_{m,n} V_{mn} \exp\left(i\left(m\Phi + ns\right)\right)}_{V_s}. \qquad (7.27)
\end{aligned}
$$

This Hamiltonian is simple and has a periodicity of 2π in the s variables. I will use calculations on it to represent in its simplest form, the kind of Hamiltonian perturbation one might encounter in astronomy or other fields. It is also at the root of standard Hamiltonian treatments in accelerator physics, see Ref. [4] for example and also Chap. 8. Anyway, the most important thing is that a simple "s" dependence renders it amenable to the standard tools of Hamiltonian perturbation theory. Here I will let an s-dependent Lie map act on it in the hope of turning it into a generator of rotations.

$$
\begin{aligned}
H_s^{new} &= \exp\left(: \varepsilon F_s :\right) H_s + \varepsilon \sum_{n=1}^{\infty} \frac{: \varepsilon F_s :^{n-1}}{n!} \frac{\partial F_s}{\partial s} \qquad (7.28) \\
&\approx H_s + \varepsilon \left[F_s, \nu J\right] + \varepsilon \frac{\partial F_s}{\partial s} + \text{Order}\left(\varepsilon^2\right) \\
&= \nu J + \varepsilon \left\{ \nu \frac{\partial F_s}{\partial \Phi} + \frac{\partial F_s}{\partial s} + V_s \right\} + \cdots. \qquad (7.29)
\end{aligned}
$$

For the purpose of this section, I compute F_s to leading[7] order. Writing F_s as

$$
F_s = \sum_{m,n} F_{mn} \exp\left(i\left(m\Phi + ns\right)\right). \qquad (7.30)
$$

I can solve for F_s to leading order and get rid of V_s

$$
F_{mn} = \frac{i}{\nu m + n} V_{mn}. \qquad (7.31)
$$

The function F_s is just given by

$$
F_s = -\frac{J}{4} \left\{ \frac{1}{2\nu + 1} \sin\left(s + 2\Phi\right) + \frac{1}{2\nu - 1} \sin\left(2\Phi - s\right) + 2\sin\left(s\right) \right\} \qquad (7.32)
$$

and the resulting Hamiltonian is just

$$
H_s^{new} = \nu J + \cdots O(\varepsilon^2). \qquad (7.33)
$$

[7] A reader familiar with perturbation theory will recognize that Eq. (7.29) to leading order would also be correct if F_s was a mixed generating function.

The solution given by Eq. (7.31) and computed explicitly for my simple example in Eq. (7.32), is really the "natural" solution. It is natural in the sense that each term is selected for the purpose of removing a non-rotational term in the original H_s. Yet, it is not the Courant-Snyder solution. I can re-express Eq. (7.32) in terms of x and p_x,

$$F_s = \frac{1}{4\nu^2 - 1}\left\{\sin(s)\left(\left(1 - 2\nu^2\right)x^2 - \nu^2 p^2\right) + \cos(s)\,\nu\, xp\right\}. \tag{7.34}$$

and compute the matrix component A_{12} and the β function:

$$\begin{aligned} x^{new} &= x - \frac{\varepsilon}{4\nu^2 - 1}\left\{\nu\cos(s)\,x - 2\nu^2\sin(s)\,p\right\} + \cdots O(\varepsilon^2) \\ \Rightarrow A_{12} &= \frac{2\varepsilon}{4\nu^2 - 1}\sin(s) + \cdots O(\varepsilon^2) \\ \Rightarrow \beta_s &= A_{11}^2 + A_{12}^2 = 1 - \frac{2\varepsilon}{4\nu^2 - 1}\cos(s) + \cdots. \end{aligned} \tag{7.35}$$

Clearly the map of Eq. (7.35) does not obey the Courant-Snyder condition unless s is a multiple of π, and since it is accurate to order ε, we can say that the full map will not obey the Courant-Snyder condition for arbitrary values of ε. Indeed there are no reasons why it should.

I can compare the phase advance from 0 to s, which I denote by μ_{0s}, to the Courant-Snyder phase advance μ_{0s}^{cs}. The results to first-order in ε are

$$\begin{aligned} \mu_{0s} &= \int_0^s \nu ds = \nu s + O(\varepsilon^2) \\ \mu_{0s}^{cs} &= \int_0^s \frac{\nu}{\beta_s} ds = \nu s + \frac{\varepsilon 2\nu}{4\nu^2 - 1}\sin(s) + O(\varepsilon^2) \rightarrow \text{(See Eq. (2.70) for the factor of } \nu\text{)}. \end{aligned} \tag{7.36}$$

The reader will notice that if I integrate the above two phases over the period of the system, i.e., 2π, then the results are the same.

In Hamiltonian perturbation an arbitrary function $(ls + \varepsilon^n \Delta_n(s))J$ can be added to the Lie exponent at the nth-order in the calculation provided l is an integer and $\Delta_n(s)$ is a periodic function of s. If we do such thing we see that the new Hamiltonian becomes

$$H_s^{new} = \left(\nu + l + \frac{\partial}{\partial s}\Delta_1(s)\right) J + \cdots O(\varepsilon^2). \tag{7.37}$$

I can choose the function $\Delta_1(s)$ so that A_{12} vanishes. In a tracking code, this is done by a "canonisation" routine. Here I can see from Eq. (2.70) that one needs to modify F_s as follows:

$$F_s^{cs} = F_s + \underbrace{\frac{2\varepsilon\nu}{4\nu^2 - 1} \sin(s)\ J}_{\Delta_1}. \tag{7.38}$$

If I substitute $\Delta_1(s)$ from Eq. (7.38) into Eq. (7.37), then the result is

$$H_s^{new} = \nu\left(1 + \frac{2\varepsilon}{4\nu^2 - 1} \cos(s)\right) J + \cdots O(\varepsilon^2). \tag{7.39}$$

Of course, Eq. (7.37) is consistent with Eq. (2.70) because Eq. (7.39) is the ε expansion of $1/\beta$.

It is time to summarize what we have learned from a Hamiltonian treatment.

1. The fractional part of the total phase advance does not depend on the choice for canonical transformation.
2. A Hamiltonian procedure that is "natural" does not imply that the canonical transformation is a function of the one-turn map; rather it is a function of the Hamiltonian which is the infinitesimal propagator. Therefore the phase advance between matched[8] positions will depend on the form of the canonical transformation.
3. We can modify arbitrarily the canonical transformation. Thus the phase advance is defined within an integer plus a periodic function of the "lattice."

Now I will discuss the case of a phase advance defined through the map only.

7.8.2 *The Case of Map Perturbation Theory: Orbital and Spin*

In the case of a one-turn map, I can insist that if the map is the same, then the canonical transformation ought to be the same. This is not a "natural" procedure with Hamiltonians. Indeed it was much easier to impose the Courant-Snyder condition on a matrix A which normalises a one-turn matrix M than on the normalisation process applied to the simple Hamiltonian of Eq. (7.27).

Thus if the canonical transformation is defined uniquely in terms of the one-turn map, then it follows that the phase advance between matched positions will be the same. Matched positions are defined as positions with the same one-turn map. This was illustrated in an analytical example at Eq. (2.80) where I related the Courant-Snyder phase advance with my so-called "anti-Courant-Snyder" phase advance.

I will summarize the properties of a map based phase advance:

1. Generally the canonical transformation should only be a function of the one-turn map. This is my preference. It implies that the integer l in Eq. (7.37) should be zero.

[8]Matched positions have the same invariants or equivalently the same one-turn maps.

2. The phase advance between matched positions is the same because the canonical transformation depends only on the one-turn map.
3. The integer part of phase advances for the whole ring is also *probably* the same if induced by a canonical transformation based on the one-turn map only. (Actually I am not 100 % certain of this statement and thus the unmathematical term "probably.")

In orbital dynamics alone, the variable "s" describes what mathematicians called the base space of the "fibre bundle." The one-turn map at position s is simply the map from s to $s + C$ where C is the circumference[9] of the ring. All canonical transformations should be periodic in C. In a tracking code, this quantity C is an integer corresponding to the total number of integration steps[10] around the ring.

Spectator spin can more complex!

In the case of spin, there is an additional ambiguity introduced by the "spectator" nature of spin dynamics. In this book, I adopted the view that the spin also has a natural one-turn map of periodicity C, then everything I said about the orbital map also applies to the spin-orbit map

The one-turn periodicity of the spin map on the closed orbit is obvious. The one-turn periodicity of a full spin-orbit map in the Stern-Gerlach case, when spin affects the orbit, is just an extension of the orbital dynamics with an additional degree of freedom. However in the spectator case, the one-turn periodicity is also true if I look "globally" at the orbit-dependent spin-orbit map; this is what I did in this book.

However I could also consider, in the integrable spectator case, that the spin map exists on a given N-torus. Thus the base space for the spin becomes the variable s and the orbital N-torus, in other words an $N + 1$-torus; N takes the values 1, 2 or 3. The N Liouvillian phases Φ_i and the distance s parametrise the base space of the spin motion. This motion is not really periodic but actually quasi-periodic. There is no reason mathematically to fix our attention on a single turn. In that extended base space perspective, it can be shown, that in addition to the integer l of Eq. (7.37), the total spin tune can be augmented, for example, by the non-integer constant $\mathbf{m} \cdot \boldsymbol{\nu}(J)$ where $\boldsymbol{\nu}(J)$ are the J-dependent orbital tunes on the N-torus. The interested reader can consult Ref. [5].

7.8.3 Description of the Phase Advance Loop of Appendix M

I extract from the code `spin_phase_advance_isf.f90` the part that is a pure phase advance loop. This code fragment reproduces exactly the loop displayed in Fig. 7.1. I remove, as the ellipses show, part of the code that computes "useful" stuff. The fragment shows only Fig. 7.1.

The first step, as always on map based perturbation theory, is to compute the one-turn map around the closed orbit. The closed orbit is found at line (3). It is added to

[9] It certainly does not have to be the circumference: see Chap. 8.

[10] It can also include, amongst other things, discrete operations such as rotations and translations that are used to position a magnet.

the identity map and fed into the object XS of type probe_8. This object contains the six orbital coordinates and the spin directions. For example, the code PTC computes the image of the spin basis $(\mathbf{e}_1, \mathbf{e}_2, \mathbf{e}_3)$. At line (4), the one-turn map is computed by propagating the identity map.

```
call find_orbit(als,closed,1,state,c_1d_5) ! (3)

xs0=closed
id_s=1
xs=xs0+id_s
.
.
.
!!!! get spin polymorphic probe after one turn
call propagate(als,xs,+state,fibre1=1) ! (4)
!!!!!!!!!!!!!!!!!!!!!!!!!!!!!!!!!!!!!!!!!!!!!!!

! copy probe_8 into a complex damap
c_map=xs ! (5)
!!!!!!!!!!!!!!!!!!!!!!!!!!!!!!!!!!!!!!!!!!!!!!!
call c_normal(c_map,c_n,dospin=my_true) ! (6) .
.
.
.

u=c_n%as*c_n%a_t ! non-descript u exiting normal form ! (8a)

call c_full_canonise(u,u_c,d,f,a,b,r,phase,nu_spin) ! (8b)

phase(1)=0.d0 ;phase(2)=0.d0 ; ;phase(3)=0.d0; nu_spin=0.d0;

xs=xs0+u_c ! (8c)

p => als%start

do i=1,als%n

call propagate(als,xs,+state,fibre1=i,fibre2=i+1) ! (9a)

xs0=xs ! saving orbit ! (9b)
u=xs ! copying in map ! (9c)

! u = u_c o r = d o f o a o b o r
```

```
call c_full_canonise(u,u_c,d,f,a,b,r,phase,nu_spin) ! (10)
.
.
.
xs=xs0+u_c ! (11)

p=>p%next
enddo
```

After one turn, the probe object XS is turned into a *bona fide* Taylor map at line (5) and normalised at line (6). The full canonical transformation U is created at line (8a) and it is then "canonised" at line (8b). This procedure could be modified to satisfy the taste of the user. In accelerator physics, the Courant-Snyder-Teng-Edwards parametrisation, whose merits are evoked in Sect. 7.4, is universally chosen. For the fixed point map F, the nonlinear part b and the spin map D, there are no universally accepted choices: my choices are described in Sects. 7.3, 7.5–7.6.

On line (9a), I simply track the probe from position i to $i+1$. This could be a full magnet, a step of integration or even some operator whose purpose is to misalign a magnet. It could even be an approximate fringe field effect. In any event, it does not really matter what the object at position i is: that is the magic of the code and the analysis attached to it.

At line (9b), the closed orbit at the exit is saved. At line (9c) the map is produced. This map U obeys Eq. (7.2), i.e., $U = U_c \circ R$. The map U_c is extracted in line (10) as well as the rotation R which is the phase advance. This map R is a spin-orbit rotation.

In this example code, which is "drifting" in the third plane (temporal or longitudinal), the transverse phase advances are in `phase(1:2)` and the time slip is in `phase(3)`. The total spin phase advance is in `nu_spin`.

The final step of the algorithm is to "reboot" the ray (or probe) XS as the closed orbit plus the canonised form U_c. In doing so, I insure that the next call to `c_full_canonise` computes only the incremental phase advance. This resetting is done at line (11).

7.8.4 *Example of Something Being Computed in the Courant-Snyder Loop*

This section contains a few examples from an infinite number of possible selections. What a person computes depends on his needs and unfortunately also on his knowledge and competence. I will illustrate the generic power of the universal Courant-Snyder algorithm with some quantities, fully aware that my own knowledge can lead me to inefficient and even incorrect calculations. The "Code" is always right, only I, a poor fool, can be wrong.

7.8.4.1 The Average of x^2: The "Beta" Functions

Any function written as a Taylor series can be averaged and the result depends on the invariants which themselves are computable by normal form. I described this in Sect. 2.2.1 at the beginning of this book in an attempt to justify normal form, i.e., the harmonic representation of a map.

If one is not concerned with speed, it is trivial to compute the linear lattice functions just by computing phase averages of moments using a generic averaging algorithm. For example, the following code fragment deals with $\langle x^2 \rangle$:

```
if(no>1) then
 fonction =2*(1.d0.cmono.1)**2 ! 2*x**2 (ea)
 call c_average(fonction,u_c,fonction_floquet) ! (eb)
endif

write(mft,*)"position, element", i, p%mag%name

betax_1=(u_c%v(1).sub.'1000')**2+(u_c%v(1).sub.'0100')**2 !(ec)
betax_2=(u_c%v(1).sub.'0010')**2+(u_c%v(1).sub.'0001')**2 !(ed)
write(mft,*) "ripken beta_x_1 beta_x_2",betax_1,betax_2
write(mft,*) "2< x^2 >"
 if(no>1) then
 call print(fonction_floquet,mft)
 else
 write(mft,*) "unfortunately no information if no = 1"
 endif
```

If the order of the calculation is greater than or equal to two, then the process of averaging x^2 can be carried out. For example, with $n_o = 2$, the quantity `fonction_floquet` will be the polynomial

$$2\langle x^2 \rangle = \beta_{x1} u\bar{u} + \beta_{x2} v\bar{v} + \cdots . \tag{7.40}$$

By increasing the order of the calculation one can get amplitude dependent beta functions as well as energy and parameter dependent beta functions. Just to illustrate the point of "special purpose" algorithms, for a more efficient Courant-Snyder calculation—like those used by the Twiss command of a standard lattice design code—one derives a special formula for the lattice functions as shown on lines (Ec) and (Ed). These are the formulae uncovered in Sect. 2.2.4 and more generally in Sect. 4.1.3.

7.8.4.2 Spin Lattice Functions

The spin lattice functions are essentially the orbital dependence of the ISF **n** and also, if needed, the basis vectors **l** and **m** defined after Eq. (6.24) on p. 158 and also on p. 174 during the discussion on the spin resonance normal form. In the example code of Appendix M, I concentrate on the computation of the ISF **n**.

Inside the lattice function loop—the Courant-Snyder or Twiss loop in the parlance of accelerator physicists—I compute and print the ISF in polynomial form using Eq. (6.21), namely $\mathbf{n} = \widetilde{D}e_y$. This is done after every elements. This can be seen in this code fragment at line (Fb) where `D` is actually $\widetilde{D}$ obtained by the canonisation routine at line (10) (see Sect. 7.8.3).

```
ISF=2   ! (Fa)
ISF=D%s*ISF ! (Fb)

Write(mft,*)"ISF vector n"
call print(ISF,mft)
```

In case the reader wonders, line (Fa) stuffs the vector $\mathbf{e}_y$ into `ISF` of type c_spinor. The ISF is then printed in the file containing the Courant-Snyder data.

At the exit of the Courant-Snyder loop, I check Barber's condition for the ISF: $S\mathbf{n} = \mathbf{n} \circ m$. It is done in the code fragment:

```
call kanalnummer(mfisf,"checking_isf.txt")

write(mfisf,*) "!!!! Exploring the ISF at the end
of the lattice !!!!"
Write(mfisf,*) "Testing Barber's Equation S ISF = ISF o m"

S_ISF = c_map%s*ISF    !(12a)
ISFoM = ISF*c_map      !  (12b)

Write(mfisf,*) " |S ISF-  ISF o m|/ |S ISF|"

do i=1,3
dISF%v(i)=S_ISF%v(i)-ISFoM%v(i)
write(mfisf,*)i,full_abs(dISF%v(i)),full_abs(dISF%v(i))/
full_abs(S_ISF%v(i)) !(12c)
enddo
```

The one-turn map is in `c_map`. In line (12a), I apply the spin matrix on the ISF. On line (12b), I substitute the orbital one-turn map into the ISF. Finally in the little "do-loop," at line (12c), the two spin vectors are compared via a norm. The result is printed in the file `checking_isf.txt`.

The next and final code fragment concerns stroboscopic averaging. This is a method discovered by Heinemann and Hoffstätter [6]. I give my own "normal form based" explanation of this technique in supplemental Chap. 12.

```
x=0.d0
x(1)=0.001d0 ;x(3)=0.001d0 ;     ! (13a)
xs1=closed+x

xst=0
cray%x=0.d0
cray%x(1:6)=x
do i=1,3
 n_isf(i) = ISF%v(i).o.cray  ! (13b)
enddo

Write(6,*)" Stroboscopic Average 5000 turns : patience "
nturn=5000
kp=1000
call stroboscopic_average(als,xs1,xst,1,STATE,nturn,kp, ISF_strobo,mfisf) ! (14c)

Write(mfisf,*)" Stroboscopic Average "

write(mfisf,*);
write(mfisf,'(a19,4(1x,g20.13),a19,i4)')" ISF  for x(1:4) = "&
,x(1:4)," number of turns = ", nturn
write(mfisf,'(a24,3(1x,g20.13))')" Stroboscopic average ", ISF_strobo
write(mfisf,'(a24,3(1x,g20.13))')" From the normal form ",n_isf
write(mfisf,'(a4,20x,3(1x,g20.13))')" n0 ", real(ISF%v(1).sub.'0'), &
real(ISF%v(2).sub.'0'),real(ISF%v(3).sub.'0')
```

At lines (13a) and (13b), I evaluate the ISF computed by normal form which is a collection of three polynomials. At line (14c), stroboscopic average is used to compute numerically the ISF $\mathbf{n}$. The results are compared and printed. The vector $\mathbf{n}_0$ is printed for comparison sake: to convince myself and the reader that I am not just reproducing the trivial $\mathbf{n}_0$.

7.8.4.3 Evaluating a Formula : Analytical Formula for $\langle x \rangle$

N.B. For this calculation to be performed, the parameter `thin` *of the example program must be set to true. If misalignments are turned off, then the agreement will be perfect. If not, there will be a small disagreement because of coupling which the formula of Eq.* (7.41) *does not take into account. See its derivation in supplemental* Chap. 13. *For a thick sextupole, the present code would have to look inside each integration steps. This is possible but not permitted in the program of Appendix A.*

In supplemental Chap. 13, I compare two Hamiltonian methods on the calculation of $\langle x \rangle$ for a simple 1-d-f distribution of sextupoles.

The first method, in Sect. 13.1, is the Green's function approach which can be connected directly to the map methods preferred by the code. This means that if the Hamiltonian of the calculation is *exactly* that of the code, then the results of the "turn-the-crank" map techniques and the results of the analytical perturbation theory of Sect. 13.1 will agree to *machine precision.*

The second method, which is carefully analysed to first-order in Sect. 13.2 of supplemental Chap. 13, requires a Fourier transform in some longitudinal variable along the beam line. The individual Fourier modes do not, in any way, correlate to an object of the code, such as a magnet propagator. It is the technique furthest from the methods of the code but nevertheless accessible as explained in Chap. 8. These Fourier methods are accessible to the Code precisely because the Code has the faithful map methods at its disposal. Read other authors and notice that they never tell you how to proceed on the realistic Hamiltonian of their "Code" assuming they have one!

In Sect. 13.2, I show that if you can sum all the Fourier series in the final result for $\langle x \rangle$, then the two approaches do agree indeed. In general, it is clear that the methods have different aims and domain of application. Anyway, this book favours the Green's function method not on fundamental grounds, but on "political" ground: it matches the "Code" and thus ought to be learnt first. Unfortunately it is not the case in accelerator pedagogy.

$$\begin{aligned}\frac{\partial \langle x_s \rangle}{\partial (2J)} &= \frac{\partial \langle x_s \rangle}{\partial u \overline{u}} \\ &= \beta_s^{1/2} \times \frac{1}{4\left(1-\cos\left(\mu\right)\right)} \oint_0^C \left(-\sin\left(\mu_{s\sigma}\right)+\sin\left(\mu_{s\sigma}-\mu\right)\right) \beta_\sigma^{3/2} k_S(\sigma) d\sigma \,.\end{aligned} \tag{7.41}$$

The formula of Eq. (7.41) is certainly computable in a code with TPSA and an analysis package. For that not to be the case, ordinary sextupoles would have to be absent from the code's zoo of magnet models. I chose it so that the reader can compare with the exact result from a TPSA calculation which is available in the example code PTC.

In general a formula like Eq. (7.41) could involve something outside the code like a magnet or a fringe effect not implemented in the code. It could also involve high order multipole terms too high for TPSA to handle. In such a case the standard techniques of Sect. 13.2 produce formulae like that of Eq. (7.41): integrals around the ring. Finally last but not least, it can involve quantities outside the realm of single particle dynamics. The Green's function or map method is *perfectly* suited for approximate beam-beam calculations which are extremely localised and poorly described by Fourier transforms in the time-like variable s. It can also emanate from other types of collective effects such as intra-beam scattering.

I will now show you the code fragment which you can easily test using the FPP and PTC libraries.

```
If(thin) then
 if((p%mag%name(1:2)=="SF".or.p%mag%name(1:2)=="SD")) then
 DX_AVERAGE_DCS=(betax_1)**1.5_DP*p%mag%BN(3)/4.0_DP &  ! (Fa)
  *(-SIN(PHASE(1)*TWOPI)+SIN((PHASE(1)-c_n%TUNE(1))*TWOPI)) &
    /(1.0_DP-COS(c_n%TUNE(1)*TWOPI)) + DX_AVERAGE_DCS
 endif
endif
write(mft,*)"!!!!!!!!!!!!!!!!!!!!!!!!!!!!!!!!!!!!!!!!!!!!!
!!!!!!!!!!"

XS=XS0+U_c ! (11)
endif

p=>p%next
enddo

if(thin) then ;

 DX_AVERAGE_DCS=DX_AVERAGE_DCS*SQRT(betax_1) ! (Fb)

 write(mfa,'(a11,F20.13,a20)')' d<x>/dCS', dx_average_dCS,"
 < ---- analytical "

 if(no==2.or.no==3) then

 fonction =(1.d0.cmono.1) ! x
 write(mfa,*);write(mfa,*)"Full <x> "
  call C_AVERAGE(fonction,U_c,fonction_FLOQUET) ! (Fc)
  call print(fonction_FLOQUET,mfa)
  write(mfa,*);write(mfa,*)"Full x-dispersion  "
  call print(F%v(1),mfa)
 endif
endif
```

Line (Fa) is exactly the second factor in Eq. (7.41). All the inputs come from the lattice function loop discussed previously. At line (Fb), the factor of $\beta_s^{1/2}$ at the point of observation multiplies the sum that was computed at line (Fa).

The result is compared with the TPSA calculation at line (Fc). It should agree if the sextupoles are thin lenses and the lattice does not contain any errors.

References

1. D. Edwards, L. Teng, IEEE Trans. Nucl. Stud. **NS-20**(3) (1973)
2. D.P. Barber, Pre-existing betatron motion and spin flipping with RF fields in storage rings, talk at Juelich, SPIN2010, 30 Sept 2010
3. E. Forest, M. Berz, J. Irwin, Part. Accel. **24**, 91 (1989)
4. S.Y. Lee, *Accelerator Physics* (World Scientific Publishing, Singapore, 2004)
5. D.P. Barber, J.A. Ellison, K. Heinemann, Phys. Rev. ST Accel. Beams **7**, 124002 (2004)
6. K. Heinemann, G.H. Hoffstätter, Phys. Rev. E **54**, 4240 (1996)

Chapter 8
Deprit-Guignard Perturbation Theory Faithful to the Code

Abstract I discuss how it is possible to Fourier transform the Hamiltonian of the ring with the output of the code only. With the help of normalisation theory applied on the one-turn map, the general Courant-Snyder loop of Chap. 7 and the logarithm of supplemental Chap. 11, I construct the Hamiltonian of the ring expanded in Fourier modes in the variable s. I also present a fully functional prototype of Deprit-Guignard normalisation on the s-dependent Hamiltonian.

Keywords s-Dependent Hamiltonian perturbation theory · Deprit-Guignard perturbation theory · Logarithm of maps

8.1 How About Hamiltonian Perturbation Theory?

In this book I have emphasized the hierarchical nature of perturbation theory in accelerator physics. This hierarchy is not a figment of my imagination but a reality coming from the ubiquitous supremacy of your Code which ever it might be, with all its imprecision and conceptual mistakes. As I said, the Code is always right; perturbation theory must slavishly serve the Code unless Big Brother says otherwise.

If you have a code, you have brute force tracking. If you have TPSA, you have approximate Taylor maps. And if you have normal form, you have approximate canonical transformations and phase advances. Finally, if you have a tool to get the logarithm of a magnet map, you have a pseudo-Hamiltonian for this magnet, and therefore you can produce a Fourier transform around the lattice. The algorithm for the logarithm of a nonlinear map is explained in supplemental Chap. 11.

There are a few reasons why you might want to analyse the motion using Hamiltonian perturbation theory on the equations of motion rather than on maps.

1. You want the elusive smooth Hamiltonian that is omnipresent in an accelerator text book but whose actual construction is never explicitly given. Well you came to the right place to see it.
2. You might have access to some tools which permits time-dependent normalisation. These tools are common in astronomy and they rely on a Fourier expansion of the motion in some time-like variable "s." These tools require the Hamiltonian as

E. Forest, *From Tracking Code to Analysis*,
DOI 10.1007/978-4-431-55803-3_8

a function of time in terms of sine and cosine functions. I will provide a prototype of this normalisation for your perverse enjoyment.

3. Sending a ring to a colleague to use in some simple code is messy: sending a function written in terms of simple functions—Fourier modes—is perhaps a pleasing thing to do.
4. Perhaps you have a ring with a mild symmetry breaking. In such a case, the maps for each super period is most relevant to the dynamics: they correspond to specific Fourier terms in the s-dependent Hamiltonian. You may want to analyse the ring using these Fourier terms.
5. …

To realise the above items, the most important construction is that of the smooth Hamiltonian in linear Floquet variables. Its construction requires *the map tools discussed in previous chapters* to fully normalise the linear problem.

I will discuss two such Hamiltonians. I will start, in Sect. 8.3, with a straight forward application of the Courant-Snyder-Teng-Edwards linear transformation and its effect on the Hamiltonian.

8.2 Definining a Time-Like Variable

Of course my starting point is the "Code." Therefore I will assume that the ring is made of "N" steps[1] of integration. These maps can be anything: integration steps, coordinate transformations for displacing magnets, fringe fields kicks, or even entire beam lines. Basically I do not care what the steps are. Of course if there is only *one* step, I am doing perturbation theory on the one-turn map. So it is useful for the reader to imagine a large number of steps otherwise this chapter would be truly useless.

These steps are between N distinct locations in the ring where location 0 is the same as location N. Furthermore, I can construct a one-to-one function from the integer N to the angles:

$$
\begin{array}{c}
0 \longrightarrow 0 \\
\cdot \\
\cdot \\
\cdot \\
j \longrightarrow \theta_j \\
\cdot \\
\cdot \\
\cdot \\
N \longrightarrow 2\pi \,.
\end{array}
\tag{8.1}
$$

[1] In my previous book [1], I defined the mathematical ring as a collection of ordered maps. I demonstrated that the usual textbook Hamiltonian is a ring with an infinite number of maps.

These angles θ_j can be proportional to a design length or some other length, but this, as it turns out, is irrelevant to the basic mechanics of Hamiltonian perturbation theory although it affects the details. The code does not care how you choose the mapping in (8.1). But, as we will see, you have an interest in choosing θ judiciously if some symmetries are (almost) present in the ring.

8.3 Using a Courant-Snyder Type Phase Advance

Temporarily, I will consider the linear part of the maps and assume our ability to normalise them within the code.

Thus imagine that I compute the one-turn linear map at $j = 0$ and normalise it:

$$m_0 = a_0 \circ r \circ a_0^{-1}. \tag{8.2}$$

The map a_0 is defined solely in terms of the one-turn map following a canonisation routine. For example a_0 can be the Courant-Snyder-Teng-Edwards transformation.

Next I construct my usual lattice function loop as described in Eq. (2.55):

$$r_{j\ j+1} = a_{j+1}^{-1} \circ m_{j\ j+1} \circ a_j. \tag{8.3}$$

Here a_{j+1} is canonised into the desired form. Now I apply Eq. (8.3) to the *full* non-linear map of the code. I use capital letters for the full nonlinear maps and keep lower case for linear maps.

$$N_{j\ j+1} = a_{j+1}^{-1} \circ M_{j\ j+1} \circ a_j. \tag{8.4}$$

The map $N_{j\ j+1}$ is to leading order a simple rotation (i.e., $r_{j\ j+1}$) but also contains higher order terms not yet normalised. If it represents a small part of the ring, say a magnet or a single integration step, then it is near the identity! Therefore I can apply the technique of Chap. 11 and take its logarithm. In other words,

$$\begin{aligned} \text{if } N_{j\ j+1} &= \exp\left(\Delta s_j \mathbf{G} \cdot \nabla\right) I \\ &= \exp\left(-\Delta s_j :H_j:\right) I \end{aligned}$$

$$\Longrightarrow \log\left(N_{j\ j+1}\right) = \Delta s_j \mathbf{G} \Rightarrow -\Delta s_j H_j. \tag{8.5}$$

The step Δs_j is connected by (8.1) to the angles which are chosen here to be proportional to a distance s,

$$\Delta s_j = s_{j+1} - s_j = \frac{\theta_{j+1} - \theta_j}{2\pi} C \tag{8.6}$$

where C is the largest value taken by s. In other words, the variable s runs from $s = 0$ to $s = C$. Obviously, when comparing to standard accelerator Hamiltonian theory, C will correspond to the total circumference of the so-called design orbit. It is important to realise that this correspondence need not be true. However for the sake of comparing to known problems, I make sure that it is true in the example program of Appendix N!

The local (pseudo-)Hamiltonian H_j has the form:

$$H_j = \frac{\Delta \mu_{j;1}}{\Delta s_j} J_1 + \frac{\Delta \mu_{j;2}}{\Delta s_j} J_2 + V_j . \tag{8.7}$$

If I assume that the Hamiltonian for the full ring can be expanded in Fourier modes in θ,

$$H\left(\mathbf{z}; \theta(s)\right) = \sum_{n=-M,M} \exp\left(i\ n\theta(s)\right)\ h_n, \tag{8.8}$$

then the Fourier coefficients h_n can be easily computed using the output from the "phase advance" loop i.e., Eq. (8.5). The answer is

$$\begin{aligned} h_n &= -\frac{1}{2\pi} \sum_{j=0}^{N-1} \exp\left(-in\theta_{j+1}\right) \log\left(N_{j\ j+1}\right) \underbrace{\frac{\Delta\theta_j}{\Delta s_j}}_{=2\pi/C} \\ &= -\frac{1}{C} \sum_{j=0}^{N-1} \exp\left(-in\theta_{j+1}\right) \log\left(N_{j\ j+1}\right) . \end{aligned} \tag{8.9}$$

Equation (8.9) gives the coefficient of the s-dependent Hamiltonian: Eq. (8.8) must be integrated from $s = 0$ to $s = C$ to give us the one-turn map. Clearly if you use θ for the variable of integration, the corresponding Hamiltonian is just:

$$\begin{aligned} H\left(\mathbf{z}; \theta\right) &= \frac{C}{2\pi} \sum_{n=-M,M} \exp\left(i\ n\theta\right)\ h_n \\ &= \sum_{n=-M,M} \exp\left(i\ n\theta\right)\ \widetilde{h}_n \end{aligned} \tag{8.10}$$

$$\text{where } \widetilde{h}_n = -\frac{1}{2\pi}\sum_{j=0}^{N-1} \exp\left(-in\theta_{j+1}\right)\log\left(N_{j\ j+1}\right). \tag{8.11}$$

It is easy to check that the quadratic part of h_0 is a constant Hamiltonian which produces the total phase advance along the ring. I will check that on an example in Sect. 8.5.

It is also true that even the linear part contains terms which are θ dependent, in other words, the quadratic part of $H(\theta)$ is

$$H_2 = \left\{ \underbrace{\nu_1 J_1 + \nu_2 J_2}_{\text{quadratic part of } h_0} + \sum_{n=-M,M} e^{in\theta}\left\{\nu_{1;n} J_1 + \nu_{2;n} J_2\right\} \right\}. \tag{8.12}$$

This is to be expected because the phase advance is not a uniform function of the variable θ. In fact for the usual quadratic Hamiltonian, the phase advance is the integral of $\frac{1}{\beta}$ over the design distance s as shown in Sect. 2.4.3. This is certainly *not* a constant function of the variable s or θ. The reader will see this fact numerically demonstrated in Sect. 8.5 on Fig. 8.1.

In the next section I show how one can insure that the quadratic Hamiltonian has exactly a constant phase advance. The technique is documented in standard textbooks, for example in Lee's book [2]. Here however, I use only tools and techniques available to the "Code."

Fig. 8.1 Fourier transform of the Hamiltonian $H = \frac{1}{\beta} J$

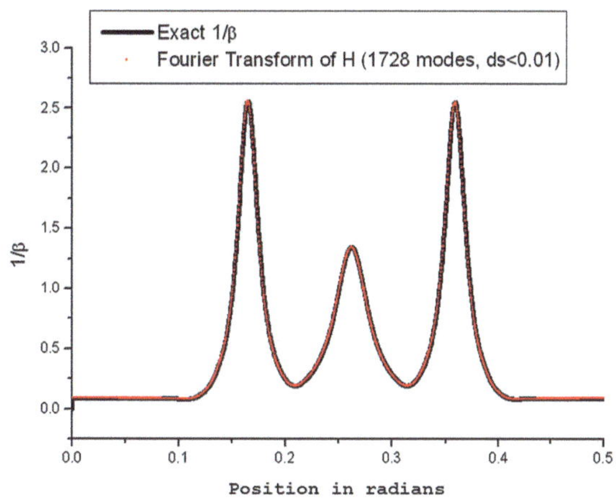

8.4 Using a Constant Phase Advance

In Eq. (8.3), the canonical transformation is a strict function of the map at positions j and $j+1$. The process of "canonisation" sets a_{j+1} in terms of the transfer map $m_{j\ j+1}$ and a_j. The phase advance emerges from the resulting $r_{j\ j+1}$. The phase advance is not an input, it is computed and it depends strongly on the choice of a_j. Please look at Sect. 2.4.3 or Chap. 7 if you need a reminder.

In this section, I will do the reverse: the phase advance will be an input from which the canonical transformation a_{j+1} is computed. The trick used by accelerator physicists is to first compute the total linear phase advance using a canonised recipe and write the following Hamiltonian

$$K = \nu_1 J_1 + \nu_2 J_2 \tag{8.13}$$

where the tunes ν_i's contain the integer part. Accelerator physicists[2] reinterpret Eq. (8.3) as an equation *not* for the phase advance $r_{j\ j+1}$ *but* for the canonical transformation a_{j+1}.

The trick in a code-based-theory is to decree *prima facie* that $r_{j\ j+1}$ is

$$r_{j\ j+1} = \exp\left(\Delta\theta_j : -\{\nu_1 J_1 + \nu_2 J_2\} :\right) I. \tag{8.14}$$

The equation for the canonical transformation a_{j+1} is given by

$$a_{j+1} = m_{j\ j+1} \circ a_j \circ r_{j\ j+1}^{-1}. \tag{8.15}$$

which can be rewritten with Lie exponents:

$$a_{j+1} = \underbrace{\exp\left(\Delta\theta_j : \{\nu_1 J_1 + \nu_2 J_2\} :\right)}_{\text{Lie map for } r_{j\ j+1}^{-1}} \quad \underbrace{m_{j\ j+1} \circ a_j}_{\text{tracked transformation}} \quad . \tag{8.16}$$

Equation (8.15) (or (8.16)) defines an acceptable transformation if it leads to a transformation a_j that is periodic. That is actually easy to prove by connecting a_j to the starting point a_0:

[2]The reader can consult Lee's book [2] on nonlinear resonances for an exposition of the methods using standard Hamiltonian theory.

$$
\begin{aligned}
a_1 &= m_{0\,1} \circ a_0 \circ r_{0\,1}^{-1} \\
a_2 &= m_{1\,2} \circ a_1 \circ r_{1\,2}^{-1} \\
&\vdots \\
a_{j+1} &= m_{j\;j+1} \circ a_j \circ r_{j\;j+1}^{-1} \\
&\vdots \\
a_{N-1} &= m_{N-2\;N-1} \circ a_{N-2} \circ r_{N-2\;N-1}^{-1} \\
a_N &= m_{N-1\;N} \circ a_{N-1} \circ r_{N-1\;N}^{-1} \\
&\Downarrow \\
a_N &= \underbrace{m_{N-1\;N} \circ m_{N-2\;N-1} \circ \cdots \circ m_{23} \circ m_{12} \circ m_{01}}_{m_{0N}} \circ a_0 \circ \underbrace{r_{01}^{-1} \circ \cdots \circ r_{N-2\;N-1}^{-1} \circ r_{N-1\;N}^{-1}}_{r_{0N}^{-1}}
\end{aligned}
$$

$$
\begin{aligned}
\Rightarrow a_N &= m_{0N} \circ a_0 \circ r_{0N}^{-1} \\
&= a_0 \circ \underbrace{a_0^{-1} \circ m_{0N} \circ a_0}_{r_{0N}} \circ r_{0N}^{-1} = a_0 r_{0N}^{-1} r_{0N} = a_0 .
\end{aligned} \tag{8.17}
$$

The reader must notice that all the above arguments do not depend on the original a_0. It is natural, being an accelerator physicist, to select a_0 to be the Courant-Snyder-Teng-Edwards transformation, but it is totally irrelevant to the procedure described in this section. In fact, if I select a_0 to satisfy the Courant-Snyder condition, $A_{12} = 0$, it is unlikely to be satisfied anywhere else around the ring except between super periods.

The rest of the calculation proceeds as in Sect. 8.3, from Eq. (8.4) to Eq. (8.9). The real change comes at Eq. (8.12) which I rewrite here:

$$
H_2 = \nu_1 J_1 + \nu_2 J_2 + \underbrace{\sum_{n=-M,M} e^{in\theta} \left\{ \nu_{1;n} J_1 + \nu_{2;n} J_2 \right\}}_{\approx 0} . \tag{8.18}
$$

If our code has an infinite number of steps and no discontinuous operators (fake fringe fields or misalignment operators), then the second term in Eq. (8.18) will vanish if the angle θ is defined to be proportional to the integration length.

However, we can make it vanish exactly by choosing the angle mapping of (8.1) judiciously. The idea is to select the phase advance within each step to be identical. In the case where the ring has a definite symmetry, it is worthwhile to make sure that the N locations are distributed identically in each super-periods. This may not be easy or convenient because some super-periods may differ in dynamically insignificant ways. For example, BPMs can be positioned asymmetrically in the lattice. There might be insertion devices that break the symmetry. If one is careless, the second term in Eq. (8.18) will not be exactly zero. This does *not* mean that the computed Hamiltonian is wrong, but it is not exactly what we naively expected.

8.5 Code Example for Sect. 8.3: Using the Courant-Snyder Phase Advance

Before discussing the code in Appendix N, it is interesting to look at Fig. 8.1. On this figure I simply plotted $\frac{1}{\beta}$ around the ring as well as a piece of the Hamiltonian evaluated by Fourier transform. I chose the 1-d-f case with no errors and used an angle θ strictly proportional to the standard distance s along the so-called design orbit. I also set the maximum distance between integration kicks to be 0.01. We expect the Hamiltonian to be

$$H = \frac{1}{\beta} J + \cdots = \frac{1}{2\beta} u\bar{u} + \cdots . \tag{8.19}$$

This is confirmed on Fig. 8.1, showing the first cell of the ALS, where the Hamiltonian matches $\frac{1}{\beta}$ nearly perfectly. This will be explained in this section.

The actual lattice is described in Appendix A, but it suffices to see that it has the following structure

```
ALS = 11*sup1+supb;
```

where the beam line `supb` is identical to `sup1` except for the presence of a thin lens cavity.

The code of Appendix N implements completely Sect. 8.3. I will first present a fragment which implements the logical core of this algorithm, namely from Eq. (8.4) to Eq. (8.9):

```
if(used_ds_ave) then
 circ=twopi
else
 call GET_LENGTH(als,circ)
endif

if(int_step) then              ! (10)
 ns=als%t%n
else
 ns=als%n
endif

allocate(bet(0:ns),theta(0:ns),ht(0:ns))
bet=0.d0
theta=0.d0
ds_ave=twopi/ns                ! (11)
ds=0.d0
```

```
if(icase==1) U_c=U_c.cut.2

s=0.d0
ray=U_c+ray_closed;          ! (12)

p=>als%start                 ! (13a)
t=>p%t1                      ! (13b)

do i=1,ns

if(mod(i,100)==0) then
  write(6,*) ns-i,"steps remaining"
endif

if(used_ds_ave) then
   ds=ds_ave                  ! (14a)
else
  if(int_step) then
   ds=twopi*t%s(5)/circ       ! (14b)
  else
   ds=twopi*p%mag%p%ld/circ ! (14c)
  endif
endif


if(int_step) then
 call propagate(als,RAY,state,node1=i,node2=i+1)    ! (15a)
else
 call propagate(als,RAY,state,fibre1=i,fibre2=i+1) ! (15b)
endif

  ray_closed=ray ! Saving orbit
  U=ray ! copying in map

  U_c=U

bet(i)=1.d0/((u_c%v(1).sub.'10')**2+(u_c%v(1).sub.'01')**2)
 ! (15c)


if(icase==1) U_c=U_c.cut.2      ! (16a)
call  c_canonise(U_c,U_c,f,A,b) ! (16b)
U=U_c**(-1)*U  ! (16c)

  Gh=log(U)        ! (17a)
  h=getpb(Gh)      ! (17b)


! Checking convergence of the logarithm
a=exp(-(fh.cut.2),(U.sub.1))   ! (18)
do k=1,c_%nd2
 if(abs(full_abs(a%v(k))-1)>1.d-5) then
```

```
    call print(a,6)
    write(6,*);write(6,*) "Log failed at element",i, p%mag%name
    stop
 endif
enddo

s=s+ds;  theta(i)=s
do k=0,n_mode
 hn(k)=hn(k)-exp(-i_*k*theta(i))*h/circ ! (19)
enddo
```

Lines (15) and (16) implement Eq. (8.4) which I rewrite here:

$$N_{j\ j+1} = \underbrace{\underbrace{a_{j+1}^{-1}}_{\text{U\_c line (16b)}} \circ \underbrace{M_{j\ j+1} \circ a_j}_{\text{U line (15)}}}_{\text{line (16c)}} . \tag{8.20}$$

Lines (17a) and (17b) contain the novel part of the algorithm, which to my knowledge, has not been described anywhere else.

$$\log\left(N_{j\ j+1}\right) = \underbrace{\Delta s_j \mathbf{G}}_{\text{line (17a)}} \Rightarrow -\underbrace{\Delta s_j H_j}_{\text{line (17b)}} . \tag{8.21}$$

Line (18) is a check on the internal consistency of the logarithm of a map. If the map is too "large", namely if the tune is close to 90° or beyond it, I do not expect my algorithm to work. Thus the code stops.

If all is fine, the contribution to the Fourier integral is incremented at line (19) as described in Eq. (8.9).

$$h_n = \sum_{j=0}^{N-1} \underbrace{-\frac{1}{C} \exp\left(-in\theta_{j+1}\right) \log\left(N_{j\ j+1}\right)}_{\text{line (19)}} . \tag{8.22}$$

In running the example program, I will suggest a few inputs. First here is the code fragment from the program of Appendix N which describes the various input:

```
Write(6,*)"Random errors -> t, no errors -> f"
read(5,*) mis

1 write(6,*) "Choose the state"
if(.not.mis) write(6,*) "only_2d0  -> 1-d-f map if no errors ->
 type 1"
write(6,*) "only_4d0   -> 2-d-f map                 -> type 2"
write(6,*) "delta0     -> 2-d-f map + delta         -> type 3"
write(6,*) "nocavity0 -> 3-d-f map if no cavity -> type 4"
write(6,*) "default0   -> 3-d-f map if    cavity -> type 5
<- SLOW"
```

```
read(5,*) i
                                   .
                                   .
                                   .
                                   .
                                   .
  write(mf,*) "L MAX "
  write(mf,*) "0.3"         ! (1) all ds of the order  0.3 metre
  write(mf,*) "CUTTING ALGORITHM"
                                   .
                                   .
                                   .
write(6,*) "Do you want to break the symmetry of the lattice"
write(6,*) "by mispowering a single sextupole ? "
write(6,*) "Yes -> t    No -> f "
read(5,*) asym

if(asym) then
 p=>als%start; call move_to(ALS,p,"SF",pos);  ! (2a)
 call add(p,3,1,10d0); !call add(p,2,1,.01d0);     ! (2b)
endif
                                   .
                                   .
                                   .
write(6,*) "int_step t or f"
read(5,*) int_step        ! (3)
write(6,*) "Uniform ds -> t"
read(5,*) used_ds_ave     ! (4)

write(6,*) "case 1,2 "
write(6,*) "case = 1 -> linear transformation"
write(6,*) "case = 2 -> full nonlinear transformation: not
interesting"
read(5,*) icase           ! (5)

write(6,*) "Number of Fourier modes: make it 12 or more"
read(5,*) n_mode
allocate(hn(0:n_mode))      ! (6)

map_order=4                 ! (7)
```

First, random errors can be put in the lattice. Secondly the dimension of the calculation is read. Line (1) is not an input but the reader can obviously modify it before compilation. It affects the size of the integration steps. The parameter asym powers a single sextupole, and, if you reintroduce the second call of line (2b), it also puts a linear break of the symmetry. Line (3) allows the maps to be that of individual integration steps rather than magnets. This, with a small integration step, allows the calculation of the code to approach the results of "exact" Hamiltonian perturbation theory when relevant. Remember that the "Code" is always right! Therefore the type of calculation performed here is correct even when the individual steps describe maps totally outside the realm of continuous Hamiltonians: misalignments for example. Line (4) is not useful when trying to reproduce a Courant-Snyder type result:

it redefines θ to be constant at each step whether integration steps, misalignment operators or integrated fringe kicks. Fringe kicks are used in models of bends[3] for example.

The example I will show here is the "extreme" case of Hamiltonian dynamics. First the reader must change line (1) from 0.3 to 0.01. This will trigger a very small cutting of all magnets so that ds is about 1 centimetre typically. The rest of the input is shown below.

```
  Random errors -> t, no errors -> f
f
  Choose the state
 only_2d0   -> 1-d-f map if no errors -> type 1
 only_4d0   -> 2-d-f map              -> type 2
 delta0     -> 2-d-f map + delta      -> type 3
 nocavity0 -> 3-d-f map if no cavity -> type 4
 default0   -> 3-d-f map if    cavity -> type 5    <- SLOW
1
  Do you want to break the symmetry of the lattice
  by mispowering a single sextupole ?
  Yes -> t    No -> f
f
                    .
                    .
                    .
  int_step t or f
t
  Uniform ds -> t
f
  case 1,2
  case = 1 -> linear transformation
  case = 2 -> full nonlinear transformation: not interesting
2
  Number of Fourier modes: make it 12 or more
1728
           22004  steps remaining
                    .
                    .
                    .
               4  steps remaining
```

The results are in file `guignard_hamiltonian.txt` and I list here the non-vanishing Fourier modes:

```
  Results for 12-fold symmetric ring
  2 x Hamitonian of the ring Guignard style


             0
```

[3] A well-known approximation for a parallel face bend consists in using a sector bend sandwiched between two thin lens quadrupoles. See [1] or any other accelerator cookbook available in any decent pharmacy.

```
 Properties, NO =    4, NV =    2, INA =  181
 *********************************************

   2   14.36779999999980        0.000000000000000       1  1
   4  -1491.899920542154        0.000000000000000       2  2

          1
            .
            .
            .
            .
            .

         12

 Properties, NO =    4, NV =    2, INA =  193
 *********************************************

   2  -6.485176171637776       0.1124135382493850E-01   1  1
   4   1576.285541487028      -1.290731929174263        2  2

         13
            .
            .
            .
            .
            .

       1727

 Properties, NO =    4, NV =    2, INA = 2118
 *********************************************

  Complex Polynomial is zero

       1728

 Properties, NO =    4, NV =    2, INA = 2059
 *********************************************

   2 -0.1878555919484628E-03  0.4424503035230257E-03   1  1
   4   143.1056868536650      -40.57186418794430        2  2
```

The perfect symmetry implies that only Fourier modes that are multiples of 12 can appear. This perfect symmetry extends to the nonlinear tune since the sextupolar distribution of this lattice is also 12-fold symmetric. Of course the phase advance is a function of θ since I use the Courant-Snyder definition for the canonical transformation. Therefore the second term Eq. (8.12) that contains the θ-dependent terms cannot be zero. The result for the coefficient in Eq. (8.19) is confirmed to be $\frac{1}{\beta}$ by the graphs in Fig. 8.1.

The reader is invited to reduce the number of steps by setting ds back to 0.3 in line (1) and try other inputs. In particular the reader is invited to run the 3-d-f examples. The case with a cavity takes a longer time to run: please note that my synchrotron tune, above transition, is negative.

8.6 Code Example for Sect. 8.4: Using a Constant Phase Advance

The code of Appendix O used here is very similar to that used in Sect. 8.5. As I said the difference lies in the way the phase advanced is computed. In standard accelerator theory, formulae like $\mu \approx \int ds/\beta$ emerge from our choice of canonical transformation which is the source of my so-called "canonisation" process. One fixes the canonical transformation, preferably a function of the one-turn map only, and here "pops" a phase advance.

In Hamiltonian perturbation theory presented in Sect. 8.4, the reverse is done. The phase advance, integer part included, is precomputed using your favourite holy canonised version. Then you choose a mapping between θ and a variable s of your own taste; it can be the standard s along the design orbit. But generally, if you have misalignments and discrete maps representing approximate fringes (for example), it should perhaps be something else.

The point of this exercise is to force the linearly[4] normalised physics to be that of a constant and trivial Hamiltonian, i.e.,

$$K = \nu_1 J_1 + \nu_2 J_2 \tag{8.23}$$

in 2-d-f.

I will show the major differences with the code of Sect. 8.5. They all relate to the fact that the phase advance is given by Eq. (8.14) and is thus fixed. There is no "canonisation" to determine it. Rather the reverse is true: the phase advance is known and the transformation is extracted from it. This is shown in some code fragments from the program of Appendix O. First I construct the vector field operator corresponding to Eq. (8.23):

```
call c_normal(one_turn_map,normal_form)  ! (8c)
U=normal_form%A_t     ! att=c_n%A_t*c_n%As

intp=0
if(normal_form%nres==0) then
```

[4]One can partially normalise the Hamiltonian to say second or third-order, but for the sake of simplicity, I keep this out of the discussion of this book.

```
 intp(1)=14
 intp(2)=8
 intp(3)=0
endif

! id_s is a rotation
id_s=U**(-1)*one_turn_map*U ! (8d)

call extract_linear_from_normalised(id_s,b,a,f_lin,f_non,intp)
 ! (8e)

call  c_canonise(normal_form%A_t ,U_c,f,A,b) ! (9a)
id_s=U_c                                     ! (9b)

 if(icase==1) then
   h_left=f_lin
 else
   h_left=f_non+f_lin
 endif
h_left=(1.d0/twopi)*h_left
```

In array `intp(1:3)` I stored the integer part. Notice that it could be computed in the program itself, but I did not bother here since we know the answer from the various ALS-based codes in the appendices. On line (8e), I extract the linear and nonlinear parts of the normalised vector field. I then divide it by 2π. The result, in `h_left` will permit me to construct the map for the phase advance, namely the map of Eq. (8.14). The polynomial `h_left` will turn out to be the Poisson bracket operator $: -\{\nu_1 J_1 + \Delta\nu_2 J_2\}:$ of Eq. (8.14) as expected.

I will now display the relevant loop to see the actual computation of the canonical transformation.

```
do i=1,ns

if(mod(i,100)==0) then
  write(6,*) ns-i,"steps remaining"
endif

if(used_ds_ave) then
   ds=ds_ave                  ! (14a)
else
  if(int_step) then
   ds=twopi*t%s(5)/circ       ! (14b)
  else
   ds=twopi*p%mag%p%ld/circ ! (14c)
  endif
endif

if(int_step) then
 call propagate(als,RAY,state,node1=i,node2=i+1)    ! (15a)
else
 call propagate(als,RAY,state,fibre1=i,fibre2=i+1) ! (15b)
endif
```

```
  ray_closed=ray ! Saving orbit  ! (16a)
  U=ray ! copying in map  ! (16b)

  U_c=exp(-ds*h_left,U)          !  (16c)
   if(icase==1) U_c=U_c.cut.2    !  (16d)
  U=U_c**(-1)*U                  !  (16e)

  Gh=log(U)        ! (17a)
  h=getpb(Gh)      ! (17b)

! Checking convergence of the logarithm
a=exp(-(fh.cut.2),(U.sub.1))  ! (18)
do k=1,c_%nd2
 if(abs(full_abs(a%v(k))-1)>1.d-5) then
    call print(a,6)
    write(6,*);write(6,*) "Log failed at element",i, p%mag%name
    stop
 endif
enddo

s=s+ds;
do k=0,n_mode
 hn(k)=hn(k)-exp(-i_*k*s)*h/circ ! (19)
enddo

ray=ray_closed+U_c

if(int_step) then   ! (20)
 t=>t%next
 p=>t%parent_fibre
else
 p=>p%next
endif

enddo
```

The real difference here is line (16c) which implements Eq. (8.16). Line (16e) recomputes the rotation $r_{j\ j+1}$ unless `icase==1`. In that case, which is the most interesting, the map `U` at line (16e) will be a nonlinear map whose linear part, and *only* its linear part, is normalised. Lines (17a, b) compute the contribution to the vector field and Hamiltonian. Line (19) performs once more the Fourier transform.

It is time for some examples. In the first test, I completely normalise the Hamiltonian, including the nonlinear part, but I use the "design distance" s. Since there are no errors, I expect a simple result. But first here are the inputs:

```
  Random errors -> t, no errors -> f
f
  Choose the state
 only_2d0   -> 1-d-f map if no errors -> type 1
 only_4d0   -> 2-d-f map              -> type 2
 delta0     -> 2-d-f map + delta      -> type 3
```

```
 nocavity0 -> 3-d-f map if no cavity -> type 4
 default0  -> 3-d-f map if    cavity -> type 5   <- SLOW
1
  Do you want to break the symmetry of the lattice
  by mispowering a single sextupole ?
  Yes -> t    No -> f
f
                         .
                         .
                         .
  int_step t or f
t
  Uniform ds -> t
f
  case 1,2
  case = 1 -> linear transformation
  case = 2 -> full nonlinear transformation: not interesting
2
  Number of Fourier modes: make it 12 or more
12
        2984  steps remaining
                         .
                         .
                         .
          84  steps remaining
```

Notice that I did not select a "`uniform ds`" but the ds of standard accelerator theory. The results for the Fourier modes are:

```
  Results for 12-fold symmetric ring
  1/pi x Hamitonian of the ring Guignard style


           0


 Properties, NO =    4, NV =    2, INA =  305
 *********************************************

   2   14.36779999999997       0.000000000000000       1  1
   4  -1400.051094947328       0.000000000000000       2  2
                         .
                         .
                         .

          11


 Properties, NO =    4, NV =    2, INA =  429
 *********************************************

  Complex Polynomial is zero

          12
```

```
Properties, NO =     4, NV =     2, INA =  366
*********************************************

  2  0.2238180754947364E-02 -0.3440156050595830E-01    1  1
  4 -0.2180965352641198        3.352214156897517       2  2
```

By selecting a nonlinear normalisation, i.e., `case = 2`, I "vaporised" all terms in the Hamiltonian. Actually by doing so, I select the rotation of Eq. (8.14) to include nonlinear terms as well:

$$r_{j\ j+1} = \exp\left(-\Delta\theta_j : \nu_1 J_1 + \nu_2 J_2 + \frac{1}{2}\nu_{11} J_1^2 + \frac{1}{2}\nu_{22} J_2^2 + \nu_{12} J_1 J_2 + \cdots :\right) I. \tag{8.24}$$

Since this is a perfect normalisation and the phase advance is supposed to be constant, then all the Fourier modes should vanish except the zeroth-order mode. It is not the case in the above run because the variable s is discrete and irregular. Of course if we decrease ds, the size of the higher order modes will decrease. This can be tested by changing line (1) of the code of Appendix O to a smaller ds. For example I can use $ds = 0.01$ instead of $ds = 0.3$. Then the results are:

```
Results for 12-fold symmetric ring
1/pi x Hamitonian of the ring Guignard style


         0


Properties, NO =     4, NV =     2, INA =  329
*********************************************

  2   14.36779999999740       0.000000000000000       1  1
  4  -1491.899920542967       0.000000000000000       2  2

                    .
                    .
                    .

        12


Properties, NO =     4, NV =     2, INA =  327
*********************************************

  2   0.000000000000000     -0.9174340199695377E-04    1  1
  4 -0.2006042085544238E-05  0.9526314186243319E-02    2  2
```

We can see that the Fourier mode $n = 12$ is now greatly reduced. The reader has noticed that the second-order tune shift with amplitude has changed. The coefficient of the term of $(h\overline{h})^2$ has changed from -1400.05 to -1491.89. This is because the sextupoles are thick and their influence changes with the number of integration steps.

Now, we can rerun the example with $ds = 0.3$ but force every integrations step to carry the same phase advance. This is done with the input

```
   Uniform ds -> t
f
```

and the result for the Hamiltonian is

```
  Results for 12-fold symmetric ring
  1/pi x Hamitonian of the ring Guignard style


            0


 Properties, NO =    4, NV =    2, INA =  366
*********************************************

   2   14.36779999999974       0.000000000000000       1  1
   4  -1400.051094957094       0.000000000000000       2  2

                       .
                       .
                       .

           12


 Properties, NO =    4, NV =    2, INA =  364
*********************************************

  Complex Polynomial is zero
```

One can see that now everything disappears and only the tune shifts are left. Of course this is not interesting. The real application is to create a Hamiltonian which can be used for non-trivial analysis. Therefore the degree of the canonical transformation should be less than the degree of the calculation. In the code of Appendix O (or N) I only permit a linear transformation, so let us repeat the preceding calculation with `icase==1`:

```
  Results for 12-fold symmetric ring
  1/pi x Hamitonian of the ring Guignard style


            0


 Properties, NO =    4, NV =    2, INA =  299
*********************************************
```

```
  2    14.36779999999974         0.000000000000000        1  1
  3  0.8263498722229337         0.457463936778557O        3  0
  3    46.77748306707147         6.960293949960596        2  1
  3    46.77748306707147        -6.960293949960596        1  2
  3  0.8263498722229337        -0.4574639367785570        0  3
  4 -0.1159118703440558        -0.3923617105875322E-01   4  0
  4    20.47647718428394         6.255117771490995        3  1
  4    46.09742446009528         0.000000000000000        2  2
  4    20.47647718428394        -6.255117771490995        1  3
  4 -0.1159118703440558         0.3923617105875322E-01   0  4

                          .
                          .
                          .


            12


 Properties, NO =     4, NV =     2, INA =  297
 ********************************************

  3    6.365220492977832         1.939002861890301        3  0
  3   -38.96913428458305        -0.5542307133387245       2  1
  3   -62.56589696763955         18.06013546866444        1  2
  3   -11.88680943171767         7.728575486524896        0  3
  4    2.812212005606214         1.325106178487980        4  0
  4   -10.18376486615540        -1.748430061899771        3  1
  4   -40.40135069325562         5.376782363158791        2  2
  4   -24.52952547124339         11.19777711815500        1  3
  4   -1.809148764522738         1.620874278368010        0  4
```

The result is the Hamiltonian, linearly normalised, as a function of θ. Since I did not introduce any asymmetry, only the modes which are multiple of twelve survive. If I rerun once more this example with

```
Do you want to break the symmetry of the lattice
  by mispowering a single sextupole ?
  Yes -> t     No -> f
t
```

then I obtain the following Hamiltonian:

```
  Results for an asymmetric ring
  1/pi x Hamitonian of the ring Guignard style


            0


 Properties, NO =     4, NV =     2, INA =  429
 ********************************************
```

```
  2   14.36779999999974          0.000000000000000         1  1
  3  0.8940931373272247          0.4777866096934882        3  0
  3   46.98860011460854          6.980868928718808         2  1
  3   46.98860011460854         -6.980868928718808         1  2
  3  0.8940931373272247         -0.4777866096934882        0  3
  4 -0.1251935202802885         -0.4303629588136917E-01    4  0
  4   20.43713113412903          6.247375081478787         3  1
  4   46.03728209715597          0.000000000000000         2  2
  4   20.43713113412903         -6.247375081478787         1  3
  4 -0.1251935202802885          0.4303629588136917E-01    0  4

         1


Properties, NO =     4, NV =     2, INA =   366
*********************************************

  3  0.7032858744489333E-01  0.7486707513945572E-02    3  0
  3  0.2112918567786044     -0.1869461878138229E-01    2  1
  3  0.2037063352393425     -0.5913938193594349E-01    1  2
  3  0.6283608526862743E-01 -0.3246209157659385E-01    0  3
  4 -0.9823097918915750E-02 -0.2024035938903566E-02   4  0
  4 -0.4009904556361971E-01 -0.3570101578719038E-03   3  1
  4 -0.5911169224381812E-01  0.1108655310767991E-01    2  2
  4 -0.3724449380244228E-01  0.1486299407094169E-01    1  3
  4 -0.8422079282043460E-02  0.5445966835407864E-02    0  4

                 .
                 .
                 .
              etc.
```

The reader will appreciate that these polynomials can now be read into *any* standard program for *time dependent* Hamiltonian perturbation theory including my own FPP prototype as shown in the coming sections.

8.7 Normalising the $\theta(s)$-Dependent Equations of Motion: Deprit-Guignard Approach

Dear Reader, in this book, I illustrated the theory using a real code (PTC) and more importantly, a real analysis package (FPP). It would difficult for me to disprove the contention that this book is just a disguised attempt to promote FPP and supplement its manual…I just hope that you will believe me when I say that there was nary a package which I could have used that is as complete as FPP. So I shamelessly peddled my code. I am only interested in the promotion of the theoretical and practical techniques and not in promoting my own faulty software. In fact I take greater pride in the existence of "gifted imitators" like my colleague Lingyun Yang[5] than in the existence of "users." Dr. Yang not only wrote a TPSA package, thus emulating Martin Berz, but also wrote an analysis routine like my own FPP.

In this section, I state up front that the software I will use was hacked together for the sole purpose of this book. It can perhaps inspire someone to write a decent set of tools but it certainly does *not* qualify as a nice set of tools. Do I feel guilty? Absolutely not.

The normalising algorithm I will use here is more than you can find in *any* book on accelerator physics. Rather it is common for authors to pontificate about Hamiltonian theory and, in the end, crack a peanut with the sledge hammer theory they poorly sell to the reader. I will not try to sell anything but, at least, I will show you how one might program a real tool for the purpose of Hamiltonian theory.

For me, the theoretical "battle fatigue" is epitomised by the following quote from an introductory textbook by Edwards and Syphers [3] who, after recracking a previously cracked peanut, conclude with a defeatist statement that makes the French army of 1940 look overly aggressive:

> Only the most basic methods of dynamics have been used thus far, because we feel that the physics at work is most transparently illustrated in that way. But much of accelerator physics makes use of one form or another of higher dynamics. The Hamiltonian approach is the method most frequently encountered in the literature.
>
> In this section, we review the Hamiltonian form of dynamics, and then recast much of the material of the earlier discussions in this language. No new physics is introduced, but the generality obtained may be helpful to the reader who wishes to pursue this approach further.

I will also recrack the same peanut…However I will provide tangible proof that it is possible to go further thanks to TPSA by actually delivering an algorithm to the reader. Moreover, the basic methods mentioned by Edwards and Syphers were actually first-order map methods. The reader by now must know that they are part of a powerful hierarchical set of methods. They are as much a part of "higher dynamics" as the Hamiltonian methods of Guignard. In fact, as this chapter illustrates, they underpin any serious attempt at using Guignard's methods.

8.7.1 Transforming the Equations of Motions

As you all know by now, any decent package should use general vector fields. Therefore, I ought to Fourier transform the force field, i.e., the plain logarithm of the map as in Eq. (8.21). Then Eq. (8.22) simply reads as

[5] I asked Dr. Yang to write the foreword of this book precisely because I wanted to de-emphasize my own software!

$$\mathbf{G}^k = \sum_{j=0}^{N-1} \underbrace{-\frac{1}{C} \exp\left(-ik\theta_{j+1}\right) \log\left(N_{j\ j+1}\right)}_{\text{line (18) in Appendix } P} \tag{8.25}$$

and the θ dependent force is given by the inverse transform:

$$\mathbf{G} = \sum_{k=-M,M} \exp\left(i\ k\theta\right)\ \mathbf{G}^k. \tag{8.26}$$

The vector field $\mathbf{G}$ is a θ dependent vector field. It turns out that it transforms under a canonical transformation like the Hamiltonian of Eq. (7.28). Assuming that the canonical transformation is

$$\begin{aligned} \mathscr{A} &= \exp\left(\mathbf{F} \cdot \nabla\right) \\ \text{where } \mathbf{F} &= \sum_{k=-M,M} \exp\left(i\ k\theta\right)\ \mathbf{F}^k, \end{aligned} \tag{8.27}$$

then the new force $\mathbf{K}$ is given by:

$$\mathbf{K} = \exp\left(:\mathbf{F}:\right)\mathbf{G} - \sum_{n=1}^{\infty} \frac{:\mathbf{F}:^{n-1}}{n!} \frac{\partial \mathbf{F}}{\partial \theta} \tag{8.28}$$

$$\approx \mathbf{G} + \left\langle \mathbf{F}, \mathbf{G}_1^{k=0} \right\rangle - \frac{\partial \mathbf{F}}{\partial \theta} + \text{Order}\left(F^2\right). \tag{8.29}$$

The attentive reader has probably noticed the minus sign between the two terms in Eq. (8.28) while there is a plus sign in Eq. (7.28) for the Hamiltonian. This is due to the fact that $\mathbf{G} \cdot \nabla =: -H :$ where H is the Hamiltonian function.

The vector function $\mathbf{G}_1^{k=0}$ is the linear part of the constant Fourier mode. Here I have assumed that the linear part of the map has been normalised following the methods of Sect. 8.4. For example, in 1-d-f, the operator $\mathbf{G}_1^{k=0} \cdot \nabla$ is given by:

$$\mathbf{G}_1^{k=0} \cdot \nabla = -i\nu \left\{ u\frac{\partial}{\partial u} - \overline{u}\frac{\partial}{\partial \overline{u}} \right\} \quad \text{where } u = x + ip. \tag{8.30}$$

Finally it is interesting to give the actual expression for the Lie bracket of two operators such as $\mathbf{G}$:

$$\begin{aligned} \langle \mathbf{G}, \mathbf{H} \rangle &= \left\langle \sum_{k=-M,M} \exp\left(i\ k\theta\right)\ \mathbf{G}^k,\ \sum_{p=-M,M} \exp\left(i\ p\theta\right)\ \mathbf{H}^p \right\rangle \\ &\approx \sum_{|p+k| \le M} \exp\left(i\ \{k+p\}\theta\right) \left\langle \mathbf{G}^k, \mathbf{H}^p \right\rangle. \end{aligned} \tag{8.31}$$

Equation (8.31) will permit me to evaluate the two terms of Eq. (8.28). The derivative with respect to θ is trivial:

$$\frac{\partial \mathbf{F}}{\partial \theta} = i \sum_{k=-M,M} k \exp\left(i\ k\theta\right)\ \mathbf{F}^k. \tag{8.32}$$

Now onwards with the normalisation!

8.7.2 The Actual Deprit-Guignard Normal Form

In this section I will perform a complete normalisation. I am cracking the same peanut as before for the sake of comparison. I hope it will convince the reader of the power of the tools, and perhaps, he can come up with different nuts[6] to crack with Hamiltonian perturbation theory.

What follows is certainly intelligible on its own. However the uncertain reader can consult Sect. 3.4 and in particular Sect. 3.4.4.2 where I normalise a time independent vector field, namely the logarithm of the standard map. Here the Hamiltonian represents the ring and is therefore time (or θ) dependent.

Let me assume, as is customary in perturbation theory, that the Hamiltonian has been normalised up to order $o-1$. Thus we have

$$\mathbf{G}_{o-1} = \exp\left(:\mathbf{F}_{o-1}:\right)\mathbf{G} - \sum_{n=1}^{\infty} \frac{:\mathbf{F}_{o-1}:^{n-1}}{n!} \frac{\partial \mathbf{F}_{o-1}}{\partial \theta:} \tag{8.33}$$

$$= \mathbf{K}_{<o} + \mathbf{K}_o + \mathbf{K}_{>o}$$

$$= \mathbf{G}_1^{k=0} + \mathbf{K}_{2\to o-1} + \mathbf{K}_o + \mathbf{K}_{>o}. \tag{8.34}$$

To continue the process of normalisation, I perform an additional transformation on Eq. (8.34):

$$\exp\left(:\mathbf{f}_o:\right)\mathbf{G}_{o-1} = \mathbf{G}_1^{k=0} + \mathbf{K}_{2\to o-1} + \underbrace{\left\langle \mathbf{f}_o, \mathbf{G}_1^{k=0}\right\rangle - \frac{\partial \mathbf{f}_o}{\partial \theta} + \mathbf{K}_o}_{=\text{ Tune shifts}} + \cdots. \tag{8.35}$$

In Eq. (8.35) the vector field $\mathbf{f}_o$ is chosen to clean $\mathbf{K}_o$ of all non-tune-shift terms. To order o the canonical transformation becomes:

$$\mathbf{F}_o = \mathbf{F}_{o-1} + \mathbf{f}_o. \tag{8.36}$$

[6] θ-dependent methods provide a natural way to study systems with slightly broken symmetries although maps of finite θ can be used as well. But I will refrain in this book from discussing topics I did not have time to investigate myself. I leave it to the reader! See Sect. 9.2.1.

To proceed further I simply need to increase the index from o to $o+1$ and return to Eq. (8.33).

I will finish this section by solving Eq. (8.35) as I did in Sect. 3.4.4.2 at Eq. (3.124). The only difference is the presence of a time dependence whose influence we will now discover. First I expand the operator $\mathbf{K}_o$ in phasors and Fourier modes, limiting myself to 2-d-f for notational simplicity. Here is the ath component of the vector $\mathbf{K}_o$:

$$K_o^a = \sum_{k=-M,M} \exp(i\, k\theta)\; K_{o;m}^{k;a} u^{m_1} \overline{u}^{m_2} v^{m_3} \overline{v}^{m_4}. \tag{8.37}$$

Solving for f_o^a, I get for $a = 1, 2, 3, 4$:

$$\begin{aligned}
f_{o;m}^{k;a=1} &= \frac{-K_{o;m}^{k;a=1}}{i\,\{\nu_1(m_2-m_1+1)+\nu_2(m_4-m_3)-k\}} \\
f_{o;m}^{k;a=2} &= \frac{-K_{o;m}^{k;a=2}}{i\,\{\nu_1(m_2-m_1-1)+\nu_2(m_4-m_3)-k\}} \\
f_{o;m}^{k;a=3} &= \frac{-K_{o;m}^{k;a=3}}{i\,\{\nu_1(m_2-m_1)+\nu_2(m_4-m_3+1)-k\}} \\
f_{o;m}^{k;a=4} &= \frac{-K_{o;m}^{k;a=4}}{i\,\{\nu_1(m_2-m_1)+\nu_2(m_4-m_3-1)-k\}}.
\end{aligned} \tag{8.38}$$

The reader will notice that if $k \neq 0$, it is always possible to perform the inversion of Eq. (8.38). In that case, the final result is a θ-independent generator of rotation. This is precisely the normalisation described in Sect. 8.3 if the canonical transformation a is nonlinear.[7] For $k = 0$, some terms cannot be removed and they correspond to the tune shift terms also computable from a map based theory. Actually the map based theory gets these terms exactly. I apologise for not providing a new use for the Deprit-Guignard methods: my goal is to show that if there is a will on the part of the reader, he will be able to program new tools based on the Deprit-Guignard approach.

As in the map based theory of Sect. 5.4, it is possible to leave a single resonance in the Hamiltonian or vector field. Then it is also possible to go into a co-moving frame from which an invariant valid all over phase space is computable. I will not discuss any of this in this book. I will simply show some prototypical tools which prove that my hierarchical approach is indeed more powerful than any other approach when realistic calculations are needed.

[7] If a is computed to the order of truncation of the TPSA package.

8.7.3 Example Code for a Deprit-Guignard Normalisation

For this example, I created a new Fortran 90 type to represent the vector field found in Eq. (8.26).

```
type c_vector_field_fourier
 integer :: n=0
 type (c_vector_field), allocatable :: f(:)
end type c_vector_field_fourier
```

For example, the kth Fourier mode $\mathbf{G}^k$ of the vector field $\mathbf{G}$ (of type c_vector_field_fourier) is represented as:

$$\mathbf{G}^k \rightarrow G\%f(k). \tag{8.39}$$

The actual main program is in Appendix P. I limited myself to a 1-d-f example and, for the sake of accuracy, I use a single cell of the ALS lattice by invoking the optional parameter `onecell=.true.` when creating the lattice:

```
call build_lattice_als(ALS,mis,exact=.false.,onecell=.true.) ! (0)
```

The main program differs very little in structure from the program of Sect. 8.6: a few flags are hard-wired for the convenience of the reader. In particular a constant phase advance is used and the Courant-Snyder normalisation is limited to the linear map as one would normally do in a Guignard style normalisation.

However, there is one big difference: I perform a Fourier transform of the vector field rather than the Hamiltonian. This is more general and consistent with the general philosophy of the FPP package. Here is the corresponding code fragment:

```
 call propagate(als,ray,state,node1=i,node2=i+1)    ! (14)


  ray_closed=ray ! Saving orbit  ! (15a)
  U=ray ! copying in map  ! (15b)

  U_c=exp(-ds*h_left,U)         !  (15c)
  U_c=U_c.cut.2                 !  (15d)
  U=U_c**(-1)*U                 !  (15e)

  logN=log(U)       ! (16)

! Checking convergence of the logarithm
a=exp(-(logN.cut.2),(U.sub.1))  ! (17)
do km=1,c_%nd2
 if(abs(full_abs(a%v(km))-1)>1.d-5) then
    call print(a,6)
    write(6,*);write(6,*)"Log failed at element",i, p%mag%name
    stop
 endif
enddo
```

```
s=s+ds;
do km=-n_mode,n_mode
 G%f(km)=G%f(km)+(exp(-i_*km*s)/circ)*logN ! (18)
enddo
```

Line (18) performs the computation of Fourier modes and thus implements Eq. (8.25). The actual normalisation is done at lines (21) and also, using a faster method, at line (22) as shown in the following fragment.

```
U=from_phasor()
call transform_vector_field_fourier_by_map(G,G,u) ! (20)

prec=1.d-5
write(mf,*); write(mf,*) "Results of Guignard Normalisation"
write(mf,*) "One exponent k=0 with",n_fourier, "modes"
call normalise_vector_field_fourier(G,F,K)      ! (21)
call c_clean_vector_field_fourier(K,K,prec)
call print(K%f(0),mf)

write(mf,*) "Factored k=0 with",n_fourier, "modes"
call normalise_vector_field_fourier_factored(G)     ! (22)
call c_clean_vector_field_fourier(G,K,prec)
call print(K%f(0),mf)
```

At line (20), the vector field **G** is transformed into the phasors' basis. This is a time-independent transformation and therefore the formula of Eq. (3.60) can be applied to each Fourier mode of **G** in Eq. (8.26). This is done at line (20) by a subroutine call. The reader is invited to look at this routine in Ci_tpsa.f90 of the FPP library.

The next step is the normalisation. I will describe `normalise_vector_field_fourier`[8] which implements exactly the normal form described in Sect. 8.7.2. Again the reader can look at Ci_tpsa.f90. Because of its importance, I reproduce its core fragment here:

```
    i1=2 ;if(present(F1)) i1=1 ;nl=0; nl=n_extra;

    do o=i1,no

        ht=H1
       IF(O>1) call exp_vector_field_fourier(F,Ht,Ht)    ! (2)

    do m=-n_fourier,n_fourier

      do ki=1,n
```

[8]The subroutine `normalise_vector_field_fourier_factored` is also a normal form. In that routine the canonical transformation is factorised à la Dragt-Finn. I did not bother to save the transformation, so the routine accepts the force field and only returns its normalised form. I invite the reader to look at the routine in Ci_tpsa.f90. I left in, commented out, the calls from `normalise_vector_field_fourier` for easier comparison.

```
    j=1
     do while(.true.)
       temp=ht%f(m)%v(ki).sub.o
       call  c_cycle(temp,j,v ,je); if(j==0) exit;

  if(m/=0) then
     removeit=.true.
  else
       call check_kernel(ki,n,je,removeit)
  endif
        if(removeit) then
          lam=-i_*m                     ! (3a)
          je(ki)=je(ki)-1
            do l=1,n
            if(coast(l)) cycle
              lam=lam-eg(l)*je(l)       ! (3b)
            enddo
          je(ki)=je(ki)+1
           F%f(m)%v(ki)=F%f(m)%v(ki)-(v.cmono.je)/lam   ! (4)
         endif

     enddo  ! over monomials
    enddo  ! over vector index
   enddo ! over fourier mode
   IF(o==1) THEN
     call exp_vector_field_fourier(F,Ht,H1,nlin=nl)
     F1=F; F=0;NL=0;
   ENDIF
 enddo  ! over order o
  ht=H1
    call exp_vector_field_fourier(F,Ht,Ht)    ! (5)
  K=ht
```

At line (1), I extract the linear frequencies which enter in the normalisation as shown explicitly at Eq. (8.38). The present algorithm assumes that the linear part produces a θ-independent rotation. As we have seen and as it is explained in standard textbooks [1], it is always possible to create such a Hamiltonian (or force field). The denominators of Eq. (8.38) are created at lines (3a, b) while line (4) is actually the implementation of Eq. (8.38). At line (2), I compute the normalised force field at that point in the iteration. Line (5) is not used in this example: it is necessary if the linear part has some s-dependence. For example, if the linear part is "Courant-Snyder", i.e., $H_2\left(J;\theta\right)=\frac{C}{2\pi}\frac{1}{\beta(s(\theta))}J$, then the position dependence of the rotation is removed at line (4). This contributes to a linear term in the computation of the Lie operator `F`. Its effect on the Hamiltonian is computed at line (5). This requires an infinite sum: in practice a sum until convergence is reached. Thus it slows down the algorithm. At line (6), the final normalised force is computed from the original force using the *total* canonical transformation.

The subroutine exp_vector_field_fourier is critical. It cannot be done by a simple substitution because of the θ-dependence. Again, I show here a code fragment from the subroutine exp_vector_field_fourier which implements Eq. (8.28):

```
subroutine exp_vector_field_fourier(s1,h0,hr)
                      .
                      .
                      .
    call ddt_vector_field_fourier(F,dF_dt) ! (1)

    s3=H ; t=H ; dhs=dF_dt;
    fac=1.0_dp;     fac1=1.0_dp;

    do i=1,no+nl    ! extra terms for cheap convergence
     fac=1.0_dp/i

      call bra_vector_field_fourier(F,t,t) ! (2a)
      call mulc_vector_field_fourier(t,fac,t)  !  t=fac*t
      call add_vector_field_fourier(s3,t,s3) ! (2b)
      call bra_vector_field_fourier(F,dF_dt,dF_dt) ! (3a)
      fac1=1.0_dp/(i+1)
      call mulc_vector_field_fourier(dF_dt,fac1,dF_dt)
      call add_vector_field_fourier(dhs,dF_dt,dhs) ! (3b)
    enddo
      fac=-1.d0
      call add_vector_field_fourier(s3,dhs,s3,fac) ! (3c)
     K=s3
```

This subroutine contains two calculations: the usual exponentiation is done at lines (2a, b). The second term in Eq. (8.28), the so-called integrated exponential, is done at lines (3a, b, c). This term is non-zero if the derivative of the operator K computed at line (1) is non-zero. I called this subroutine "exp_vector_field_fourier" obviously assuming that it is an exponentiation of the operator F found in its calling argument. Actually, if one extends phase space to include the time-like variable θ, then one sees that it is an exponentiation. In other words, the operator

$$\mathbf{G}^{ext} \cdot \nabla = \mathbf{G} \cdot \nabla + \frac{\partial}{\partial \theta} \tag{8.40}$$

transforms under the operator $\exp(\mathbf{F} \cdot \nabla)$ as:

$$\mathbf{K}^{ext} \cdot \nabla = \exp(\mathbf{F} \cdot \nabla) \left\{ \mathbf{G} \cdot \nabla + \frac{\partial}{\partial \theta} \right\} \exp(-\mathbf{F} \cdot \nabla) . \tag{8.41}$$

I leave it to the reader[9] to show that $\mathbf{K}^{ext}$ is really $\mathbf{K}$ of Eq. (8.28).

[9]Equation (8.41), if applied to a mixed variable generating function, will reproduce the simple result $k^{ext} = k \circ a + \frac{\partial f}{\partial \theta}$ where f is the mixed variable generator. Lie methods are more convenient in

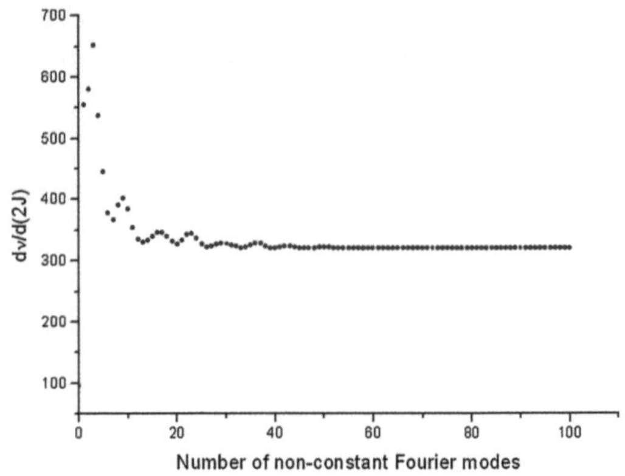

Fig. 8.2 Tune shift from a Guignard normalisation

I will stop here and present the numerical results. The reader is invited to dig deeper in the routines associated to this Guignard normal form; they do not contain anything surprising.

The graph of Fig. 8.2 was done with the "commented-out" section in Appendix P. Without a doubt, the algorithm converges to the correct result obtained from the one-turn map normalisation.

8.7.4 *Numerical Example* $\langle x \rangle$*: Analytical, Guignard and Map*

Here I redirect you to (13.3) of Chap. 13. Of course feel free to read the entire chapter. There I perform the analytical calculation for $\langle x \rangle$ two different ways. The program in Appendix Q, described in Sect. 13.3, uses the analytical formula, the map method and two different Fourier representation of the force field for the normalisation. Notice that I have already described the implementation of the formula in Sect. 7.8.4.3.

Please take a look at all the relevant sections!

8.7.5 *The Final Strategy: Be Prepared to Mix Everything!*

This section is not written. Please look at Sect. 9.2.1 to see what ought to have been written and discussed here. I take the unorthodox liberty to leave the section blank!

(Footnote 9 continued)
standard perturbation because they are explicit. In the Hamiltonian case, mixed variable functions have some great theoretical advantages.

References

1. E. Forest, *Beam Dynamics: A New Attitude and Framework* (Harwood Academic Publishers, Amsterdam, The Netherlands, 1997)
2. S.Y. Lee, *Accelerator Physics* (World Scientific Publishing, Singapore, 2004)
3. D.A. Edwards, M.J. Syphers, *An Introduction to the Physics of High Energy Accelerators, Wiley Series in Beam Physics and Accelerator Technology* (Wiley, New York, 1993)

Chapter 9
Here Is the Conclusion of This Book

Abstract I give a final summary of this book. It is as much an introduction as it is a conclusion since it describes what I wanted to achieve and what I have actually achieved. I try here to show pictorially using Fig. 9.1 what I call the universal Courant-Snyder loop. It ties, in the code, map perturbation theory and standard Hamiltonian theory.

Keywords Hierarchy · Summary · Schematic flow chart · Missing parts

9.1 Conclusion

I began with the statement that the "Code" is the arbiter of truth in accelerator theory. Indeed most accelerator physicists would agree with me that the easiest and safest way to "understand" the motion of non-interacting particles is simply to write an integrator. It is a brute force approach but we think that it is more reliable than Taylor series maps or worse, simplistic analytical formulae. The accelerator represented inside the code can have misalignment errors, powering errors, can include radiation, spin, complex fringe fields, etc. Therefore it does not seem to be easily amenable to perturbation theory.

In standard books we see formulae, we see a Fourier expansion of some ideal Hamiltonian and various concepts introduced on the basis of the existence of this expansion. However no one is ever shown a realistic calculation, or even a semi-realistic one, from beginning to end. The reason for this lack of realism is simply that the theoretical tools and computational tools exist side by side without any special thoughts given on how they should be integrated.

This state of affair is the result of the natural evolution of this highly technological field. The computers of the 1950s were primitive by today's standard. However the calculations performed in this book are *real* since you can download and run them. They would have been only dreams in the 1950s. In fact, when I entered the field in 1985, I can tell the reader that the type of calculations done here were barely feasible. But they were not sitting at a distant horizon, and therefore, with the help of other scientists, Alex Dragt, Martin Berz, Richard Talman, Ronald Ruth, I began

E. Forest, *From Tracking Code to Analysis*,
DOI 10.1007/978-4-431-55803-3_9

to put together all the necessary tools. Recently with the help of Desmond Barber, I added also spin to the collection of objects which can be analysed by the perturbation theory tools of the library FPP. I did not implement a Fourier decomposition of the spin-orbit motion in FPP, but that is possible. In fact, as this book is being finalised, I am extending the logarithm of a map to include spin which is the first step towards the Guignard-Deprit perturbation scheme of Chap. 8.

The important message, which I seriously documented in this book, is that the hierarchy presented in this book, is the only reliable and self-consistent approach to the merging of simulation and perturbation theory. It is worth depicting it once more.

The structure in Fig. 9.1 is, in my view, universal. Therefore I encourage people who use other packages to reproduce part of this structure.

The central piece is the "Code" in dark blue. Its unadulterated output are rays which I called *probes*. These probes are the *propagata*. They contain the material that is "trackable" by the code, by your code. Through the magic of polymorphic TPSA, I can say that the code can track Taylor series, ordinary rays being series of degree zero. These Taylors series, if correctly initialised, can be elevated to the status of approximate maps: they go from *propagata* to *propagators*.

The logic and content of this book is summarized on Fig. 9.1. First there is Big Brother: this is the person or team of persons that created or even uses the Code. It or he or they decided what the models are, what the approximations are, etc. Of course Big Brother can modify the code as he sees fit. But, generally, the users simply use the Code without attempting to modify it. Sometimes, as designers, we are allowed to choose the lattice and the models from the buffet selection within the Code. Sometimes we become Big Brother and change the Code itself. In this book we did not play Big Brother. So the reader will notice that I avoided discussing what the code PTC was, is or should be. For the purpose of this book, which is a description of the algorithms schematically depicted by Fig. 9.1, I did not need to know what the Code truly is. Of course when comparing with analytical formulae, a part of me needed to assume the role of an inner party member to use Orwell's language. Thus, in Sect. 7.8.4.3, for example, I displayed a knowledge of the model used by the code and, in fact, coerce the code to use a model that would agree perfectly with the analytical formula derived for $\langle x \rangle$.

So I have a code with Taylor polymorphism, what can I do with it? On Fig. 9.1, we can first follow the blue lines. We get tracking and, out of tracking, we get one-turn maps in the blue pentagon. This is trivial and contains zero accelerator theory besides what is in the code. In fact that statement would be true if the code described spectrometers, light sources, electron microscopes or even the trajectory of a bullet out of a rifle! This trivial usage of the code was described in Sect. 1.7 of the introduction.

The real fun started when I assumed that the map is that of a circular accelerator and computed a normal form. A normal form is similar in spirit to a diagonalisation of the map. Therefore it invokes a numerical library: FPP. I purposely coloured FPP

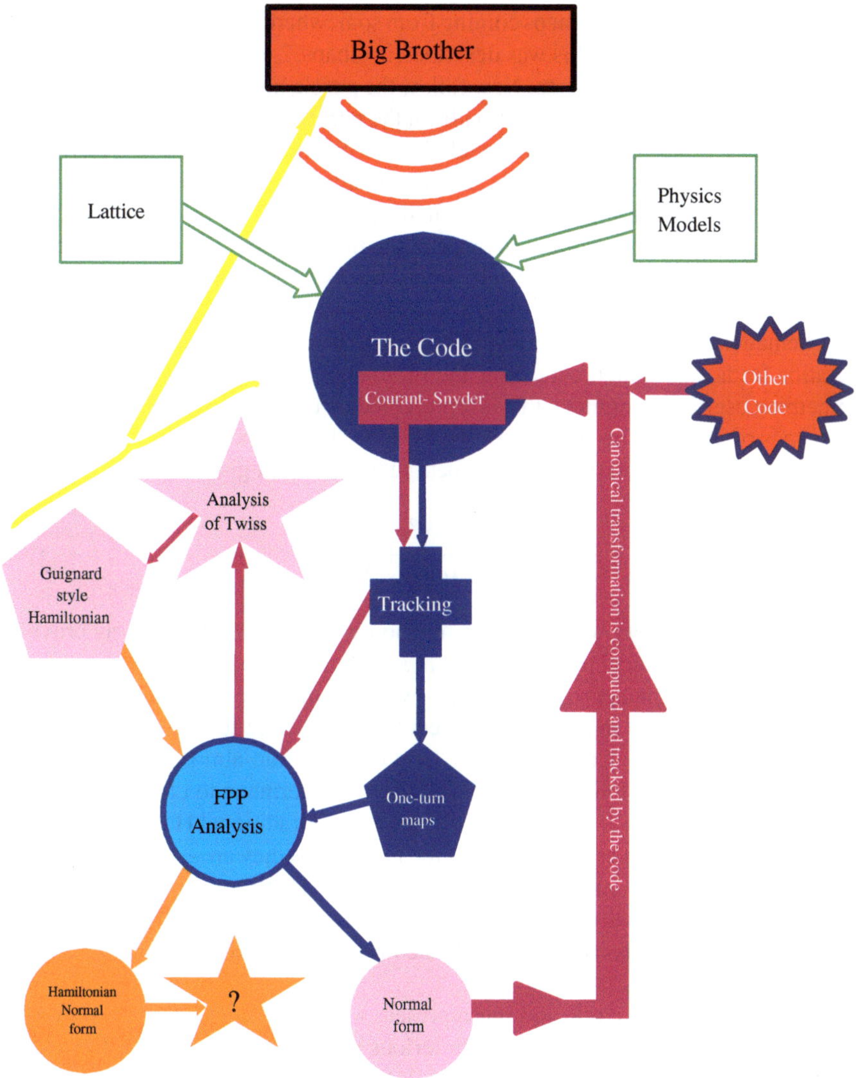

Fig. 9.1 The hierarchy consistent with the code

in a different shade of blue. I want to emphasize that while my own FPP is an integral[1] part of PTC, this is an accident of my poor intellect. It is quite possible to ship a map computed by PTC to the analysis tools of say Cosy-Infinity or to the library of my colleague Lingyun Yang. So I really want to emphasize the "indepedent library"

[1]You can run FPP as a standalone and glue it to a code written in Fortran 90. It is a library. You cannot run PTC with another library easily. However you could analyse the maps of PTC with a different library. For example my complex library is *not* the TPSA library used by PTC!

aspect of FPP. I can analyse maps coming from somewhere else and it can be replaced by something else! All of this was described in Chaps. 2–6.

So we have the one-turn map. Now follow the purple line from the normal form to the code: the canonical transformation from the normal form is tracked by the code. This result is what I call the "universal Courant-Snyder loop." It was described in Sect. 2.4 in 1-d-f both with the code and analytically. I then explained in great detail in Chap. 7 the concept of phase advance in the presence of nonlinearities and spin. I also included a discussion of the advance of the time slip first discussed in Sect. 4.2.

The Courant-Snyder loop by itself simply tracks a canonical transformation relevant[2] to the one-turn map. It is what you do with it that has meaning. For example, as schematically displayed on Fig. 9.1, the purple arrow of the Courant-Snyder computation feeds data from the tracking algorithm into the analysis package FPP. Then something useful is computed; this is represented by the star shaped algorithm called "Analysis of Twiss." The simplest example is the "beta" function. In 1-d-f, if I track the canonical transformation, then the beta function is just given by

$$\beta(s) = \left(\frac{\partial x_s}{\partial x_0}\right)^2 + \left(\frac{\partial x_s}{\partial p_0}\right)^2. \tag{9.1}$$

In Eq. (9.1), I purposely wrote the beta function in terms of the tracked ray provided the input is the canonical transformation at $s = 0$; of course derivatives produce the Jacobian matrix—the approximate map—which is the object from which the beta function was extracted in previous chapters (see Sect. 2.2.2 for example).

The analysis of the Courant-Snyder loop can generate almost anything: lattice functions, spin lattice functions, radiative quantities germane to electron rings, etc. But, as shown in Chap. 8, it can also be the input for a traditional Hamiltonian calculation. This calculation uses the maps of the magnets as they are—errors and bizarre fringe fields included—and computes the s-dependent Hamiltonian as a Fourier series in s. This is schematically represented by the light purple pentagon. It is a result of a Courant-Snyder analysis which I indicated by an arrow from the star to the light purple pentagon.

The orange arrows, circle and star represent work that is not yet seriously implemented. Nonetheless I developed tools serious enough to illustrate the theory: the simple normalisation algorithm of Sect. 8.7. Of course the Fourier transform of the Hamiltonian can be the input of a normalisation code written by astronomers.

The idea that the "Code" can produce an input for another type of code, because I am too dumb to write my own or too lazy, can also be reversed. This is represented by the "other code" on Fig. 9.1. This "code" can send a map, a canonical transformation, etc. For example, you might have designed a damping ring with your favourite tool. From that code of yours, you extract information about the beam, say for example, the equilibrium beam sizes. You then want to inject that beam into a machine modelled

[2]For a transfer line or a linear accelerator, the canonical transformation usually describes the moments of the input distribution. I did not discuss this topic because it is a trivial generalisation: the initial canonical transformation is computed externally.

in *my code* and see if the beam behaves properly. One way is simply to diagonalize the envelope of your beam (see footnote 2) using a canonical transformation and then track it inside my own code. You can also use this canonical transformation to produce matched tracking input and do *bona fide* tracking of rays.

Finally there is the yellow line on Fig. 9.1. It represents information being fed back to Big Brother by the thought police: this is called fitting. It can imply a modification of the lattice parameters, a modification of the model, a modification of the lattice, and in some cases, purging the code itself and replacing it by another code. In fact there are ways to use PTC in order to discard PTC in account of inadequacy. Then you find a better code that suits your purpose. The yellow line brings fear in the heart of the code[3] programmer. But so is life under the Code!

I hope that I have given the reader a small glance of the hierarchy of perturbation theory which I consider to be the most effective and self-consistent expression of perturbation theory in accelerator physics. As I often said, nothing is lost, even Hamiltonian perturbation theory can be regained in an even more complete and faithful manner.

9.2 Exclusion

In this book I wanted to avoid "bull…" as much as possible. Every topic addressed was documented with an example program. The example programs, as well as the code PTC itself, can be molested by the user. After all the reader of this book is Big Brother while he attempts to understand its delusional prose. However there are some topics which I kept out because I feel that I do not have the best tools in my FPP arsenal. They are excluded due to present lack of software tools and therefore I am unable to support them by software evidence that the user can personally examine.

9.2.1 A Deeper Discussion About Guignard Normalisation

The first obvious exclusion concerns Chap. 8. I did show a prototype of time dependent perturbation theory. I ended up cracking a peanut with a sledge hammer which is against my religious beliefs. However I did so with a piece of software which, I hope sincerely, can convince the reader that real powerful and general tools can be written.

There is an important topic which I did not explore and it is that of the empty Sect. 8.7.5. The attentive reader will notice that in Chap. 8 I state that the relationship

[3]There is another option: the code can absorb other codes. The code BMAD of Cornell has literally swallowed everything in sight. So in BMAD you are most likely to change the model or ask its programmer to increase its dominion.

between the time-like variable θ and some true distance s along the machine is arbitrary.

This implies that it is possible to mix finite map theory with Hamiltonian theory à la Guignard. For example, imagine a ring like the LHC: hundreds of cells with various interaction regions. It would not be feasible or desirable to use enough modes to describe the cellular structure of that lattice. However one can easily imagine representing arcs with a single map and the interaction region with a θ-dependent Hamiltonian. In a ring like KEKB, where the interaction regions are the main source of nonlinearities, this would seem sensible at first glance. For two reasons, no one has ever done that: the tools did not exist and the experts at Hamiltonian theory mostly pontificate with no software. Moreover they do not understand a map based theory.

I hope that the reader can see that it is possible and indeed desirable to mix both theories if necessary. The hierarchical scheme of this book allows this but I did not provide any convincing example. I do not feel guilty because most textbooks do not provide any realistic examples. The fanatical reader will see how one can modify the example of Chap. 8 so as to mix finite maps and θ-dependent force fields.

9.2.2 Synchro-Betatron Effects

The second exclusion is a more serious issue. It concerns the longitudinal or temporal plane. If you read this book carefully, you will notice that there is nothing fundamentally different between the two transverse planes and the temporal plane. Indeed the examples of this book allow for a full 3-d-f calculation with longitudinal focussing. The entire radiation calculation of Sect. 4.4 depended on this.

However the attentive reader has perhaps noticed a problem with the logic presented in this book with respect to the RF cavity. At page 84, I affirm that the standard map with a small tune is very well modelled by a pendulum. I also said that the purpose of a normal form is to simplify away—sweep under the carpet—what we already understand. Therefore, if we print a map, can we extract the pendulum[4] out of the map accurately? The answer is *no* for a technical reason that has been known for a long time by many scientists such as Professor Kaoru Yokoya of KEK in the context of spin normal form.

A normal form on the longitudinal map alone can be done to a high order and the result is very close to the tracking code. On the other hand, normal form on the transverse map is basically "pure garbage" if carried out to high order. The solution to this problem is simple in theory: we need TPSA tools which allow us to produce Taylor maps which discriminate between planes. In other words, we need high order in the longitudinal plane while retaining low order in the transverse variables. This is the only way to keep the calculation manageable, but it requires a TPSA package which is not what exists in FPP.

---

[4]In reality this map could include nonlinear time slip terms not present in the classical pendulum map, but this is a tiny detail that can easily be included.

The same applies to perturbation theory on the Hamiltonian. I explain in Chap. 8 that one can Fourier transform the Hamiltonian using a phase which is uniformly distributed and linear around the ring. It is proposed in standard treatments of accelerator physics. This implies in 3-d-f that the phase advance of the unperturbed problem (see Eq. (8.14)) should be

$$r_{j\ j+1} = \exp\left(\Delta\theta_j : -\{\nu_1 J_1 + \nu_2 J_2 + \nu_3 J_3\} :\right) I. \tag{9.2}$$

But Eq. (9.2) is incorrect if we insist that the longitudinal plane be as close as possible to a pendulum-like oscillator. The correct answer should be

$$r_{j\ j+1} = \exp\left(\Delta\theta_j : -\left\{\nu_1 J_1 + \nu_2 J_2 + \int_0^{J_3} \nu_3(J_3')dJ_3'\right\} :\right) I \tag{9.3}$$

if the entire theory is able to deal with high orders in the cavity plane. Because of this technical inability, I refrained from talking about synchro-betatron resonances. This lacuna also affects spin resonances involving the longitudinal plane.

It is important to realise that the lacuna is technical: we lack the tools. Perhaps a clever reader will develop the TPSA package(s) necessary for this type of analysis.

In conclusion, I listed things that, within the internal logic of this book, should have been included but were not for the reasons I just discussed. Of course there might be other things which should be documented and were not. For example, it would have been interesting to make a large catalogue of quantities that are computed numerically and analytically in accelerators, and then, to provide examples of their computation in my appendices. Hopefully the readers can contribute with suggestions that could enter in a future edition. I wanted to keep this book short and to the point.

All's Well That Ends Well
—*William Shakespeare*—

Chapter 10
Phasors Basis: Why Do I Reject Symplectic Phasors?

Abstract In this supplementary chapter I discuss with examples why I chose a non-symplectic phasors' basis. This can only be justified if I examine the implication of my choice in the nonlinear regime.

Keywords Phasors' basis · Complex symplectic transformation

Since FPP is a complex package, would it be more pleasing to use a complex symplectic transformation instead of Eq. (2.21)? In that case, the Poisson bracket would be one. Moreover the rules for changing the Hamiltonian would be the usual substitution rule. So why did I pick a non-symplectic definition for the phasors?

The answer is connected to the tune shifts and the damping decrements. In the linear case, it is always true that the tune is a pure imaginary number and that the damping is a real number. If we choose the phasors of Eq. (2.21), this results extends to the nonlinear tunes and the amplitude dependent damping: they are respectively polynomial with pure imaginary coefficients and real coefficients at all orders. This is not true if a symplectic phasor transformation is used. For example, I can use the following transformation:

$$\zeta = \frac{x+ip}{1-i} \quad \text{and} \quad \overline{\zeta} = \frac{x-ip}{1-i} \quad [\zeta, \overline{\zeta}] = 1 \tag{10.1}$$

I substitute Eq. (10.1) into a simple Hamiltonian with amplitude dependent tune shift:

$$H = \omega \frac{x^2+p^2}{2} + \Delta\left(x^2+p^2\right)^2 \rightarrow \; -i\omega\zeta\overline{\zeta} - 4\Delta\left(\zeta\overline{\zeta}\right)^2. \tag{10.2}$$

We can easily compute the change of ζ with respect to time by taking the *usual* Poisson bracket:

$$\frac{d\zeta}{dt} = [\zeta, H] = -i\omega\zeta - 8\Delta\left(\zeta\overline{\zeta}\right)\zeta = \underbrace{\left\{-i\omega - 8\Delta\left(\zeta\overline{\zeta}\right)\right\}}_{\text{tune}}\,\zeta. \tag{10.3}$$

E. Forest, *From Tracking Code to Analysis*,
DOI 10.1007/978-4-431-55803-3_10

Equation (10.3) displays a tune which has an imaginary linear tune and a "real" tune shift with amplitude. Let us do the same calculation with the non-symplectic phasors,

$$h = x + ip \quad \text{and} \quad \overline{h} = x - ip \quad [h, \overline{h}] = -2i \tag{10.4}$$

and the result is

$$H = \omega \frac{x^2 + p^2}{2} + \Delta\left(x^2 + p^2\right)^2 \rightarrow \frac{\omega}{2} h\overline{h} + \Delta\left(h\overline{h}\right)^2 \tag{10.5}$$

and the tune is given by

$$\frac{dh}{dt} = -2i\,[h, H] = -i\omega h - 4i\Delta\left(h\overline{h}\right)h = \underbrace{\left\{-i\omega - 4i\Delta\left(h\overline{h}\right)\right\}}_{\text{tune}}\,h. \tag{10.6}$$

And we then substitute h for ζ in Eq. (10.3), we get:

$$\text{Tune} = -i\omega - 8\Delta\zeta\overline{\zeta} = -i\omega - 8\Delta\frac{h\overline{h}}{(1-i)^2} = -i\omega - 4i\Delta h\overline{h}. \tag{10.7}$$

So, as we can see, the result is the same but the use of the non-symplectic phasors facilitates the identification of the tunes and the damping decrements: tunes are imaginary and damping decrements are real. We must simply remember that if the Poisson bracket is used, it involves a factor of $-2i$ as I explained near Eq. (3.74). This is not a big deal especially within the context of FPP where the Poisson bracket is *never* used internally so as to accommodate non-symplectic maps and non-canonical variables.

Chapter 11
The Logarithm of a Map

Abstract In this supplementary chapter I discuss the brute force extraction of a Lie exponent by using the Taylor series of the logarithm. This is not totally trivial because I lack a matrix representation. It must be pointed out that in general matrices can have complex logarithms just like numbers, however I am interested here in real maps having real Lie operators as logarithm. It could include spin.

Keywords Logarithm of map · Lie operator · Nonlinear logarithm · Single exponent

If a matrix M is near the identity, the following series converges:

$$\log(M) = \sum_{n=1}^{\infty} (-1)^{n+1} \frac{(M-1)^n}{n}. \tag{11.1}$$

Of course this applies trivially to a linear map of phase space. What about a Lie map? Consider a nonlinear map m and let us assume that it can be approximated by a Lie exponent:

$$m = \exp(F \cdot \nabla) I \approx I + F \cdot \nabla I + \cdots \approx I + F + \cdots . \tag{11.2}$$

The Lie map $\exp(F \cdot \nabla)$ acts on the space of functions and thus it is possible to write a matrix for it using a basis made of monomials. For example, in 1-d-f, for polynomials of degree n_0, the space of polynomials is of dimension $\frac{(n_0+2)!}{n_0!2!}$. Therefore the matrix for $\exp(F \cdot \nabla)$ is of dimension $\left(\frac{(n_0+2)!}{n_0!2!}\right)^2$. This matrix, as I pointed out in Ref. [1], is the transpose of the matrix which propagates the moments. In the linear case, this was confirmed in Sect. 2.3.1 namely at Eq. (2.3.8).

Here I do not have matrices in the nonlinear case, so I must be a little more resourceful. First one notices that Eq. (11.2) gives us a trivial approximation of the vector field:

$$F \approx m - I. \tag{11.3}$$

E. Forest, *From Tracking Code to Analysis*,
DOI 10.1007/978-4-431-55803-3_11

I proceed further by assuming that the vector field F is already known to some order and that the result is F^k. I can write:

$$\exp\left(-F^k \cdot \nabla\right) m = I + t. \tag{11.4}$$

This algorithm starts with $t = m - 1$. At the end we expect t to be zero. My goal is to find a relatively fast algorithm. The first step is to find a vector field which can reproduce the map $I + t$. To third-order in t, I can write:

$$\begin{aligned} &\exp(\,\tau_3 \cdot \nabla\,)\, I = I + t \\ &\quad \text{where} \;\; \tau_3 = t + \varepsilon_2 + \varepsilon_3 . \end{aligned} \tag{11.5}$$

Solving for ε_2 and ε_3, I get:

$$\varepsilon_2 = -\frac{1}{2} t \cdot \nabla t \quad \text{and} \quad \varepsilon_3 = -\frac{1}{2}\varepsilon_2 \cdot \nabla t - \frac{1}{6} t \cdot \nabla \varepsilon_2 . \tag{11.6}$$

I can then rewrite Eq. (11.5) using Eq. (11.6):

$$\begin{aligned} \exp\left(-F^k \cdot \nabla\right) m &= \exp(\tau_3 \cdot \nabla)\, I \\ &\Downarrow \\ m &= \exp\left(F^k \cdot \nabla\right) \exp(\tau_3 \cdot \nabla)\, I . \end{aligned} \tag{11.7}$$

Now I can apply the Baker-Campbell-Hausdorff formula (CBH) to go to the next step in the iteration:

$$\begin{aligned} F^{k+1} = {} & F^k + \tau_3 + \frac{1}{2}\left\langle F^k, \tau_3\right\rangle + \frac{1}{12}\left(\left\langle F^k, \left\langle F^k, \tau_3\right\rangle\right\rangle - \left\langle \tau_3, \left\langle F^k, \tau_3\right\rangle\right\rangle\right) \\ & - \frac{1}{24}\left\langle F^k, \left\langle \tau_3, \left\langle F^k, \tau_3\right\rangle\right\rangle\right\rangle + \text{higher order} \cdots . \end{aligned} \tag{11.8}$$

The Lie bracket in Eq. (11.8) was defined in Eq. (3.39).

Recursive application of Eq. (11.8) will produce a series which convergences at a rate faster than linear. The map must be close to the identity; however it is possible to change the algorithm and obtain the logarithm of maps even when the series of Eq. (11.1) doesn't converge. But this is a topic for another book.

Finally, the efficacy of the algorithm proposed here can be compared with a pure linear algorithm where the CBH formula is completely ignored. This is done at the end of the program of Appendix F in the following code snippet:

```
!!!!!!!!!!!!!!!!!!!!   Illustrating the Logarithm    !!!!!!!!!!!!!!!!!!!
!

           lielib_print(3)=1    ! printing iterates
write(6,*);write(6,*) "Testing the logarithm of a map";write(6,*);
```

```
write(6,*);write(6,*) "First Case";write(6,*);

          vf=log(M)


!!!!  forcing linear behaviour until 10^-15 is reached
write(6,*); write(6,*) "Second Case";write(6,*);

          vf=log(M,epso=1.d-14)

write(6,*); write(6,*) "Third Case";write(6,*);
extra_terms_log=.true.
          vf=log(M)
```

Here is the abbreviated results of this code:

```
Testing the logarithm of a map


First Case

Norm of linear iteration #    1 0.16420059355772
Norm of linear iteration #    2 0.13412592351757E-01
Norm of   CBH  iteration #    4 0.24126056691115E-03
Norm of   CBH  iteration #    5 0.26360240866976E-04
Norm of   CBH  iteration #    6 0.40916713300836E-08
Norm of   CBH  iteration #    7 0.17461427983157E-11
Norm of   CBH  iteration #    8 0.17462221669734E-14
Norm of   CBH  iteration #    9 0.11539838746724E-15
iteration    8

Second Case

Norm of linear iteration #    1 0.16420059355772
Norm of linear iteration #    2 0.13412592351757E-01
Norm of linear iteration #    3 0.24126056691115E-03
Norm of linear iteration #    4 0.26036930351810E-04
Norm of linear iteration #    5 0.62740960903693E-05
Norm of linear iteration #    6 0.17486020329745E-05
                               .
                               .
                               .
Norm of   CBH  iteration #   33 0.81461400443222E-14
Norm of   CBH  iteration #   34 0.53481867860725E-14
Norm of   CBH  iteration #   35 0.12822022828854E-15
Norm of   CBH  iteration #   36 0.11306162475990E-15
iteration   35

Third Case

Norm of linear iteration #    1 0.16420059355772
Norm of linear iteration #    2 0.13412592351757E-01
Norm of   CBH  iteration #    4 0.24126056691115E-03
Norm of   CBH  iteration #    5 0.26356657389704E-04
Norm of   CBH  iteration #    6 0.55200509961495E-10
```

```
Norm of   CBH  iteration #    7 0.90784481929196E-15
Norm of   CBH  iteration #    8 0.11148873415683E-15
iteration     7
```

The first call to `log(M)` jumps to the application of the CBH formula after the first iteration. Eight steps are necessary. The reader can see that the convergence is not quadratic but not too far. The second call (`log(M,epso=1.d-14)`) delays the application of the CBH formula to a point when the vector field is within 10^{-15} of the final result. It now takes 35 steps to reach convergence. If another layer of higher order terms is added to the CBH formula of Eq. (11.8), we achieve a marginally better convergence as 7 steps are needed.

Reference

1. E. Forest, *Beam Dynamics: A New Attitude and Framework* (Harwood Academic Publishers, Amsterdam, The Netherlands, 1997)

Chapter 12
Stroboscopic Average for the ISF Vector n

Abstract In this supplementary chapter I discuss the numerical extraction of the invariant spin field using a method first proposed by Heinemann and Hoffstätter. As you might expect, I explain it via the potential existence of a normal form for the spin.

Keywords Numerical average · Stroboscopic methods · Spin

The "stroboscopic" average of spin motion is a novel technique (see Ref. [1], p. 142 and also [2]) first proposed by Heinemann and Hoffstätter for computation of the invariant spin axis **n**. I put the word stroboscopic in quotation marks because the expression is highly unfortunate for a person dealing primarily with a map based theory. In a map based theory, a stroboscopic average would involved recording data not at every turn but once in a while. We assume that Heinemann and Hoffstätter were thinking about the equations of motion when they coined the term; their paper [2] confirms the prejudice. A true stroboscopic average is rarely discussed in accelerator theory; for example dispersion can be defined rigorously in the presence of 3 tunes as a stroboscopic skipping over p turns if the temporal tune ν_3 is nearly equal to q/p where q and p are integers.[1] I retain the term "stroboscopic average" by respect for the originators and also to avoid undue confusion.

I will start with the idea that a normal form, whether we can compute it analytically or not, gives us a way to average quantities. Namely the idea is that an average over turns is actually an average over the invariant tori.

I start with the expression for the one-turn map expressed as a normal form. It is given by Eq. (6.20). I get for the j-turn map:

$$
\begin{aligned}
T^j &= \left(I, \widetilde{D}\right) \circ \left(m^j, e^{j\theta\circ a^{-1} L_y}\right) \circ \left(I, \widetilde{D}^{-1}\right) \\
&= \left(m^j, \widetilde{D} \circ m^j \; e^{j\widetilde{\theta} L_y} \; \widetilde{D}^{-1}\right) \quad \text{where } \widetilde{\theta} = \theta(\mathbf{J}) \circ a^{-1} \equiv \theta(\mathbf{I}). \qquad (12.1)
\end{aligned}
$$

[1] The generalised dispersions at Eq. (4.33) based on the de Moivre representation of the one-turn matrix are stroboscopic averages in the sense I just mentioned.

E. Forest, *From Tracking Code to Analysis*,
DOI 10.1007/978-4-431-55803-3_12

If I apply this spin-orbital map to an arbitrary spin $\mathbf{s}_0$ at location $\mathbf{z}_0$ in orbital space, I get for $\mathbf{s}_j$ at turn j:

$$\mathbf{s}_j = \widetilde{D}_{\mathbf{z}_j}\, e^{j\widetilde{\theta} L_y}\, \widetilde{D}_{\mathbf{z}_0}^{-1}\, \mathbf{s}_0 . \tag{12.2}$$

However in Eq. (12.2), we can see that the role of the initial position is special unlike the case of the normalised space. Indeed in actual space there is no guaranty that the spin axis at some position $\mathbf{z}_j$ bears any resemblance to the spin axis at the initial point $\mathbf{z}_0$. Therefore, I must invert Eq. (12.2):

$$\mathbf{s}_0 = \widetilde{D}_{\mathbf{z}_0} e^{-j\widetilde{\theta} L_y}\, \widetilde{D}_{\mathbf{z}_j}^{-1}\, \mathbf{s}_j . \tag{12.3}$$

I now sum and take the limit:

$$\prod = \lim_{j\to\infty} \frac{1}{N} \sum_{j=1}^{N} \widetilde{D}_{\mathbf{z}_0} e^{-j\widetilde{\theta} L_y}\, \widetilde{D}_{\mathbf{z}_j}^{-1} . \tag{12.4}$$

Equation (12.4) is a little tricky. The problem resides in the index j appearing in the matrix $\widetilde{D}_{\mathbf{z}_j}^{-1}$. To perform the average I use the assumption of an orbital normal form and express the matrix $\widetilde{D}_{\mathbf{z}_j}^{-1}$ in terms of the initial action angle variables $(\boldsymbol{\psi}, \mathbf{I})$:

$$\begin{aligned} \widetilde{D}_{\mathbf{z}_j}^{-1} &= \sum_{\mathbf{m}} \Gamma_{\mathbf{m}}^{0}(\mathbf{I})\ \exp\left(i\mathbf{m}\cdot\{\boldsymbol{\psi} + j\boldsymbol{\mu}\}\right) \\ &= \sum_{\mathbf{m}} \Gamma_{\mathbf{m}}(\boldsymbol{\psi}, \mathbf{I})\ \exp\left(ij\mathbf{m}\cdot\boldsymbol{\mu}\right) . \end{aligned} \tag{12.5}$$

The matrices $\Gamma_{\mathbf{m}}$ in Eq. (12.5) can be rewritten as three columns vectors:

$$\Gamma_{\mathbf{m}} = \left(\boldsymbol{\gamma}_1^{\mathbf{m}}, \boldsymbol{\gamma}_2^{\mathbf{m}}, \boldsymbol{\gamma}_3^{\mathbf{m}}\right) . \tag{12.6}$$

The matrix Π becomes

$$\Pi = \widetilde{D}_{\mathbf{z}_0} \sum_{\mathbf{m}} \lim_{N\to\infty} \frac{1}{N} \sum_{j=1}^{N} \left(e^{-j\widetilde{\theta} L_y}\boldsymbol{\gamma}_1^{\mathbf{m}}, e^{-j\widetilde{\theta} L_y}\boldsymbol{\gamma}_2^{\mathbf{m}}, e^{-j\widetilde{\theta} L_y}\boldsymbol{\gamma}_3^{\mathbf{m}}\right) \exp\left(ij\mathbf{m}\cdot\boldsymbol{\mu}\right) . \tag{12.7}$$

In Eq. (12.7), each vector $\boldsymbol{\gamma}_k^{\mathbf{m}}$ is made of 3 components which transform differently under the effect of $e^{-j\widetilde{\theta} L_y}$. First the second component $\boldsymbol{\gamma}_{k;2}^{\mathbf{m}}$ is left invariant by the rotation. The first and third component can be expressed using spin phasors:

$$\boldsymbol{\sigma}_{\pm} = \begin{pmatrix} \pm i \\ 0 \\ 1 \end{pmatrix} \quad \text{where } e^{-j\widetilde{\theta} L_y}\boldsymbol{\sigma}_{\pm} = e^{\mp ij\widetilde{\theta}}\boldsymbol{\sigma}_{\pm} . \tag{12.8}$$

Therefore we have:

$$\boldsymbol{\gamma} = \frac{\gamma_3 - i\gamma_1}{2}\boldsymbol{\sigma}_+ + \frac{\gamma_3 + i\gamma_1}{2}\boldsymbol{\sigma}_- + \gamma_2 \begin{pmatrix} 0 \\ 1 \\ 0 \end{pmatrix}. \tag{12.9}$$

The vector in Eq. (12.9) represents any of the three column vectors of Eq. (12.7).

$$\begin{aligned} e^{-j\tilde{\theta}L_y} e^{ij\mathbf{m}\cdot\boldsymbol{\mu}} \boldsymbol{\gamma} &= e^{-ij(\tilde{\theta}+\mathbf{m}\cdot\boldsymbol{\mu})} \frac{\gamma_3 - i\gamma_1}{2}\boldsymbol{\sigma}_+ + e^{-ij(-\tilde{\theta}+\mathbf{m}\cdot\boldsymbol{\mu})} \frac{\gamma_3 + i\gamma_1}{2}\boldsymbol{\sigma}_- \\ &+ e^{-ij\mathbf{m}\cdot\boldsymbol{\mu}} \gamma_2 \begin{pmatrix} 0 \\ 1 \\ 0 \end{pmatrix}. \end{aligned} \tag{12.10}$$

From Eq. (12.10), we see that the infinite series of Eq. (12.7) will converge to zero for all terms such that $\mathbf{m} \neq 0$, provided we are not sitting on some spin-orbit resonance.

$$\begin{aligned} \Pi &= A_{\mathbf{z}_0} \begin{pmatrix} 0 & 0 & 0 \\ \boldsymbol{\gamma}_{1;2}^{\mathbf{m}=0} & \boldsymbol{\gamma}_{2;2}^{\mathbf{m}=0} & \boldsymbol{\gamma}_{3;2}^{\mathbf{m}=0} \\ 0 & 0 & 0 \end{pmatrix} \\ &= \left(\boldsymbol{\gamma}_{1;2}^{\mathbf{m}=0}\, \mathbf{n} \mid \boldsymbol{\gamma}_{2;2}^{\mathbf{m}=0} \mathbf{n} \mid \boldsymbol{\gamma}_{3;2}^{\mathbf{m}=0} \mathbf{n} \right). \end{aligned} \tag{12.11}$$

In conclusion, the process of stroboscopic averaging leads potentially to three copies of the ISF $\mathbf{n}$. In most cases, the direction closest to $\mathbf{n}_0$ dominates. For example if $\mathbf{n}_0 \approx \mathbf{e}_y$, then "at infinity" $\boldsymbol{\gamma}_{2;2}^{\mathbf{m}=0}\mathbf{n}$ gives the best answer for $\mathbf{n}$.

References

1. K. Heinemann, G.H. Hoffstätter, Phys. Rev. E **54**, 4240 (1996)
2. G.H. Hoffstaetter, *High Energy Polarized Proton Beams—A Modern View* (Springer, New York, USA, 2006)

Chapter 13
Hierarchy of Analytical Methods

Abstract In this supplementary chapter I compare a simple first-order analytic calculation done two different ways. I first compute it using the Green's function or map method which most closely parallels the code. I then redo the calculations using the Fourier mode methods discussed in standard textbooks as well as in Sect. 8.7. These involve infinite sums which in this simple case can be carried out exactly. It should be clear that the hierarchy emerges from the extreme locality of the perturbation in accelerator physics; it is not a fundamental mathematical ordering.

Keywords Analytical methods · Map methods · Deprit-Guignard methods · Hamiltonian · Fourier expansions · Dirac delta function expansions · Infinite sums · Perturbation theory

I will perform a simple *first-order* nonlinear calculation on the Hamiltonian used by the lattice hardwired in Appendix A. The calculation will be performed by two different analytical methods. The Taylor methods of the code for this calculation are shown in Sect. 7.8.4.3. I will show that the Taylor method of the code matches the Green's function method (analytical map method actually). Therefore, in my way of thinking, methods based on Fourier transforms are further down the hierarchy and therefore appear in Chap. 8 at the end of this book. They are further away from the tracking code; but that does not mean that they are useless! In some cases, they can outperform the methods based on Green's functions, but it is not true in the trivial example of this chapter.

My simple example will be that of a tracking code with a sextupole as the perturbation for which we want analytical results. The Hamiltonian looks like this:

$$H = \underbrace{\frac{p^2}{2} + k_Q(s)\frac{x^2}{2}}_{H_0} + k_S(s)\frac{x^3}{3}. \tag{13.1}$$

The particular result I will derive is the average of the position, i.e., $\langle x \rangle$.

E. Forest, *From Tracking Code to Analysis*,
DOI 10.1007/978-4-431-55803-3_13

13.1 Green's Function Method

In Ref. [1], I lists the rules of analytical perturbation theory as they apply to map based methods. What if I introduce a certain potential around the ring and want to evaluate a quantity to order κ in that potential? The following rules must be followed in the map based method.

13.1.1 *The Rules of Analytical Perturbation Theory with Maps*

Rules for κth-order analytical perturbation theory on maps.

1. Go into the Floquet variables of the problem you already know how to solve: all the unperturbed maps should be rotations.
2. Transform the perturbations by the canonical transformation which gave you the rotations of item 1.
3. Insert κ distinct perturbations in the ring labelled from one to κ in order of appearance. The one-turn map at some arbitrary point will be made of a product of $2 * \kappa + 1$ maps.
4. Compute the one-turn map by "phase advancing" all the perturbations at the beginning or end of the ring using the zeroth-order propagator of item 1; generally rotations.
5. Lump all the perturbations into one exponent using the BCH theorem.
6. Compute the quantity of interest, the average of x for example.
7. Sum/integrate over the actual distribution of the perturbations using the index j labelling the perturbations ($1 \le j \le \kappa$) as a time-ordering label.

13.1.2 *The Actual Calculation with Maps*

I will apply each one of these rules on the Hamiltonian of Eq. (13.1). First, I assume that the linear one-turn map, generated by H_0, can be normalised at any position s. So the relevant matrices obey

$$M_s = A_s R A_s^{-1}. \tag{13.2}$$

For A_s I choose the Courant-Snyder definition:

$$A_s = \begin{pmatrix} \sqrt{\beta_s} & 0 \\ -\frac{\alpha_s}{\sqrt{\beta_s}} & \frac{1}{\sqrt{\beta_s}} \end{pmatrix}. \tag{13.3}$$

I know from general principles, that if I apply Eqs. (13.3) to (13.1), the result will be:

$$H = \omega_s J + \beta_s^{3/2} k_S(s) \frac{x^3}{3}. \tag{13.4}$$

The form of ω_s is not important in an actual calculation since the phase advances will be computed by the code. Of course for Eq. (13.1), with A_s defined by Eq. (13.3), the answer is actually

$$\omega_s = \frac{1}{\beta_s}. \tag{13.5}$$

Next I apply rule 2 to compute the map of a single pertubation at some point s. The answer, in terms of Lie maps, is just:

$$\mathscr{P}_s = \exp\left(-ds : \beta_s^{3/2} k_S(s) \frac{x^3}{3} :\right). \tag{13.6}$$

Now I construct the map according to rule 3. Since we are looking at a first-order calculation, I need only one map $\mathscr{P}_s$ in our "abstract" ring. The one-turn map at some position s is given by:

$$M_s = \mathscr{R}_{ss_1} \mathscr{P}_{s_1} \mathscr{R}_{s_1 s} \tag{13.7}$$

Remark: If we were interested in a second order calculation, then the equivalent of Eq. (13.7) would be:

$$M_s = \mathscr{R}_{ss_1} \mathscr{P}_{s_1} \mathscr{R}_{s_1 s_2} \mathscr{P}_{s_2} \mathscr{R}_{s_2 s}. \tag{13.8}$$

This is necessary in the calculation of the tune shift due to a sextupole distribution.

Now I phase advance the perturbation in Eq. (13.7) as demanded by rule 4:

$$\begin{aligned} M_s &= \mathscr{R}_{ss_1} \mathscr{P}_{s_1} \mathscr{R}_{s_1 s} \\ &= \mathscr{R}_{ss_1} \mathscr{P}_{s_1} \mathscr{R}_{ss_1}^{-1} \underbrace{\mathscr{R}_{ss_1} \mathscr{R}_{s_1 s}}_{\mathscr{R}} \end{aligned}$$

$$= \exp\left(-ds\mathscr{R}_{ss_1} : \beta_{s_1}^{3/2} k_S(s_1)\frac{x^3}{3} : \mathscr{R}_{ss_1}^{-1}\right)\mathscr{R}$$
$$= \exp\left(-ds : \beta_s^{3/2} k_S(s_1)\frac{\mathscr{R}_{ss_1}x^3}{3} :\right)\mathscr{R}$$
$$= \exp\left(-ds : \beta_{s_1}^{3/2} k_S(s_1)\frac{\left(\mathscr{R}_{ss_1}x\right)^3}{3} :\right)\mathscr{R}. \tag{13.9}$$

Rule 5 is not relevant to a first-order calculation because Eq. (13.9) has only one factor containing the perturbation.

We have reached the most complex step. We need to compute the "quantity of interest" which in my case is the average of the position x. To do this, I need to compute the part of the canonical transformation which depends on the perturbation. The purpose of this transformation is to turn the perturbed map into a rotation so that I can perform the average.

$$\mathscr{R}_1 = \exp(: F :)\exp\left(-ds : \beta_{s_1}^{3/2} k_S(s_1)\frac{\left(\mathscr{R}_{ss_1}x\right)^3}{3} :\right)\mathscr{R}\exp(: -F :)$$
$$= \exp(: F :)\exp\left(-ds : \beta_{s_1}^{3/2} k_S(s_1)\frac{\left(\mathscr{R}_{ss_1}x\right)^3}{3} :\right)\exp(: -\mathscr{R}F :)\mathscr{R}$$
$$= \exp\left(: F - \mathscr{R}F - ds\beta_{s_1}^{3/2} k_S(s_1)\frac{\left(\mathscr{R}_{ss_1}x\right)^3}{3} + \cdots :\right)\mathscr{R}. \tag{13.10}$$

I am left with the following possibility:

$$(I - \mathscr{R})\,F - ds\beta_{s_1}^{3/2} k_S(s_1)\frac{\left(\mathscr{R}_{ss_1}x\right)^3}{3} \stackrel{?}{=} 0. \tag{13.11}$$

Can I choose F so that Eq. (13.11) is satisfied? In general I can choose F so as to leave only powers of the action J in the exponent. These produce tune shifts. However, as is well known about sextupoles, they produce no amplitude dependent tune shifts in leading order. Thus Eq. (13.11) can be satisfied by a proper selection of F. I first express x^3 in the eigenphasors basis of Eq. (2.21):

$$\left(\mathscr{R}_{ss_1}x\right)^3 = \frac{1}{8}\left(\mathscr{R}_{ss_1}\Phi_+ + \mathscr{R}_{ss_1}\Phi_-\right)^3$$
$$= \frac{1}{8}\left(e^{-i\Delta}\Phi_+ + e^{i\Delta}\Phi_-\right)^3$$

$$= \frac{1}{8}\left(e^{-i3\Delta}\Phi_+^3 + e^{i3\Delta}\Phi_-^3 + 3e^{-i\Delta}\Phi_+^2\Phi_- + 3e^{i\Delta}\Phi_+\Phi_-^2\right) \tag{13.12}$$

where $\Delta = \mu_{ss_1}$ $\leftarrow$ phase advance from s to s_1.

The answer for F is thus:

$$\begin{aligned} F &= ds\beta_{s_1}^{3/2} k_S(s_1)(I-\mathscr{R})^{-1}\frac{\left(\mathscr{R}_{ss_1}x\right)^3}{3} \\ &= \frac{ds}{24}\beta_{s_1}^{3/2} k_S(s_1)\left(\frac{e^{-i3\Delta}}{1-e^{-i3\mu}}\Phi_+^3 + \frac{e^{i3\Delta}}{1-e^{i3\mu}}\Phi_-^3 + \frac{3e^{-i\Delta}}{1-e^{-i\mu}}\Phi_+^2\Phi_- + \frac{3e^{i\Delta}}{1-e^{i\mu}}\Phi_+\Phi_-^2\right). \end{aligned} \tag{13.13}$$

I can now find the average of x by transforming x with the full canonical transformation:

$$\begin{aligned} \widetilde{x} &= \exp(:F:)\mathscr{A}_s x \\ &= (1+:F:+\cdots)\mathscr{A}_s x \\ &= \sqrt{\beta_s}x + \sqrt{\beta_s}\,[F,x] + \cdots \\ &= \sqrt{\beta_s}\left(\frac{\Phi_+ + \Phi_-}{2} + \frac{[F,\Phi_+] + [F,\Phi_-]}{2}\right) \\ &= \sqrt{\beta_s}\left(\frac{\Phi_+ + \Phi_-}{2} + i\left\{\frac{\partial F}{\partial \Phi_-} - \frac{\partial F}{\partial \Phi_+}\right\}\right). \end{aligned} \tag{13.14}$$

Because the phasors in Eq. (13.14) move on circles, I can easily average $\widetilde{x}$ as explained in Sect. 2.2.1:

$$\begin{aligned} \langle\widetilde{x}\rangle &= i\sqrt{\beta_s}\left\langle\frac{\partial F}{\partial \Phi_-} - \frac{\partial F}{\partial \Phi_+}\right\rangle \\ &= \sqrt{\beta_s}2\ \mathrm{Im}\left\langle\frac{\partial F}{\partial \Phi_+}\right\rangle \\ &= \frac{ds}{2}\beta_s\beta_{s_1}^{3/2}k_S(s_1)\Phi_+\Phi_-\ \mathrm{Im}\left(\frac{e^{-i\Delta}}{1-e^{-i\mu}}\right) \\ &= \frac{ds}{2}\beta_s^{1/2}\beta_{s_1}^{3/2}k_S(s_1)\frac{-\sin(\Delta)+\sin(\Delta-\mu)}{1-\cos(\mu)}J\,. \end{aligned} \tag{13.15}$$

Finally we apply rule 7. We sum/integrate over the entire ring.

$$\frac{\partial\langle\widetilde{x}\rangle}{\partial J} = \frac{\beta_s^{1/2}}{2(1-\cos(\mu))}\oint_0^C\left(-\sin(\mu_{s\sigma}) + \sin(\mu_{s\sigma}-\mu)\right)\beta_\sigma^{3/2}k_S(\sigma)d\sigma. \tag{13.16}$$

Notice: we could have applied rule 7 first to the polynomial F. Indeed F is given by:

$$F = \oint_0^C \frac{1}{24}\beta_\sigma^{3/2} k_S(\sigma) \left(\frac{e^{-i3\mu_{s\sigma}}}{1 - e^{-i3\mu}} \Phi_+^3 + \frac{e^{i3\mu_{s\sigma}}}{1 - e^{i3\mu}} \Phi_-^3 \right.$$
$$\left. + \frac{3e^{-i\mu_{s\sigma}}}{1 - e^{-i\mu}} \Phi_+^2 \Phi_- + \frac{3e^{i\mu_{s\sigma}}}{1 - e^{i\mu}} \Phi_+ \Phi_-^2 \right) d\sigma. \quad (13.17)$$

In first-order perturbation theory, it suffices to examine the effect of a thin perturbation to get the complete result. This is why I refer to this method as a Green's function method.

13.2 Fourier Mode Calculations with the Hamiltonian

I will describe how calculations are done on the Hamiltonian using a Fourier transform in the time-like position s. This is the traditional approach in accelerator physics as championed by Guignard and others. It is, of course, in Lee's book [2]. A lot of material presented in the traditional way can be found in Wiedemann's book as well [3]. I devoted Chap. 8 to the serious computation of the Fourier transformed Hamiltonian in realistic and arbitrarily complex lattices, something, I am sad to say, you will not find in any reference that advocates Guignard-like computations. Here I concentrate on a purely analytical calculation.

13.2.1 Changing the Time-Like Variable into a Phase Advance

It is best to change the "time" variable s into the actual phase advance of the linear part of H.

$$H = \omega_s J + \beta_s^{3/2} k_S(s) \frac{x^3}{6}$$
$$\Downarrow$$
$$K = \nu J + \frac{\nu}{\omega_s} \beta_\theta^{3/2} k_S(\theta) \frac{x^3}{6} \quad (13.18)$$
$$\text{where } \theta\nu = \int_0^s \omega_\sigma d\sigma. \quad (13.19)$$

2-d-f or more: In several degrees of freedom, the trick of Eq. (13.19) does not work alone. In order to get a form like say,

$$K = \nu_1 J_1 + \nu_2 J_2 + V, \tag{13.20}$$

the other phases must be modified. This amounts to a change of the phase advances, or equivalently, a change of the linear matrix A_s. The perturbation receives extra phases so that in the end everything still applies as described here although I confine myself to the 1-d-f case for simplicity. See Chap. 8 for a full explanation with an example from the Code.

13.2.2 The Fourier Method Approach: Guignard

The s or θ dependence of Eq. (13.18) is quite arbitrary. In accelerator physics, we can assume very little about the linear and the nonlinear parts besides a general Liouville-Arnold structure near the origin, i.e., we have doughnuts. Faced with this conundrum, we are forced to expand the perturbation into some set of known functions. The standard approach of accelerator physicists, which is furthest from the code, is to expand Eq. (13.18) in a Fourier series in the periodic variable θ.

$$\frac{\nu \beta_\theta^{3/2} k_S(\theta)}{3\omega_s} = \sum_{k=-\infty,\infty} V_k e^{ik\theta} \tag{13.21}$$

$$V_k = \frac{1}{2\pi} \int_0^{2\pi} \frac{\nu \beta_{\theta'}^{3/2} k_S(\theta')}{3\omega_s} e^{-ik\theta'} d\theta', \tag{13.22}$$

and the Hamiltonian becomes:

$$K = \nu J + \sum_{k,m} \frac{V_k}{8} e^{ik\theta} |m\rangle \quad \text{where} \quad \begin{array}{ll} |3\rangle & = \Phi_+^3 \\ |-3\rangle & = \Phi_-^3 \\ |1\rangle & = 3\Phi_+^2 \Phi_- \\ |-1\rangle & = 3\Phi_+ \Phi_-^2 . \end{array} \tag{13.23}$$

In general, the sum over k is infinite and must be truncated. Since our magnets are nearly Dirac delta functions, it is not clear where we need to stop. In the case of astronomy, where this theory originates, the number of harmonics is small if you put the Sun at the centre: this is why Copernicus almost beats Ptolemy and Galileo avoided the stake!

My goal is to compare the Fourier mode methods to the map based methods of Eq. (7.41). So I will keep all the Fourier terms, all the way to infinity! I must first start with the transformation of a time (or θ) dependent Hamiltonian. It is given by

$$K_\theta^{new} = \exp\left(: F_\theta :\right) K_\theta + \sum_{n=1}^{\infty} \frac{: F_\theta :^{n-1}}{n!} \frac{\partial F_\theta}{\partial \theta} \tag{13.24}$$

$$\approx K_\theta + [F_\theta, K_\theta] + \frac{\partial F_\theta}{\partial \theta} + \text{Order}\left(|F_\theta|^2\right) \cdots . \tag{13.25}$$

Equation (13.24) is the exact effect of a canonical transformation represented by the Lie transform $\exp(: F_\theta :)$ on the Hamiltonian K_θ. Equation (13.25) is a first-order expansion of Eq. (13.24) which is sufficient for our needs. To proceed further I expand F_θ in a Fourier series as well,

$$F_\theta = \sum_{k,m} F_k^m e^{ik\theta} \left|m\right\rangle, \tag{13.26}$$

and substitute the series in Eq. (13.25):

$$K^{new} = \nu J + \sum_{k,m} F_k^m e^{ik\theta} \left[\left|m\right\rangle, \nu \frac{\Phi_+ \Phi_-}{2} \right] + \sum_{k,m} ik F_k^m e^{ik\theta} \left|m\right\rangle + \sum_{k,m} \frac{V_k}{8} e^{ik\theta} \left|m\right\rangle$$

$$= \nu J + \sum_{k,m} \left\{ i\left(k - m\nu\right) F_k^m + \frac{V_k}{8} \right\} e^{ik\theta} \left|m\right\rangle + \cdots . \tag{13.27}$$

Because I have an actual functional form for the Hamiltonian, albeit an infinite series, I can solve for F:

$$F_k^m = i \frac{V_k}{8\left(k - m\nu\right)} \quad \Longrightarrow F_\theta = \sum_{k,m} \frac{i V_k}{8\left(k - m\nu\right)} e^{ik\theta} \left|m\right\rangle \tag{13.28}$$

The Lie methods implementation used here was discovered by Hori [4, 5] and the actual algebraic manipulations were pioneered by André Deprit [6]. All of this goes back to the nineteenth century under the name of secular perturbation theory with mixed variable generating functions instead of Lie operators.

The final showdown: I want to show that the answer in Eq. (13.28) is the same as the map result of Eq. (13.17). Therefore I must keep all the terms, all the way to infinity, in the Fourier transform! First I substitute the expression for V_k in Eq. (13.28):

$$F_\theta = \sum_{k,m} \frac{i \frac{1}{2\pi} \int_0^{2\pi} \frac{\nu \beta_{\theta'}^{3/2} k_S(\theta')}{24\omega_s} e^{-ik\theta'} d\theta'}{\left(k - m\nu\right)} e^{ik\theta} \left|m\right\rangle$$

$$
\begin{aligned}
&= \int_0^{2\pi} i \frac{1}{2\pi} \frac{\nu \beta_{\theta'}^{3/2} k_S(\theta')}{24\omega_s} d\theta' \sum_{k,m} \frac{e^{-ik\theta'}}{(k-m\nu)} e^{ik\theta} \left|m\right\rangle \\
&= \int_0^{2\pi} \frac{\nu \beta_{\theta'}^{3/2} k_S(\theta')}{24\omega_s} d\theta' \sum_m \underbrace{\frac{i e^{-im\nu(\theta'-\theta)}}{2\pi} \sum_k \frac{e^{-i(k-m\nu)(\theta'-\theta)}}{(k-m\nu)}}_{G_{mk}} \left|m\right\rangle . \qquad (13.29)
\end{aligned}
$$

The four terms in Eq. (13.17) must be equal to G_{mk} of Eq. (13.29). There is no easy way to sum this horror directly. However, since I know that this can be easily derived by Green's functions, i.e., the rules of Sect. 13.1.1, I am tempted to take the derivative of G_{mk} with respect to $\tau = \theta' - \theta$, hoping that a Dirac-delta function will emerge:

$$
\begin{aligned}
\frac{\partial G_{mk}}{\partial \tau} &= \frac{\partial}{\partial \tau} \frac{i e^{-im\nu\tau}}{2\pi} \sum_k \frac{e^{-i(k-m\nu)\tau}}{(k-m\nu)} \\
&= -imG_{mk} + e^{-im\nu\tau} \underbrace{\frac{1}{2\pi} \sum_k e^{-i(k-m\nu)\tau}}_{\delta(\tau) e^{im\nu\tau}} \\
&\Downarrow \\
\frac{\partial G_{mk}}{\partial \tau} &= -imG_{mk} + \delta(\tau) . \qquad (13.30)
\end{aligned}
$$

To solve Eq. (13.30), I first solve outside the delta function:

$$
G_{mk} \propto \lambda e^{-im\nu\tau}. \qquad (13.31)
$$

To fix λ I integrate around $\tau = 0$:

$$
\text{for } \varepsilon \to 0: \quad G_{mk}(\varepsilon) - G_{mk}(-\varepsilon) = 1. \qquad (13.32)
$$

But G_{mk} must be periodic, therefore it must be true that:

$$
G_{mk}(-\varepsilon) = G_{mk}(2\pi - \varepsilon) \Longrightarrow G_{mk}(2\pi - \varepsilon) = \lambda e^{-im\nu 2\pi} = \lambda e^{-im\mu}. \qquad (13.33)
$$

Equation (13.32) is rewritten as

$$
\text{for } \varepsilon \to 0: \quad \lambda - \lambda e^{-im\mu} = 1 \qquad (13.34)
$$

and the final result is thus:

$$
G_{mk} = \frac{e^{-im\nu(\theta'-\theta)}}{1 - e^{-im\mu}} = \frac{e^{-im\Delta}}{1 - e^{-im\mu}} \qquad (13.35)
$$

$$F_\theta = \int_0^{2\pi} \frac{\nu \beta_{\theta'}^{3/2} k_S(\theta')}{24\omega_s} d\theta' \sum_m \frac{e^{-im\Delta}}{1 - e^{-im\mu}} |m\rangle$$
$$= \int_0^{C} \frac{\beta_\sigma^{3/2} k_S(\sigma)}{24} d\sigma \sum_m \frac{e^{-im\Delta}}{1 - e^{-im\mu}} |m\rangle \qquad \text{since } \frac{d\theta'}{d\sigma} = \frac{\omega_\sigma}{\nu}, \quad \text{Q.E.D.} \tag{13.36}$$

13.3 Numerical Example $\langle x \rangle$: Analytical, Guignard and Map

In this section I describe the implementation of the calculation of $\langle x \rangle$ using every method under the sky (not really…). The relevant code is in Appendix Q.

I will start with a trivial code fragment from Appendix Q describing the input:

```
call build_lattice_als(ALS,mis,exact=.false.,thin=.true.,onecell=.true.)


write(6,*)"Give integration step ds"
write(6,*) "> 3 and nothing is cut; each step is a full magnet"
write(6,*) " real fun starts around ds=0.5"
read(5,*) ts
                         .
                         .
                         .
Write(6,*) "Constant phase advance per step        ---> t"
write(6,*) "Courant Snyder phase advance per step ----> f"
read(5,*) used_ds_ave
if(used_ds_ave) then
 Write(mf,*) "Constant phase advance per step"
else
 write(mf,*) "Courant Snyder phase advance per step"
endif

write(6,*) "Enter number of Fourier modes"
read(5,*) n_mode
                         .
                         .
                         .
```

The call to `build_lattice_als` constructs one cell of the ALS and uses a thin lens model for the sextupole. This is to insure that the analytical formula of Eq. (13.16) is exact on a lattice with no errors and no coupling.

The next input concerns the typical integration step size. If you input $ds = 3.0$ or so, nothing is cut. This is because magnets in the ALS lattice are 3 m or less. If you select $ds = 0.1$ or smaller, the number of slices will start to be inversely proportional to ds. Thus you can study the effect of a fine split by reducing ds.

The next input is interesting. In the standard approach as described in Ref. [2] or Sect. 8.4, a constant phase advance is required. For this you select "t" as the input for `used_ds_ave`. If you select "f" the Courant-Snyder transformation will be used.

Finally the user can select the number of modes and see what happens as one increases that number.

```
do i=1,ns

 call propagate(als,ray,state,node1=i,node2=i+1)    ! (14a)
 call propagate(als,ray_cs_twiss,state,node1=i,node2=i+1)    ! (14b)

  ray_closed=ray ! Saving orbit  ! (15a)
  a_cs=ray_cs_twiss
  a_cs=a_cs.sub.1
  call c_full_canonise(a_cs,a_cs,phase=phase)

  U=ray ! copying in map  ! (15b)

if(used_ds_ave) then
 U_c=exp(-ds*h_left,U)          !  (16a)
 U_c=U_c.cut.2                  !  (16b)
 U=U_c**(-1)*U                  !  (16c)
else
 ds=twopi*t%s(5)/circ
 U_c=U.cut.2                    !  (16d)
 call  c_canonise(U_c,U_c)      !  (16e)
 U=U_c**(-1)*U                  !  (16f)
endif
  logN=log(U)                   !  (17)

s=s+ds;
do km=-n_mode,n_mode
 G%f(km)=G%f(km)+(exp(-i_*km*s)/twopi)*logN ! (19)
enddo

  betax_1=(a_cs%v(1).sub.'10')**2+(a_cs%v(1).sub.'01')**2  ! (20a)
 if((p%mag%name(1:2)=="SF".or.p%mag%name(1:2)=="SD").and.t%cas==case0) then
  DX_AVERAGE_DCS=(betax_1)**1.5_DP*p%mag%BN(3)/4.0_DP &     ! (20b)
    *(-SIN(PHASE(1)*TWOPI)+SIN((PHASE(1)-normal_form%TUNE(1))*TWOPI)) &
    /(1.0_DP-COS(normal_form%TUNE(1)*TWOPI)) + DX_AVERAGE_DCS
 endif

 ray=ray_closed+U_c                       ! (21)
 ray_cs_twiss=ray_closed+a_cs
 t=>t%next
 p=>t%parent_fibre
enddo
```

I removed from the above loop code anything that is not necessary for the basic understanding of the loop.

I will point the reader's attention to the main loop. The reader will notice that I track the ray twice: am I stuttering? First we look at lines (20a, b). This is the evaluation of the analytical formula from Eq. (13.16). The phase advance in that formula *must be* the Courant-Snyder phase advance. Therefore the call at line (14b) performs a normal Courant-Snyder loop with the Courant-Snyder choice in mind. Could I have avoided this call with additional contortions with the output of line (14a)? The answer is yes. However to keep things clear, rather than fast, I perform two separate calculations in this program.

Lines (16a, b, c) in conjunction with lines (17) and (19) implement the calculation of Sect. 8.4: the case of constant linear phase advance. Lines (16d, e, f) implements the case of the Courant-Snyder phase advance—the case of Sect. 8.3—which will require a linear normalisation.

Here both methods use the variable θ for the time-like variable of the Hamiltonian as explained on p. 209. Here are the normalisation calls:

```
if(used_ds_ave.and.doit) then
 call normalise_vector_field_fourier(G,F,K)      ! (23a)
else
 n_extra=50
 call normalise_vector_field_fourier(G,F,K,F1)   ! (23b)
endif
```

Please notice that the flag `doit` is set to true by default. Therefore the normal form will not compute the linear vector field $\mathbf{F}_1$ if a constant linear phase advance is used. By setting the flag to false, you can check that it is indeed zero if computed.

The total canonical transformation a is given by:

$$a = \exp(\mathbf{F} \cdot \nabla) \exp(\mathbf{F}_1 \cdot \nabla) a_{cs}. \tag{13.37}$$

This is implemented with the following fragment:

```
theta=0.d0
call c_evaluate_vector_field_fourier(f1,theta,f_non)

f_non=to_phasor()*f_non   !  Turns f_non in Cartesian basis
id_s=exp(f_non,a_cs)
call c_evaluate_vector_field_fourier(f,theta,f_non)

f_non=to_phasor()*f_non !  Turns f_non in Cartesian basis
A=exp(f_non,id_s)
```

The vector fields $\mathbf{F}_1$ and $\mathbf{F}$ are evaluated at the end of the ring ($\theta = 0 = 2\pi$) by `c_evaluate_vector_field_fourier`. The rest of the fragment is a straight implementation of Eq. (13.37). Notice that my Guignard normal form produces a result in phasors' basis; this is changed by the calls `f_non=to_phasor()*f_non` which implements Eq. (3.27) on the Fourier expanded vector field.

I invite you to run an example. Here is a slightly tedious example: $ds = 0.01$ and 100 modes. I will just compare the results of $\langle x \rangle$. First let me display the constant phase result

```
Results for <x>
         100 modes

  d<x>/dCS     120.2054785736042     < ---- analytical

 Result of map normal form

 Properties, NO =     4, NV =     2, INA =  214
*********************************************

   2    120.2054785736068        0.000000000000000       1  1
   4    189811.7154277755     -0.9038103598868474E-11    2  2

 Constant linear phase advance
 dtheta =  3.338568175972150E-003  radians
 ds <=   1.000000000000000E-002  metres

 Result of Guignard normal form

 Properties, NO =     4, NV =     2, INA =  214
*********************************************

   2    120.1946989163008       0.9271505792053573E-16   1  1
   4    182947.4935034170      -0.2726367796098808E-10   2  2
```

followed by the result with a Courant-Snyder phase

```
 Courant-Snyder linear phase advance
 ds <=   1.000000000000000E-002  metres

 Result of Guignard normal form

 Properties, NO =     4, NV =     2, INA =  214
*********************************************

   2    120.5484654655627       0.1058377837737943E-15   1  1
   4    187859.0799306934       0.5311593586168033E-11   2  2
```

The reader can see that the agreement is good with the map based theory at all orders. Since there are no errors, the leading order analytical calculation agrees with the exact result from the one-turn map.

Finally, the reader is invited to uncomment the following lines in the code of Appendix Q:

```
!  write(mf,*) "MISALIGN EVERYTHING"
!  write(mf,*) "0 0 0 0 0.1   0  5"
```

This call imposes a random rotation around the vertical axis on all magnets including the sextupoles. This error does not violate mid-plane symmetry and thus the 1-d-f system is well defined. Therefore if all goes well, the reader will discover that the one-turn map result agrees with the Fourier mode calculation; however it will not agree with the analytic calculation because the thin sextupole is not properly handled in that case. Of course that could be fixed but it requires additional work not discussed in Sect. 13.1.

References

1. E. Forest, *Beam Dynamics: A New Attitude and Framework* (Harwood Academic Publishers, Amsterdam, 1997)
2. S.Y. Lee, *Accelerator Physics* (World Scientific Publishing, Singapore, 2004)
3. H. Wiedemann, *Particle Accelerator Physics* (Springer, Berlin, 2007)
4. G. Hori, Publ. Astr. Soc. Japan **18**, 287 (1966)
5. G. Hori, Publ. Astr. Soc. Japan **23**, 567 (1971)
6. A. Deprit, Cel. Mech. **1**, 12 (1969)

Appendix A
The Hardwired ALS Lattice

```
subroutine  build_lattice_als(ALS,mis,error,exact,sl,thin,onecell)
use madx_ptc_module
use pointer_lattice
implicit none

type(layout), target :: ALS
real(dp),optional :: error(6)
logical, optional :: exact,sl,thin,onecell
real(dp) :: alpha,lbend, cut, ksd, ksf
type(fibre)  L1,L2,L3,L4,L5,L6,L7,L8,L9,L10
type(fibre)  L11,L12,L13,L14,L15,L16,L17,L18,L19,L20,CAVM
type(fibre)  L21,L22,L23,L24,L25,L26,L27,L27A,L27B,L27C,L27D,DS
type(fibre)  QF1,QF2,QD1,QD2,QFA1,QFA2,sf,sd,cav,bend,vc5,bend1
type(layout) :: sfline,sdline,sup1,supb
logical(lp) :: mis,thi=.false.,oneperiod
!----------------------------------
if(present(thin)) thi=thin

call make_states(.true.)
exact_model = .false.;oneperiod = .false.
if(present(exact)) exact_model=exact
if(present(onecell)) oneperiod=onecell
call update_states
madlength = .false.


call set_mad(energy = 1.5d0, method = 2, step = 1)

madkind2 = matrix_kick_matrix


L1  = drift("L1 ",  2.832695d0);L2  = drift("L2 ",  0.45698d0);
L3  = drift("L3 ",  0.08902d0);L4  = drift("L4 ",  0.2155d0);
L5  = drift("L5 ",  0.219d0);L6  = drift("L6 ",  0.107078d0);
L7  = drift("L7 ",  0.105716d0);L8  = drift("L8 ",  0.135904d0);
L9  = drift("L9 ",  0.2156993d0);L10 = drift("L10",  0.089084d0);
L11= drift("L11",  0.235416d0);L12= drift("L12",  0.1245d0);
L13= drift("L13",  0.511844d0);L14= drift("L14",  0.1788541d0);
L15= drift("L15",  0.1788483d0);L16= drift("L16",  0.511849d0);
L17= drift("L17",  0.1245d0);L18= drift("L18",  0.235405d0);
L19= drift("L19",  0.089095d0);L20= drift("L20",  0.2157007d0);
L21= drift("L21",  0.177716d0);L22= drift("L22",  0.170981d0);
L23= drift("L23",  0.218997d0);L24 = drift ("L24",  0.215503d0);
L25 = drift ("L25",  0.0890187d0);L26 = drift ("L26",  0.45698d0);
L27 = drift ("L27",  2.832696d0);L27a  = drift (" L27a",  0.8596d0);
L27b  = drift (" L27b",  0.1524d0);L27c  = drift (" L27c",  0.04445d0);
L27d  = drift (" L27d",  1.776246d0);ds  = drift (" DS  ", 0.1015d0);
```


E. Forest, *From Tracking Code to Analysis*,
DOI 10.1007/978-4-431-55803-3

```
QF1 = QUADRUPOLE(" QF1 ",0.344D0, K1= 2.2474D0+6.447435260914397D-03)
QF2 = QUADRUPOLE(" QF2 ",0.344D0, K1= 2.2474D0)
QD1 = QUADRUPOLE(" QD1 ",0.187D0, K1= -2.3368D0-2.593018157427161D-02);
QD2 = QUADRUPOLE(" QD2 ",0.187D0, K1= -2.3368D0);
QFA1= QUADRUPOLE(" QFA1",0.448D0, K1= 2.8856D0);
QFA2= QUADRUPOLE(" QFA2",0.448D0, K1= 2.8856D0);

!!! 1/2 mad-x value
ksf=-41.3355516397069748d0;
ksd=56.2564709584745489d0;

sf=sextupole ("sf",2.d0*0.1015d0, K2= ksf);
sd= sextupole("sd", 2.d0*0.1015d0, K2= ksd);

 VC5=marker("vc5");
ALPHA=0.17453292519943295769236907684886d0;

LBEND=0.86621d0;

BEND = RBEND("BEND", LBEND, ANGLE=ALPHA).q.(-0.778741d0)
BEND1 = RBEND("BEND1", LBEND, ANGLE=ALPHA).q.(-0.778741d0)

CAVM=MARK("CAVM");
CAV=RFCAVITY("CAV",L=0.0000d0,VOLT=-1.0d0,REV_FREQ=500.0d6)

if(thi) then
 sf=sextupole ("sf",0.d0, K2= ksf*0.203d0);
 sd= sextupole("sd", 0.d0, K2= ksd*0.203d0);
  sfline=(ds+sf+ds);
  sdline=(ds+sd+ds);
else
 sfline=1*sf;
 sdline=1*sd;
endif

SUP1=L1+L2+L3+QF1+VC5+L4+L5+QD1+L6+L7+L8+VC5+BEND+VC5+L9+sfline+L10+&
           L11+QFA1+L12+sdline+L13+ &
           L14+BEND+L15+L16+sdline+L17+ &
           QFA2+L18+L19+sfline+L20+BEND+L21+&
           L22+QD2+L23+L24+QF2+L25+ &
           L26+VC5+L27+cavm;

SUPb=L1+L2+L3+QF1+VC5+L4+L5+QD1+L6+L7+L8+VC5+BEND+VC5+L9+sfline+L10+&
           L11+QFA1+L12+sdline+L13+ &
           L14+BEND+L15+L16+sdline+L17+ &
           QFA2+L18+L19+sfline+L20+BEND1+L21+&
           L22+QD2+L23+L24+QF2+L25+ &
           L26+VC5+L27+cav;

if(oneperiod) then
 ALS = sup1;  !11*sup1+supb;
else
 ALS = 11*sup1+supb;
endif
if(present(sl)) then
L1  = drift("L1 ",  2.832695d0);
 if( sl ) then
  Qf1 = QUADRUPOLE(" QF1 ",L=0.d0, K1= 0.01d0 ); L1  = drift("L1 ",L=0.1d0);
  ALS=L1+QF1;
 endif
endif

ALS = .ring.ALS

call survey(ALS)
```

```
if(mis) then
 sig=1.d-5; cut=4.d0;
 if(present(error)) sig=error
 call MESS_UP_ALIGNMENT(ALS,SIG,cut);
endif
end subroutine build_lattice_als
```

Appendix B
Program for one_turn_orbital_map

```
program program_one_turn_map
use madx_ptc_module
use pointer_lattice
use c_TPSA
implicit none

    interface
       subroutine build_lattice_als(ALS,MIS,error,exact,sl,thin,onecell)
         use madx_ptc_module
         use pointer_lattice
         implicit none
         type(layout), target :: ALS
         logical(lp) mis
         real(dp),optional :: error(6)
         logical, optional :: exact,sl,thin,onecell
       end subroutine build_lattice_als
    end interface


type(layout), pointer:: ALS
real(dp) prec,closed_orbit(6),mat(6,6)
type(internal_state),target :: state
logical(lp) :: mis=.true.
type(c_damap)  one_turn_map, Id
type(real_8) y(6)
integer i,map_order

prec=1.d-6 ! for printing
longprint=.false.

call ptc_ini_no_append
call append_empty_layout(m_u)
ALS=>m_u%start

call build_lattice_als(ALS,mis)

state=nocavity0

map_order=1
call init_all(state,map_order,0)
call alloc(one_turn_map,id)
call alloc(y)

call find_orbit_x(als,closed_orbit(1:6),STATE,1.e-5_dp,fibre1=1)      ! (1)

id=1   ! map is set to identity                                       ! (2)
```


E. Forest, *From Tracking Code to Analysis*,
DOI 10.1007/978-4-431-55803-3

```
write(6,*) " id%v(1) "
call print(id%v(1),6)
! map is added to closed orbit and put into the 6 polymorphs
y(1:6)=closed_orbit(1:6)+id                                        ! (3)
write(6,*) " Y(1) = closed_orbit(1)+id%v(1) "
call print(y(1),6)
call propagate(als,y(1:6),state,fibre1=1)                          ! (4)

one_turn_map=y(1:6) ! Six polymorphs are promoted to Taylor maps   ! (5)

call print(one_turn_map,6,prec)

mat=one_turn_map                                                   ! (6)
closed_orbit=y                                                     ! (7)

write(6,*) " ";
write(6,*) "  Constant part of the map and linear part (matrix) ";
 write(6,'(a16,6(1x,g12.5))') " closed orbit = ",closed_orbit(1:6)
do i=1,6
 write(6,'(a5,i1,a5,6(1x,g12.5))') " row ",i," --> ",mat(i,1:6)
enddo

call ptc_end(graphics_maybe=1,flat_file=.false.)

end program program_one_turn_map
```

Appendix C
Program one_turn_orbital_map_normal_form_2d

```
program one_turn_orbital_map_normal_form_2d
use madx_ptc_module
use pointer_lattice
use c_TPSA
implicit none

    interface
       subroutine build_lattice_als(ALS,MIS,error,exact,sl,thin,onecell)
         use madx_ptc_module
         use pointer_lattice
         implicit none
         type(layout), target :: ALS
         logical(lp) mis
         real(dp),optional :: error(6)
         logical, optional :: exact,sl,thin,onecell
       end subroutine build_lattice_als
    end interface


type(layout), pointer:: ALS
real(dp) prec,closed_orbit(6),mat(6,6),a(6,6)
complex(dp) ac(6,6),w(6)
real(dp) beta,gamma,alpha
type(internal_state),target :: state
logical(lp) :: mis=.false.
type(c_damap)  one_turn_map, Id
type(real_8) y(6)
type(c_normal_form) normal_form
type(c_taylor) e1,r2,z1,z2,z1_new,z2_new,e2
integer i,map_order
c_mess_up_vector=.true.; b_mess=-1.0_dp;
c_verbose=.false.
prec=1.d-6 ! for printing
longprint=.false.

 closed_orbit=0.d0
call ptc_ini_no_append
call append_empty_layout(m_u)
ALS=>m_u%start

call build_lattice_als(ALS,mis)

state=only_2d0

map_order=2
call init_all(state,map_order,0)
```


E. Forest, *From Tracking Code to Analysis*,
DOI 10.1007/978-4-431-55803-3

```
call alloc(one_turn_map,id)
call alloc(y)
call alloc(normal_form)
call alloc(e1,r2,z1,z2,z1_new,z2_new,e2)

call find_orbit_x(als,closed_orbit(1:6),STATE,1.e-5_dp,fibre1=1)      ! (1)

id=1   ! map is set to identity                                       ! (2)

! map is added to closed orbit and put into the 6 polymorphs
y(1:6)=closed_orbit(1:6)+id                                           ! (3)

call propagate(als,y(1:6),state,fibre1=1)                             ! (4)

one_turn_map=y(1:6) ! Six polymorphs are promoted to Taylor maps      ! (5)
closed_orbit=y                                                        ! (6)

one_turn_map=one_turn_map.sub.1                                       ! (7a)
call print(one_turn_map,6,prec)                                       ! (7b)


call  c_normal(one_turn_map,normal_form)                              ! (8)

write(6,*) " "
write(6,*) " tune = ",normal_form%tune(1)                             ! (8a)

write(6,*) " ";write(6,*) "  Constant part of the map";
 write(6,'(a16,6(1x,g12.5))') " closed orbit = ",closed_orbit(1:6)   ! (8b)

mat=one_turn_map
write(6,*) " ";write(6,*) " One-turn Map ";
do i=1,c_%nd2
 write(6,'(a5,i1,a5,6(1x,g12.5))') " row ",i," --> ",mat(i,1:c_%nd2) ! (8c)
enddo

a=normal_form%a_t
write(6,*) " ";write(6,*) "  Canonical transformation A";
do i=1,c_%nd2
 write(6,'(a5,i1,a5,6(1x,g12.5))') " row ",i," --> ",a(i,1:c_%nd2)  ! (8d)
enddo

mat=normal_form%a_t**(-1)*one_turn_map*normal_form%a_t
write(6,*) " ";write(6,*) "  Normal Form ";
do i=1,c_%nd2
 write(6,'(a5,i1,a5,6(1x,g12.5))') " row ",i," --> ",mat(i,1:c_%nd2) ! (8e)
enddo

z1=1.e0_dp.cmono.'10'                                                 ! (9a)
z2=1.e0_dp.cmono.'01'                                                 ! (9b)
z1_new=z1*normal_form%a_t**(-1)                                       ! (9c)
z2_new=z2*normal_form%a_t**(-1)                                       ! (9d)

r2=z1**2+z2**2                                                        ! (10a)
e1 = r2*normal_form%a_t**(-1)                                         ! (10b)
e2 = z1_new**2+z2_new**2                                              ! (10c)

beta = a(1,1)**2+a(1,2)**2                                            ! (11a)
gamma = a(2,1)**2+a(2,2)**2                                           ! (11b)
alpha = -a(1,1)*a(2,1)-a(1,2)*a(2,2)                                  ! (11c)
write(6,*)  " "
write(6,*) "beta = ", beta
write(6,*) "gamma = ", gamma
write(6,*) " 2 x alpha = ", 2*alpha
write(6,*)  " "
write(6,*)  " Courant-Snyder Invariant: r^2 o a^(-1) "
call print(e1,6)
```

```
write(6,*)  " Courant-Snyder Invariant : z1_new**2 + z2_new**2 "
call print(e2,6)

write(6,*) " ";write(6,*) " w_1 =",(i_-alpha)/sqrt(beta)      ! (12a)
               write(6,*) " w_2 =",-sqrt(beta)                ! (12b)

ac=from_phasor(-1) * normal_form%a_t**(-1)                    ! (13a)
ac=transpose(ac)                                              ! (13b)

write(6,*)" ";write(6,*)"Complex Canonical transformation A";write(6,*)" ";
write(6,'(a8,17x,a1,18x,a1,1/)') " column ","1","2"
do i=1,c_%nd2
 write(6,'(a5,i1,a5,6(1x,g12.5,1x,g12.5))') " row ",i," --> ",ac(i,1:c_%nd2)
enddo

call ptc_end(graphics_maybe=1,flat_file=.false.)

end program one_turn_orbital_map_normal_form_2d
```

Appendix D
Program one_turn_orbital_map_phase_ad

```
program one_turn_orbital_map_phase_ad
use madx_ptc_module
use pointer_lattice
use c_TPSA
implicit none

    interface
       subroutine build_lattice_als(ALS,MIS,error,exact,sl,thin,onecell)
         use madx_ptc_module
         use pointer_lattice
         implicit none
         type(layout), target :: ALS
         logical(lp) mis
         real(dp),optional :: error(6)
         logical, optional :: exact,sl,thin,onecell
       end subroutine build_lattice_als
    end interface


type(layout), pointer:: ALS
real(dp) prec,closed_orbit(6),mat(6,6),a(6,6),L
complex(dp) ac(6,6),w(6)
real(dp) beta,gamma,alpha
type(internal_state),target :: state
logical(lp) :: mis=.false.
type(c_damap)  one_turn_map, Id,a_1,a_2,b_2
type(real_8) y(6)
type(c_normal_form) normal_form
type(c_taylor) e2,r2,z1,z2,z1_new,z2_new,e2t,e1,phase(3)
integer i,map_order,pos
c_mess_up_vector=.true.; b_mess=-1.0_dp;
c_verbose=.false.
prec=1.d-6 ! for printing
longprint=.false.
call ptc_ini_no_append
call append_empty_layout(m_u)
ALS=>m_u%start

call build_lattice_als(ALS,mis)

state=only_2d0

write(6,*) "Write 't' for Courant-Snyder "
write(6,*) "Write 'f' for Anti-Courant-Snyder "
read(5,*) courant_snyder_teng_edwards
```


E. Forest, *From Tracking Code to Analysis*,
DOI 10.1007/978-4-431-55803-3

```
pos=2
map_order=2
call init_all(state,map_order,0)

call alloc(one_turn_map,id,a_1,a_2,b_2)
call alloc(phase)
call alloc(y)
call alloc(normal_form)
call alloc(e2,r2,z1,z2,z1_new,z2_new,e2t,e1)
z1=1.e0_dp.cmono.'10'                                                   ! (1a)
z2=1.e0_dp.cmono.'01'                                                   ! (1b)
r2=z1**2+z2**2                                                          ! (1c)
closed_orbit=0.0d0
call find_orbit_x(als,closed_orbit(1:6),STATE,1.e-5_dp,fibre1=1)       ! (2)

id=1   ! map is set to identity                                         ! (3)

! map is added to closed orbit and put into the 6 polymorphs
y(1:6)=closed_orbit(1:6)+id                                             ! (4)

call propagate(als,y(1:6),state,fibre1=1)                               ! (5)

one_turn_map=y(1:6) ! Six polymorphs are promoted to Taylor maps        ! (6)
closed_orbit=y                                                          ! (7)

call  c_normal(one_turn_map,normal_form)                                ! (8a)

write(6,'(1/,a50,1/)') " Canonical Transformation coming from Normal Form "
call print(normal_form%a_t,6)                                           ! (8b)

call c_canonise(normal_form%a_t,a_1,phase=phase);phase(1)=0.0_dp;      ! (9a)

if(courant_snyder_teng_edwards) then
 write(6,'(1/,a50,1/)') " Courant-Snyder Canonical Transformation  "
  else
 write(6,'(1/,a50,1/)') " Anti-Courant-Snyder Canonical Transformation  "
endif
call print(a_1,6,prec)                                                  ! (9b)

! map is added to closed orbit and put into the 6 polymorphs

y(1:6)=closed_orbit(1:6)+a_1                                            ! (10)

call propagate(als,y(1:6),state,fibre1=1,fibre2=pos)                    ! (11)

b_2=y(1:6)                                                              ! (12a)
call c_canonise(b_2,a_2,phase=phase)                                    ! (12b)
!!!!!!!!!!!!!!!!!!!!!!!!!  One turn map at pos = 2 !!!!!!!!!!!!!!!!!!!!
call find_orbit_x(als,closed_orbit(1:6),STATE,1.e-5_dp,fibre1=pos)     ! (13)

y(1:6)=closed_orbit(1:6)+id                                             ! (14)

call propagate(als,y(1:6),state,fibre1=pos)                             ! (15)

one_turn_map=y(1:6)                                                     ! (16)

call  c_normal(one_turn_map,normal_form)                                ! (17)
call  c_canonise(normal_form%a_t,a_2);                                  ! (18)

!!!!!!!!!!!!!!!!!!   Computes invariants at 1 and 2 !!!!!!!!!!!!!!!!

e1 = r2*a_1**(-1)                                                       ! (19a)
e2t = r2*b_2**(-1)                                                      ! (19b)
e2 = r2*a_2**(-1)                                                       ! (19c)

write(6,'(1/,a54,1/)')" Invariant at position = 1 computed from one turn map "
call print(e1,6)                                                        ! (20a)
```

```
write(6,'(1/,a54,1/)')" Invariant at position = 2  tracked from position = 1 "
call print(e2t,6)                                                        ! (20b)
write(6,'(1/,a54,1/)')" Invariant at position = 2  computed from one turn map"
call print(e2,6)                                                         ! (20c)



write(6,'(1/,a54,1/)') " Phase advance from position = 1 to position = 2      "
!!!!!!!!!!!!!!!!!!!!!!!!!!!!   Analytic results !!!!!!!!!!!!!!!!!!!!!!
if(pos==2.and.courant_snyder_teng_edwards) then
 L=2.832695d0 ! Length of first element L1
 beta=e1.sub.'02'
 alpha=(e1.sub.'11')/2

 write(6,"(a48,G20.13)") " Based on theory, the phase advance should be = ", &
                                                 atan(L/(beta-L*alpha))/twopi
else
 write(6,*) " Based on theory, the phase advance should be zero "
endif
!!!!!!!!!!!!!!!!!!!!!!!!! The code's results !!!!!!!!!!!!!!!!!!!!!!!!!

call print(phase(1),6,prec)                                              ! (21)

call ptc_end(graphics_maybe=1,flat_file=.false.)

end program one_turn_orbital_map_phase_ad
```

Appendix E
Program Pendulum

```
program pendulum
use madx_ptc_module
use pointer_lattice
use c_TPSA
implicit none
type(internal_state),target :: state
type(c_damap)  m,Id,r,a1
type(c_vector_field) vf
type(c_taylor) K,theta,p,phase,I,a_op,k_io
type(c_normal_form) N
real(dp) prec,F,dt,c4,c6,tpf
complex(dp) v
integer map_order,io,km,ji(2),ns

prec=1.d-13 ! for printing
longprint=.false.

state=only_2d0

map_order=6
call init_all(state,map_order,0)
call alloc(m,id,r,a1)
call alloc(vf);call alloc(K,theta,p,phase,I,a_op,k_io);call alloc(N);

f=1/2.d0 !  This pendulum has a period of tau=2
dt=5.d-2
theta=1.d0.cmono.1                        !(1a)
p=1.d0.cmono.2                            !(1b)

K=p**2/2+2*(2*pi*f)**2*sin(theta/2.d0)**2  !(1c)

a1%v(1)=theta/sqrt(2*pi*f)                !(2a)
a1%v(2)=p*sqrt(2*pi*f)                    !(2b)

write(6,*);write(6,*) " This is Hamiltonian in linear phasors "; write(6,*);

K=K*a1*from_phasor()                      !(2c)

       call print(K,6,prec)

      tpf=K.sub.'11'

       do io=3,map_order
```


E. Forest, *From Tracking Code to Analysis*,
DOI 10.1007/978-4-431-55803-3

```
      a_op=0.0_dp
       k_io=K.sub.io
       call c_taylor_cycle(K_io,size=ns)  !(3a)

!!!   Cycling over all monomials
     do km=1,ns
       call c_taylor_cycle(K_io,ii=km,value=v,j=ji) !(3b)
       if(ji(1)/=ji(2)) then                        !(3c)
         v=v/((ji(1)-ji(2))*(2*i_)*tpf)    !(3d)
         a_op=a_op+(v.cmono.ji)                 !(3e)
       endif
     enddo
     if(io==4) then
       write(6,*); write(6,*) " Eq 3.73 of the book is printed "
       call print(a_op,6)
     endif
       vf=cgetvectorfield( a_op )                   !(4a)
       K=exp(vf,K)                                  !(4b)

       enddo

write(6,*);write(6,*) " This is K_new directly normalised "; write(6,*);

       call print(K,6,prec)

       write(6,*);write(6,*) " From analytical calculations ";
write(6,*);
write(6,99) pi*f, "(phi+ phi-) + ",-1/16.0_dp/4," (phi+ phi-)^2 + ", &
 -1/512.0_dp/pi/f/8," (phi+ phi-)^3  "
write(6,*);
99   format(' ',(g23.16,a15,g23.16,a17,g23.16,a16))

!!!!!! Now Map Methods !!!!!!

K=p**2/2+2*(2*pi*f)**2*sin(theta/2.d0)**2  !(5)

!!! vf is the force field of the pendulum
vf=getvectorfield( -dt*K )                   !(6)
!vf%v(1)=dt*(theta.pb.K)                     !(6a)
!vf%v(2)=dt*(p.pb.K)                         !(6b)
!vf%v(1)=-dt*(K.d.2)                         !(6c)
!vf%v(2)= dt*(K.d.1)                         !(6d)

id=1                         ! (7)
m=exp(vf,id)                 ! (8)

write(6,*);
 write(6,*);write(6,*) " Normalising the map ";
write(6,*);
call  c_normal(m,n)          ! (9)

r=n%a_t**(-1)*m*n%a_t        ! (10)

r=from_phasor(-1)*r*from_phasor()           ! (11)
call print(r,6,prec)

       phase=-i_*log(r%v(2).k.(2)).cut.map_order  ! (12a)
       phase=phase/dt                             ! (12b)
       id%v(1)=id%v(1)*2.d0                       ! (12c)
       id%v(2)=1.d0                               ! (12d)
       phase=phase.o.id                           ! (12e)

write(6,*);write(6,*) " Analytical results from dK_infinity/dJ Eq.(3.22) ";
```

```
write(6,*);

       write(6,100) 2*pi*f,"+ ",-1.d0/8," J + ",-3.d0/512/pi/f," J^2   " ! (13a)

       call print(phase,6,prec)                        ! (13b)


       write(6,*); write(6,102) pi*f," theta^2 + ", 1/(4*pi*f)," p^2"
       c4=-1536*pi**3*f**3
       write(6,104) 80*pi**4*f**4/c4," theta^4 + ",-24*pi**2*f**2/c4, &
       " theta^2 p^2 + ",-3/c4 ," p^4"

       c6=2949120*pi**5*f**5
       write(6,106) 1472*pi**6*f**6/c6," theta^6 + ",2640*pi**4*f**4/c6, &
       " theta^4 p^2 + ",1620*pi**2*f**2/c6, &
       " theta^2 p^4 + ",135/c6 ," p^6"; write(6,*);     ! (14a)

 I=((((1.d0.cmono.1)**2+(1.d0.cmono.2)**2)/2.d0)*n%a_t**(-1)).cut.map_order  ! (14b)

       call print(I,6,prec)                              ! (14c)


102    format(' ',(g23.16,a11,g23.16,a4))
104    format(' ',(g23.16,a11,g23.16,a15,g23.16,a4))
106    format(' ',(g23.16,a11,g23.16,a15,g23.16,a15,g23.16,a4))
100     format(' ',(g23.16,a2,g23.16,a5,g23.16,a7))

    end program pendulum
```

Appendix F
Program standard_map

```
program standard_map
use madx_ptc_module
use pointer_lattice
use c_TPSA
implicit none
type(internal_state),target :: state
type(c_damap)  m,Id,r,a1,m_n,c,rho1,n_np1
type(c_vector_field) vf,g_io
type(c_factored_lie) f_op,t_op
type(c_taylor) K,theta,p,phase,I,f_pb,k_io,t_pb,a_op
type(c_normal_form) N
real(dp) prec,F,dt,c4,c6,tpf,beta
complex(dp) v,expmu,om
integer map_order,io,km,ji(2),ns,mf,a
logical t_o

prec=1.d-13 ! for printing
longprint=.false.
c_verbose=.false.
call kanalnummer(mf,'result_of_standard_map.txt')

state=only_2d0

map_order=8
call init_all(state,map_order,0)
call alloc(m,id,r,a1,m_n,c,rho1,n_np1)
call alloc(vf);call alloc(K,theta,p,phase,I,f_pb,k_io,t_pb,a_op);call alloc(N);
call alloc(f_op);call alloc(t_op);call alloc(g_io);



f=1/2.d0 !  This pendulum has a period of tau=2
dt=5.d-2
theta=1.d0.cmono.1                            !(1a)
p=1.d0.cmono.2                                !(1b)

! One quadratic step of integration
!  K=p**2/2+2*(2*pi*f)**2*sin(theta/2.d0)**2

 theta = theta + (dt/2) * p                                        ! (2a)
 p = p - dt * 2*(2*pi*f)**2* sin(theta/2.d0)*cos(theta/2.d0)   ! (2b)
 theta = theta + (dt/2) * p                                        ! (2c)

M%v(1)=theta    ! (3a)
M%v(2)=p        ! (3b)
```


E. Forest, *From Tracking Code to Analysis*,
DOI 10.1007/978-4-431-55803-3

```
 write(mf,*);
 write(mf,*) " Normalising the standard map using FPP software ";

call  c_normal(m,n)        ! (4)
write(mf,*) " The tune of the map is ", n%tune(1)
write(mf,*); write(mf,*) " Normalised map  ";write(mf,*);

        id=exp(n%ker)         ! (5)
        call print(id,mf,prec)

!!!!!!!!!!!!! Normalised Pseudo-Hamiltonian !!!!!!!!!!!!!!
        vf=0

        do io=1,size(n%ker%f)
         vf%v(1)=vf%v(1)+n%ker%f(io)%v(1)  ! (6a)
         vf%v(2)=vf%v(2)+n%ker%f(io)%v(2)  ! (6b)
        enddo

        k_io=-cgetpb(vf)/dt                ! (6c)

 write(mf,*); write(mf,*) " Normalised Pseudo-Hamiltonian  ";write(mf,*);

        call print(k_io,mf,prec)

!!!!!!!!!!!!!!!! Map and Vector Brackets !!!!!!!!!!!!!!!!!
write(mf,*);
write(mf,*) " Normalising the map using our software using Vector Brackets ";
write(mf,*);

        t_op=0 ;f_op=0;

        beta=(1-(pi*f*dt)**2)**(1.0_dp/2)/(2*pi*f)      ! (ia)

        a1%v(1)=sqrt(beta).cmono.1                      ! (ib)
        a1%v(2)=(1.0_dp/sqrt(beta)).cmono.2             ! (ic)

        c=from_phasor()                                 ! (iia)
        m_n=c**(-1)*a1**(-1)*m*a1*c                     ! (iib)

        rho1=m_n.sub.1                  ! (iiia)
        expmu=m_n%v(1).sub.'1'          ! (iiib)

        t_op%dir=1   ! (iva)
        f_op%dir=-1  ! (ivb)

 do io=2,map_order

         n_np1= m_n*rho1**(-1); n_np1= exp_inv(T_op,n_np1)          ! (v)
         g_io%v(1)=n_np1%v(1).sub.io      ! (via)
         g_io%v(2)=n_np1%v(2).sub.io      ! (vib)
do a=1,2
          call c_taylor_cycle(g_io%v(a),size=ns)

!!!   Cycling over all monomials
    do km=1,ns
       call c_taylor_cycle(g_io%v(a),ii=km,value=v,j=ji)    ! (vii)
       if(ji(1)-ji(2)+(-1)**a/=0) then                     ! (viiia)
         v=v/(1-expmu**(ji(2)-ji(1)-(-1)**a))              ! (viiib)
         f_op%f(io)%v(a)=f_op%f(io)%v(a)+(v.cmono.ji)    ! (viiic)
       else
         t_op%f(io)%v(a)=t_op%f(io)%v(a)+(v.cmono.ji)    ! (viiid)
       endif
    enddo
 enddo
        m_n=exp(-f_op%f(io))*m_n ;m_n=exp(f_op%f(io),m_n) ! (ix)
 enddo
```

```
        t_op%f(1)%v(1)= log(expmu).cmono.1        ! (xa)
        t_op%f(1)%v(2)=-log(expmu).cmono.2        ! (xb)

write(mf,*);write(mf,*) " Normalised map  "; write(mf,*);

        id=exp(t_op)  ! (xi)
        call print(id,mf,prec)

!!!!!!!!!!! Computing the normalised Pseudo-Hamiltonian !!!!!!!!!

        vf=0
        do io=1,size(t_op%f)
         vf%v(1)=vf%v(1)+t_op%f(io)%v(1)      ! (xiia)
         vf%v(2)=vf%v(2)+t_op%f(io)%v(2)      ! (xiib)
        enddo
     k_io=-cgetpb(vf)/dt                      ! (xiii)

write(mf,*);
write(mf,*) " Normalised Pseudo-Hamiltonian  ";
write(mf,*);
        call print(k_io,mf,prec)

!!!!!!!!!!!!!! Map and Poisson Brackets !!!!!!!!!!!!!
write(mf,*);
write(mf,*) " Normalising the map using our software using Poisson Brackets ";
write(mf,*);
        t_op=0
        f_op=0
        beta=(1-(pi*f*dt)**2)**(1.0_dp/2)/(2*pi*f)      ! (Ia)


        a1%v(1)=sqrt(beta).cmono.1                      ! (Ib)
        a1%v(2)=(1.0_dp/sqrt(beta)).cmono.2             ! (Ic)

        c=from_phasor()                                 ! (IIa)
        m_n=c**(-1)*a1**(-1)*m*a1*c                     ! (IIb)

        rho1=m_n.sub.1                  ! (IIIa)
        expmu=m_n%v(1).sub.'1'          ! (IIIb)

        t_op%dir=1  ! same as above
        f_op%dir=-1

 do io=2,map_order
         t_op%f(io)=0                   ! (IVa)
         f_op%f(io)=0                   ! (IVb)
         n_np1=m_n*rho1**(-1); n_np1=exp_inv(t_op,n_np1)            ! (V)

         g_io%v(1)=n_np1%v(1).sub.io     ! (VIa)
         g_io%v(2)=n_np1%v(2).sub.io     ! (VIb)

         k_io=cgetpb( g_io )             ! (VIc)

         f_pb=0.0_dp                    ! (VIIa)
         t_pb=0.0_dp                    ! (VIIb)

          call c_taylor_cycle(K_io,size=ns)   ! same as above

!!!   Cycling over all monomials
    do km=1,ns
       call c_taylor_cycle(K_io,ii=km,value=v,j=ji)
       if(ji(1)/=ji(2)) then                    !(VIIIa)
         v=v/(1-expmu**(ji(2)-ji(1)))           !(VIIIb)
         f_pb=f_pb+(v.cmono.ji)                 !(VIIIc)
       else
         t_pb=t_pb+(v.cmono.ji)                 !(VIIId)
       endif
```

```
    enddo
        t_op%f(io)=cgetvectorfield(t_pb)          ! (IXa)
        f_op%f(io)=cgetvectorfield(f_pb)!         ! (IXb)
        m_n=exp(-f_op%f(io))*m_n; m_n=exp(f_op%f(io),m_n) ! (X)

 enddo
        t_op%f(1)%v(1)= log(expmu).cmono.1        ! (XIa)
        t_op%f(1)%v(2)=-log(expmu).cmono.2        ! (XIb)

 write(mf,*);write(mf,*) " Normalised map  "; write(mf,*);

        id=exp(t_op)                            ! (XII)
        call print(id,mf,prec)

!!!!!!!!!!!! Computing the pseudo-Hamiltonian !!!!!!!!!!!!
write(mf,*);
  write(mf,*) " Normalising the standard map using our own little software ";
  write(mf,*) " by first computing the 'logarithm' of the map and then  ";
  write(mf,*) " normalising it using Poisson Brackets ";
write(mf,*);

        vf=log(M)                                   ! (1a)
        k=-getpb(vf)/dt                             ! (1b)

        beta=(1-(pi*f*dt)**2)**(1.0_dp/2)/(2*pi*f)  ! (2a)

        a1%v(1)=sqrt(beta).cmono.1                  ! (2b)
        a1%v(2)=(1.0_dp/sqrt(beta)).cmono.2         ! (2c)

!! This is Hamiltonian in linear phasors

K=K*a1*from_phasor()                  ! (3a)

      tpf=K.sub.'11'                  ! (3b)

       do io=3,map_order
       a_op=0.0_dp
       k_io=K.sub.io                      ! (4a)
       call c_taylor_cycle(K_io,size=ns)  ! (4b)

!!!   Cycling over all monomials
    do km=1,ns
       call c_taylor_cycle(K_io,ii=km,value=v,j=ji) ! (5a)
       if(ji(1)/=ji(2)) then                        ! (5b)
         v=v/((ji(1)-ji(2))*(2*i_)*tpf)             ! (5c)
         a_op=a_op+(v.cmono.ji)                     ! (5d)
       endif
    enddo

       vf=cgetvectorfield( a_op )                 ! (6a)
       K=exp(vf,K)                                ! (6b)

       enddo
write(mf,*);write(mf,*) " This is K_new directly normalised "; write(mf,*);

       call print(K,mf,prec)
       vf=cgetvectorfield( K )       ! (7a)
       vf=-dt*vf                     ! (7b)
       id=exp(vf)


 write(mf,*);write(mf,*) " Normalised map  "; write(mf,*);

        call print(id,mf,prec)
```

```
 !!!!!!!!!!!!!!!!!!!!!!!!!!!!!!!!!!!!!!!!!!!!!!!!!!!!!!!!!!!!
  write(mf,*)
  write(mf,*) " Normalising the standard map using our own little software ";
  write(mf,*) " by first computing the 'logarithm' of the map and then  ";
  write(mf,*) " normalising it using Vector Brackets ";


Write(6,*)"Transform vector field with formula (F.dot.a_k^-1) o a ---> type t"
Write(6,*) " Or with exp(Lie Bracket A) F ---> type f "
read(5,*) t_o
  if(t_o) then
    Write(mf,*) " Transforming vector field with formula (F.dot.a_k^-1) o a "
  else
    Write(mf,*) " Transforming vector field with exp(Lie Bracket A) F "
  endif
        f_op=0
        t_op=0

        beta=(1-(pi*f*dt)**2)**(1.0_dp/2)/(2*pi*f)        ! (1A)

        a1%v(1)=sqrt(beta).cmono.1                        ! (1B)
        a1%v(2)=(1.0_dp/sqrt(beta)).cmono.2               ! (1C)

        M=from_phasor(-1)*a1**(-1)*m*a1*from_phasor()     ! (2)

        g_io=log(M)  !                                     ! (3A)
        g_io=-(1.0_dp/dt)*g_io                             ! (3B)
        om= g_io%v(1).sub.'1'                              ! (3C)

       do io=2,map_order
!!!   Cycling over all monomials
do a=1,2
          m_n%v(a)=g_io%v(a).sub.io    ! (4)
          call c_taylor_cycle(m_n%v(a),size=ns)

!!!   Cycling over all monomials
    do km=1,ns
       call c_taylor_cycle(m_n%v(a),ii=km,value=v,j=ji)
       if(ji(1)-ji(2)+(-1)**a/=0) then                       ! (5A)
         v=v/(om*(ji(1)-ji(2)+(-1)**a))                      ! (5B)
         f_op%f(io)%v(a)=f_op%f(io)%v(a)+(v.cmono.ji)        ! (5C)
       endif
    enddo
enddo

    if(t_o) then    ! if true, use (F.dot.a_k^-1) o a
          id=exp(f_op%f(io))     ! (6A)
          a1=id**(-1)            ! (6B)
          r=0

          do ns=1,2
          do a=1,2
           r%v(ns)=g_io%v(a)*(a1%v(ns).d.a)+r%v(ns)  ! (7)
          enddo
          r%v(ns)=r%v(ns)*id                         ! (8)
          enddo

          do ns=1,2
            g_io%v(ns)=r%v(ns)
          enddo
    else       !  use a Lie Bracket operator
          g_io=exp_ad(f_op%f(io),g_io)       ! (6')
    endif

 enddo

 write(mf,*);write(mf,*) " Normalised map  "; write(mf,*);
```

```
       id=exp(g_io)                        ! (9)
       call print(id,mf,prec)

 write(mf,*);write(mf,*) " This is K_new directly normalised "; write(mf,*);

       k=cgetpb(g_io)    ! (10)
       call print(K,mf,prec)

       close(mf)

!!!!!!!!!!!!!!!!!!!!   Illustrating the Logarithm    !!!!!!!!!!!!!!!!!!!
!

          lielib_print(3)=1    ! printing iterates
write(6,*);write(6,*) "Testing the logarithm of a map ";write(6,*);
write(6,*);write(6,*) "First Case ";write(6,*);

          vf=log(M)


!!!!  forcing linear behaviour until 10^-15 is reached
write(6,*); write(6,*) "Second Case ";write(6,*);

          vf=log(M,epso=1.d-14)

write(6,*); write(6,*) "Third Case ";write(6,*);
extra_terms_log=.true.
          vf=log(M)

end program standard_map
```

Appendix G
Program one_turn_cavity_map

```
program one_turn_cavity_map
use madx_ptc_module
use pointer_lattice
use c_TPSA
implicit none

    interface
       subroutine build_lattice_als(ALS,MIS,error,exact,sl,thin,onecell)
         use madx_ptc_module
         use pointer_lattice
         implicit none
         type(layout), target :: ALS
         logical(lp) mis
         real(dp),optional :: error(6)
         logical, optional :: exact,sl,thin,onecell
       end subroutine build_lattice_als
    end interface


type(layout), pointer:: ALS
real(dp) prec,closed_orbit(6),mat(6,6),a(6,6),ai(6,6),del,error(6),mom(6,6),ray(5,6),cav_tune
real(dp) Ia(6,6,3),Sa(6,6,3),S(6,6),Ka(6,6,3),Ba(6,6,3),Ha(6,6,3),Ea(6,6,3),inv,invl,tune_s
REAL(DP) F(6,6),alpha_s,m(6,6), alphas(2),c
complex(dp) mc(6,6)
type(internal_state),target :: state
logical(lp) :: mis=.false.
type(c_damap)  one_turn_map, Id,a0,a_cs
type(real_8) y(6)
type(c_normal_form) normal_form
type(c_taylor) e(3),z_ij,ave_FLOQUET
integer expo(6),pos
character*48 command_gino,fmd
integer i,map_order,j,cas,mf,nd2a,nda,k,t
!!!!!!!!!!!!!!!!!!!!!
type(fibre), pointer :: p1
type(integration_node), pointer :: ti
!!!!!!!!!!!!!!!!!!!!!
c_verbose=.false.
prec=1.d-10 ! for printing
longprint=.false.
 del=0.d0
use_info=.true.
fmd='(1x,g17.10)'

Ia=0.d0; Sa=0.d0; S=0.d0;Ba=0.d0;Ha=0.d0;Ea=0.d0

do i=1,3
```


E. Forest, *From Tracking Code to Analysis*,
DOI 10.1007/978-4-431-55803-3

```
 S(2*i-1,2*i) = 1.d0; S(2*i,2*i-1) = -1.d0;
 Ia(2*i-1,2*i-1,i) =1.d0;  Ia(2*i,2*i,i)    = 1.d0;
 Sa(2*i-1,2*i,i)   =1.d0;  Sa(2*i,2*i-1,i) =-1.d0;
enddo

call ptc_ini_no_append
call append_empty_layout(m_u)
ALS=>m_u%start

write(6,*) " misalignments type t, otherwise f"
read(5,*) mis

If(mis) then
 write(6,'(a56,/)') " The lattice produced will have errors and thus coupling"
else
 write(6,'(a56,/)') " The lattice produced is ideal: mid-plane symmetric      "
endif

pos=15
error=0.d0
error(6)=1.d-5 ! tilt along z-axis

call build_lattice_als(ALS,mis,error,exact=.false.)


 do cas=2,0,-1

if(cas==0) then
 state=nocavity0
 nd2a=4
 nda=2
call kanalnummer(mf,"result_no_cavity.txt")
elseif(cas==1) then
 state=default0
 nd2a=6
 nda=3
call kanalnummer(mf,"result_with_cavity.txt")
elseif(cas==2) then
 state=radiation0
 nd2a=6
 nda=3
call kanalnummer(mf,"result_with_cavity_and_radiation.txt")
endif

map_order=2
call init_all(state,map_order,0)

call alloc(one_turn_map,id,a0,a_cs)
call alloc(y)
call alloc(normal_form)
call alloc(e)
call alloc(z_ij,ave_FLOQUET)

closed_orbit=0.d0; closed_orbit(5)=del;                                 ! (0)

call find_orbit_x(als,closed_orbit(1:6),STATE,1.e-8_dp,fibre1=pos)        ! (1)
del=closed_orbit(5)

id=1    ! map is set to identity                                        ! (2)

! map is added to closed orbit and put into the 6 polymorphs
y(1:6)=closed_orbit(1:6)+id                                             ! (3)

write(mf,*) " ";write(mf,*) "  Constant part of the map";
write(mf,'(a16,6(1x,g12.5))') " closed orbit = ",closed_orbit(1:6)    ! (4)
!write(mf,'(a16,6'//fmd//')') " closed orbit = ",closed_orbit(1:6)    ! (4)
```

```
call propagate(als,y(1:6),state,fibre1=pos)                                ! (5)

one_turn_map=y(1:6) ! Six polymorphs are promoted to Taylor maps         ! (6)
closed_orbit=y                                                           ! (7)

one_turn_map=one_turn_map.sub.1                                          ! (8)

write(mf,*) " "; write(mf,*) "  The linear map";

call print(one_turn_map,mf,prec)


call  c_normal(one_turn_map,normal_form)                                 ! (9)


write(mf,'(a8,3(1x,g20.13))') " tune = ",normal_form%tune(1:nda)        ! (10)
if(cas==1) tune_s=normal_form%tune(3)
write(mf,*) " ";write(mf,*) "  Constant part of the map";
 write(mf,'(a16,6(1x,g12.5))') " closed orbit = ",closed_orbit(1:6)   ! (11)

a=normal_form%a_t                                                        ! (12a)
ai=normal_form%a_t**(-1)                                                 ! (12b)

mc=from_phasor(-1)*normal_form%a_t**(-1)*one_turn_map*normal_form%a_t*from_phasor()  ! (13)

write(mf,'(/,a37,15x,a41,/)')" i  j     Real(mc(i,j))     Im(mc(i,j))", &
                              "damping  + i *   phase  ( =log(mc(i,j)) )"
do i=1,c_%nd2
do j=1,c_%nd2
if(abs(mc(i,j))>1.e-10_dp) write(mf,'(i2,1x,i2,2(5x,g12.5),5x,2(5x,g12.5))')&
                                      i,j,mc(i,j),log(mc(i,j)) ! (14)
enddo
enddo

!!!!!!! Lattice functions !!!!!!!
!! coefficient of invariant  !!

do i=1,c_%nd
 Ba(1:6,1:6,i)=matmul(matmul(a,Sa(1:6,1:6,i)),ai)  ! (15a)
 Ha(1:6,1:6,i)=matmul(matmul(a,Ia(1:6,1:6,i)),ai)  ! (15b)
 Ka(1:6,1:6,i)=-matmul(S,Ba(1:6,1:6,i))            ! (15c)
 Ea(1:6,1:6,i)=-matmul(Ba(1:6,1:6,i),S)            ! (15d)
enddo
call clean_mat(Ba,prec);call clean_mat(Ha,prec);   ! (16a)
call clean_mat(Ka,prec);call clean_mat(Ea,prec);   ! (16b)

if(cas/=2) then
  do i=1,nda
      expo=0
      expo(2*i-1)=2
     e(i)=1.d0.cmono.expo
      expo=0
      expo(2*i)=2
     e(i)=e(i)+(1.d0.cmono.expo)
     e(i)=e(i)*normal_form%a_t**(-1)                ! (17)
  enddo

  do i=1,nda
write(mf,*)" "; write(mf,*) " Invariant (q^2+p^2) o A^(-1)  in plane ",i; write(mf,*)" ";
   call print(e(i),mf,prec)
   do j=1,2*nda
    write(mf,'(6'//fmd//')')Ka(j,1:6,i)            ! (18)
   enddo
  enddo

  do i=1,2*nda
   do j=i,2*nda
```

```
      z_ij=(1.d0.cmono.i)*(1.d0.cmono.j)
      call C_AVERAGE(z_ij,normal_form%a_t,ave_FLOQUET)          ! (19a)
      write(mf,*);write(mf,'(a3,i1,a3,i1,a1)')  "<z_",i," z_",j,">"
      write(mf,'(3(5x,g12.5))') Ea(i,j,1)/2,Ea(i,j,2)/2,Ea(i,j,3)/2
      call print(ave_FLOQUET,mf,prec)
   enddo
  enddo
endif

  do i=1,3
   write(mf,*)" "; write(mf,*) " Matrix H ",i; write(mf,*)" ";
   do j=1,6
    write(mf,'(6'//fmd//')') Ha(j,1:6,i)
   enddo
  enddo

  if(cas/=0) then
   alphas(cas)=Ba(6,5,3)*sin((normal_form%tune(3))*twopi)
   write(mf,'(a33)') " Approximate time slip  "
   write(mf,'((5x,g12.5))') alphas(cas)
     write(mf,'(a42)') " H based dispersion as in Chao-Sands paper"
   do j=1,6
    write(mf,*) i, Ha(j,5,3)/Ha(5,5,3)
   enddo

  endif

   if(cas==0) then
   write(mf,'(a34)') " Standard time slip without cavity"
   write(mf,'((5x,g12.5))') real(mc(6,5))
   write(mf,'(a34)') " Time slip estimated with a cavity"
   write(mf,'((5x,g12.5))') alphas(1)
   write(mf,'(a48)') " Time slip estimated with a cavity and radiation"
   write(mf,'((5x,g12.5))') alphas(2)


!!!!  Computation by "hand" of the time slip alpha_s
call c_full_canonise(normal_form%A_t,a_cs,a0=a0)    ! (0)
    mat=one_turn_map
    f=a0                                            ! (1)

! not trusting f(6,1:4) from normal form
    f(6,1:4)=0.0_dp
    do i=1,4
    do j=1,4
     f(6,j)=f(6,j)-f(i,5)*S(i,j)                           ! (2)
    enddo
    enddo

    m(6,1:4) = mat(6,1:4) !(mat(6,1:4)-matmul(f(6,1:4),mat(1:4,1:4)))      ! (3a)
    alpha_s=mat(6,5)
    do i=1,4
     alpha_s =  m(6,i)* f(i,5)+ alpha_s ! - f(6,i)*mat(i,5) + alpha_s    !(3b)
    enddo
    write(mf,*) " Time slip alpha_s from computation and normal form"
    write(mf,*)   alpha_s,real(mc(6,5))
    call move_to(als,p1,"CAV",pos,reset=.true.)
    cav_tune=(alpha_s/2)*(p1%mag%volt*1e-3_dp/p1%mag%P%P0C)*twopi*p1%mag%freq/CLIGHT
    write(mf,*) "tune using alpha_s ", sign(acos(1.0_dp+cav_tune)/2/pi,p1%mag%volt)
    write(mf,*) "Exact temporal tune", tune_s ; write(6,*) " ";
     write(mf,'(a21)') " Standard dispersion "
    do i=1,6
     write(mf,*) i, a(i,5)
    enddo
   endif
   if(cas/=0) then
    write(mf,'(a56)') " Almost Exact Crab Angle ignoring transverse emittances "
```

```
    write(mf,'((5x,g12.5))') -Ka(5,2,3)/Ka(5,5,3)
    write(mf,'(a56)') " Approximate Crab Angle ignoring transverse emittances  "
    write(mf,'((5x,g12.5))')  Ha(1,6,3)-Ka(5,6,3)*Ha(1,5,3)/Ka(5,5,3)
  endif


!!!!!           Evaluation of the Kinetic Invariants           !!!!!!
ray(1,1:6)= (/3,4,3,6,1,3/)
ray(2,1:6)= (/3,5,2,6,1,0/)
ray(3,1:6)= (/2,4,3,6,1,1/)
ray(4,1:6)= (/3,5,3,6,1,2/)
ray(5,1:6)= (/2,3,5,8,1,0/)

 mat=one_turn_map

do t=1,10


 do k=1,5
   ray(k,1:6)=matmul(mat,ray(k,1:6))
 enddo

do i=1,6
do j=i,6
 mom(i,j)=0.d0
do k=1,5
 mom(i,j)=ray(k,i)*ray(k,j)/5+mom(i,j)
enddo
enddo
enddo

inv=0.D0


do i=1,3
inv=4*(mom(2*i-1, 2*i-1)*mom(2*i, 2*i)-mom(2*i-1, 2*i)**2)+inv
enddo
inv1=inv

inv=inv+8*(mom(1, 3)*mom(2, 4)-mom(1, 4)*mom(2, 3)) &
       +8*(mom(1, 5)*mom(2, 6)-mom(1, 6)*mom(2, 5)) &
       +8*(mom(3, 5)*mom(4, 6)-mom(3, 6)*mom(4, 5))

write(mf,'(a19,i2,a2,1x,g12.5,a18,1x,g12.5)') &
" Invariant at turn ",t,  " =",inv,' 1-d-f invariant =',inv1
enddo



call kill(one_turn_map,id,a0,a_cs)
call kill(y)
call kill(normal_form)
call kill(e)
call kill(z_ij,ave_FLOQUET)

close(mf)
 enddo ! cas

call ptc_end(graphics_maybe=1,flat_file=.false.)

end program one_turn_cavity_map

subroutine clean_mat(H,prec)
use precision_constants
implicit none
real(dp) H(6,6,3),prec
integer i,j,k
```

```
do i=1,6
do j=1,6
do k=1,3
 if(abs(H(i,j,k))<prec) H(i,j,k)=0
enddo
enddo
enddo

end subroutine clean_mat
```

Appendix H
Program radiation_map

```
program radiation_map
use madx_ptc_module
use pointer_lattice
use c_TPSA
implicit none

    interface
       subroutine build_lattice_als(ALS,MIS,error,exact,sl,thin,onecell)
         use madx_ptc_module
         use pointer_lattice
         implicit none
         type(layout), target :: ALS
         logical(lp) mis
         real(dp),optional :: error(6)
         logical, optional :: exact,sl,thin,onecell
       end subroutine build_lattice_als
    end interface


type(layout), pointer:: ALS
real(dp) prec,closed_orbit(6),mat(6,6),a(6,6),ai(6,6),del,error(6),emi(3),sij(3)
real(dp) Ia(6,6,3),Sa(6,6,3),S(6,6),Ka(6,6,3),Ba(6,6,3),Ha(6,6,3),Ea(6,6,3),xij
complex(dp) mc(6,6)
type(internal_state),target :: state,sta(2)
logical(lp) :: mis=.false.
type(c_damap)  one_turn_map, Id,a0,a_cs
type(real_8) y(6)
type(c_normal_form) normal_form
type(c_taylor) e(3),z_ij,ave_FLOQUET
integer expo(6),pos
character*48 command_gino,fmd,fmd1
integer i,map_order,j,mf,k,t,cas
type(probe) ray_closed
type(probe_8) ray
!!!!!!!!!!!!!!!!!!!!!
type(fibre), pointer :: p1
type(integration_node), pointer :: ti
!!!!!!!!!!!!!!!!!!!!!
c_verbose=.false.
prec=1.d-10 ! for printing
longprint=.false.
 del=0.d0
use_info=.true.
fmd= '(a12,1X,a3,I1,a3,i1,a4,D18.11,1x,D18.11)'
fmd1='(1X,a3,I1,a3,i1,a4,2(D18.11,1x),(f10.3,1x),a2)'
```


E. Forest, *From Tracking Code to Analysis*,
DOI 10.1007/978-4-431-55803-3

```
sta(1)=default0
sta(2)=default0 +radiation0+envelope0

Ia=0.d0; Sa=0.d0; S=0.d0;Ba=0.d0;Ha=0.d0;Ea=0.d0

do i=1,3
 S(2*i-1,2*i) = 1.d0; S(2*i,2*i-1) = -1.d0;
 Ia(2*i-1,2*i-1,i) =1.d0;  Ia(2*i,2*i,i)   = 1.d0;
 Sa(2*i-1,2*i,i)   =1.d0;  Sa(2*i,2*i-1,i) =-1.d0;
enddo

call ptc_ini_no_append
call append_empty_layout(m_u)
ALS=>m_u%start

write(6,*) " misalignments type t, otherwise f"
read(5,*) mis

If(mis) then
 write(6,'(a56,/)') " The lattice produced will have errors and thus coupling"
else
 write(6,'(a56,/)') " The lattice produced is ideal: mid-plane symmetric     "
endif

pos=15
error=0.d0
error(6)=1.d-5 ! tilt along z-axis

call build_lattice_als(ALS,mis,error,exact=.false.)

call kanalnummer(mf,"result_with_stochastic_radiation.txt")

do cas=1,2

 state=sta(cas)

map_order=1
call init_all(state,map_order,0)

call alloc(one_turn_map,id,a0,a_cs)
call alloc(y)
call alloc(normal_form)
call alloc(e)
call alloc(z_ij,ave_FLOQUET)
call alloc(ray)

closed_orbit=0.d0;                                                   ! (1)

call find_orbit_x(als,closed_orbit(1:6),STATE,1.e-8_dp,fibre1=pos)   ! (2)
if(cas==2) then
 write(mf,*) " ";write(mf,*) "  Constant part of the map";
 write(mf,'(a16,6(1x,g12.5))') " closed orbit = ",closed_orbit(1:6)
endif

ray_closed=closed_orbit      ! (3)
id=1;
! ray= closed orbit + identity map
ray=ray_closed+id;           ! (4)


call propagate(als,RAY,state,fibre1=pos)  ! (5)

! Six polymorphs and the fluctuationsare E_ij
! are promoted to Taylor maps

one_turn_map=ray                         ! (6a)
```

```
one_turn_map=one_turn_map.sub.1          ! (6b)

if(cas==2) then
 write(mf,*) " "; write(mf,*) "  The linear map";
 call print(one_turn_map,mf,prec)                        ! (7)
endif

call  c_normal(one_turn_map,normal_form)                 ! (8)

if(cas==2) write(mf,'(a8,3(1x,g20.13))') " tune = ",normal_form%tune(1:3)     ! (9)


a=normal_form%a_t                                                ! (10a)
ai=normal_form%a_t**(-1)                                         ! (10b)

mc=from_phasor(-1)*normal_form%a_t**(-1)*one_turn_map*normal_form%a_t*from_phasor()  ! (11)

if(cas==2) then
write(mf,'(/,a37,15x,a41,/)')" i  j    Real(mc(i,j))    Im(mc(i,j))", &
                             "damping  + i *   phase  ( =log(mc(i,j)) )"
do i=1,c_%nd2
do j=1,c_%nd2


if(abs(mc(i,j))>1.e-10_dp) write(mf,'(i2,1x,i2,2(5x,g12.5),5x,2(5x,g12.5))')&
                                     i,j,mc(i,j),log(mc(i,j))                     ! (12)
enddo
enddo
endif

if(cas==1) then
!!!!!!! Lattice functions !!!!!!!
!! coefficient of invariant  !!

do i=1,c_%nd
 Ba(1:6,1:6,i)=matmul(matmul(a,Sa(1:6,1:6,i)),ai)  ! (13a)
 Ha(1:6,1:6,i)=matmul(matmul(a,Ia(1:6,1:6,i)),ai)  ! (13b)
 Ka(1:6,1:6,i)=-matmul(S,Ba(1:6,1:6,i))            ! (13c)
 Ea(1:6,1:6,i)=-matmul(Ba(1:6,1:6,i),S)            ! (13d)
enddo

endif


if(cas==2) then
write(mf,*)
write(mf,'(16X,a50)') "   Equilibrium moments in Phasors Basis            "
 do i=1,6
 do j=i,6
  if(abs(normal_form%s_ijr(i,j))>1.d-20) then
   write(mf,fmd) " Phasors -> ","<x_",i," x_",j,"> = ",  &   ! (14)
                  c_clean(normal_form%s_ijr(i,j),1.d-20)
  endif
 enddo
 enddo

emi(1)=real(normal_form%s_ijr(1,2))/2        ! (15a)
emi(2)=real(normal_form%s_ijr(3,4))/2        ! (15b)
emi(3)=real(normal_form%s_ijr(5,6))/2        ! (15c)

write(mf,*)

write(mf,'(16X,a54)') "    Exact              Chao          (Exact-Chao)/Exact"
 do i=1,6
 do j=1,6
   xij=0.d0
```

```
  do k=1,3
    xij= Ea(i,j,k)*emi(k) + xij                              ! (16)
  enddo
 if(abs(normal_form%s_ij0(i,j))>1.e-15_dp) then
  write(mf,fmd1) "<x_",i," x_",j,"> = ", real(normal_form%s_ij0(i,j)), &
               xij ,abs(100*(real(normal_form%s_ij0(i,j))-xij)/       &
  real(normal_form%s_ij0(i,j)))," %"
 endif
 enddo
 enddo

 sij(1)=emi(1)*Ea(1,6,1)+emi(2)*Ea(1,6,2)+emi(3)*Ea(1,6,3)     ! (17a)
 sij(2)=emi(1)*Ea(6,6,1)+emi(2)*Ea(6,6,2)+emi(3)*Ea(6,6,3)     ! (17b)
 sij(3)=emi(1)*Ea(1,1,1)+emi(2)*Ea(1,1,2)+emi(3)*Ea(1,1,3)     ! (17c)

 write(mf,*)
 write(mf,'(a56)') " Exact Crab Angle with beam envelope                   "
 write(mf,'((5x,D18.11))') real(normal_form%s_ij0(1,6)) &
/(real(normal_form%s_ij0(6,6))-real(normal_form%s_ij0(1,1)))             ! (18)
 write(mf,'(a25)') " Chao Exact Crab Angle "
 write(mf,'((5x,D18.11))') sij(1)/(sij(2)-sij(3))                         ! (19)
 write(mf,'(a56)') " Almost Exact Crab Angle ignoring transverse emittances "
 write(mf,'((5x,D18.11))') -Ka(5,2,3)/(Ka(5,5,3)-Ka(2,2,3))              ! (20)
 write(mf,'(a56)') " Approximate Crab Angle ignoring transverse emittances  "
 write(mf,'((5x,D18.11))')  Ha(1,6,3)-Ka(5,6,3)*Ha(1,5,3)/Ka(5,5,3)      ! (21)




write(mf,*);
write(mf,*) "!!!  Raising the maps with moments to the power 2**100  !!!"

do i=1,100
  one_turn_map=one_turn_map*one_turn_map  ! (22a)
enddo

call print(one_turn_map,mf)               ! (22b)

endif


call kill(one_turn_map,id,a0,a_cs)
call kill(y)
call kill(normal_form)
call kill(e)
call kill(z_ij,ave_FLOQUET)
call kill(ray)
enddo



close(mf)

call ptc_end(graphics_maybe=1,flat_file=.false.)

end program radiation_map
```

Appendix I
Program modulated_map

```
program modulated_map
use madx_ptc_module
use pointer_lattice
use c_TPSA
implicit none

    interface
       subroutine build_lattice_als(ALS,MIS,error,exact,sl,thin,onecell)
         use madx_ptc_module
         use pointer_lattice
         implicit none
         type(layout), target :: ALS
         logical(lp) mis
         real(dp),optional :: error(6)
         logical, optional :: exact,sl,thin,onecell
       end subroutine build_lattice_als
    end interface
type(layout), pointer:: ALS
real(dp) prec,closed_orbit(6)
real(dp) L,Kq,k0,mu_mod,beta,dmu,mu_x,circ

type(internal_state),target :: state
logical(lp) :: mis=.false.
type(c_damap)  one_turn_map, drift_map, quad_map,id
type(c_normal_form) normal_form
type(c_taylor) q(4)
integer :: pos =1, nind(11)
integer i,map_order,mf,mf1,mfmap
type(probe) ray_closed
type(probe_8) ray
!!!!!!!!!!!!!!!!!!!!!


c_verbose=.false.
prec=1.d-10 ! for printing
longprint=.false.

state=only_2d0+modulation0

call ptc_ini_no_append
call append_empty_layout(m_u)
ALS=>m_u%start
```


E. Forest, *From Tracking Code to Analysis*,
DOI 10.1007/978-4-431-55803-3

```
! sl= true produces the small 1-d lattice
!  Qf1 = QUADRUPOLE(" QF1 ",0.d0, K1= 0.01d0 ); L1  = drift("L1 ",0.1d0);
! ALS=L1+QF1;

!call build_lattice_als(ALS,mis,exact=.false.,sl=.true.)

call kanalnummer(mf,"result_with_modulation.txt")
call kanalnummer(mfmap,"maps.txt")

!!!! Let us do the baby example with maps
map_order=3
call init_all(state,map_order,0)

call alloc(one_turn_map, drift_map, quad_map,id)
call alloc(q,4)
call alloc(normal_form)
call alloc(ray)

!!!!!!! Simple example !!!!!!!
!  without using PTC
do i=1,4
 q(i)=1.d0.cmono.i   ! q_1, q_2, q_3 and q_4 are created as TPSA variables
enddo
l=0.1d0; Kq=.1d0; k0=.2d0; mu_mod=twopi*0.12345d0;  ! (1a)
! Analytic calculation
beta=1.d0/sqrt(Kq)/sqrt(1.d0-Kq*L**2/4)             ! (1b)

mu_x=acos(1.d0-Kq*L**2/2)                           !  (2a)
dmu=-2*(sin(2*mu_x+mu_mod)/(1-cos(2*mu_x+mu_mod)) & !  (2b)
+sin(2*mu_x-mu_mod)/(1-cos(2*mu_x-mu_mod)))*4*(-beta*k0*L/16)**2

write(mf,*);
write(mf,*) "  Analytical tune in radians =      ",mu_x
write(mf,*) "  Analytical tune shift in radians = ",dmu
write(mf,*);

drift_map=1
quad_map=1

drift_map%v(1)=q(1)+L*q(2)     ! (3) drift
drift_map%v(2)=q(2)
drift_map%v(3)=cos(mu_mod)*q(3) + sin(mu_mod)*q(4)
drift_map%v(4)=cos(mu_mod)*q(4) - sin(mu_mod)*q(3)

quad_map%v(1)=q(1)
quad_map%v(2)=q(2)-L*(Kq+k0*q(3))*q(1) ! (4) quadrupole

! Map of system is made
one_turn_map=quad_map*drift_map   ! total map   (5)

 write(mfmap,*); write(mfmap,*) " Map hardwired in main program " ; write(mfmap,*);
 call print(one_turn_map,mfmap)

! Map is normalised
call  c_normal(one_turn_map,normal_form)       !  (6)
write(mf,*);
write(mf,*) " Result from the normal form algorithm for hardwired map ";
write(mf,*);

write(mf,*) " Normal form result for tune in radians =        ", &  !  (7)
 -aimag(normal_form%ker%f(1)%v(1).sub.'1000')
```

```
write(mf,*) " Normal form result for tune shift in radians = ", &  !  (8)
 -aimag(normal_form%ker%f(3)%v(1).sub.'1011')

! Same calculation from with PTC
! using the following little lattice
!  Qf1 = QUADRUPOLE(" QF1 ",L=0.d0, K1= 0.01d0 ); L1  = drift("L1 ",L0=.1d0);
!  ALS=L1+QF1;
call build_lattice_als(ALS,mis,exact=.false.,sl=.true.)

!!!! circ is the circumference of the ring !!!!
call get_length(als,circ)
!!!! AC_modulate.txt sets the magnet QF1 as a modulated magnet !!!!
 call kanalnummer(mf1,file="AC_modulation.txt")
  write(mf1,*) "select layout"
  write(mf1,*) 1
  write(mf1,*) " MODULATE"
  write(mf1,*) " QF1"
  write(mf1,*) "1.d0 0 0       !DC_ac,A_ac,theta_ac"
  write(mf1,*) "1.d0   2       ! D_ac,n_ac  "
  write(mf1,*) "2 0.02d0 0      ! n d_bn(n) d_an(n)  "  ! (A)
  write(mf1,*) "0  0 0 "
  write(mf1,*) " return "
 close(mf1)
 call read_ptc_command77("AC_modulation.txt")

!!!! set a modulation clock !!!!!!
ray_closed%ac%om=mu_mod/circ        ! (B1)
ray_closed%ac%x=0.d0 ;              ! (B2)
write(6,*) " Modulation tune in radians =",circ*ray_closed%ac%om

closed_orbit=0.d0;                                                  ! (C)

call find_orbit_x(als,closed_orbit(1:6),STATE,1.e-8_dp,fibre1=pos)   ! (D)

ray_closed=closed_orbit     ! (E)
id=1;
! ray= closed orbit + identity map
ray=id+ray_closed;          ! (F1)
write(mf,*); write(mf,*) " Initial value of the clock ( in type probe_8) "
write(mf,*);call print(ray%ac,mf )          ! (F2)

call propagate(als,RAY,state,fibre1=pos)  ! (G)

! Six polymorphs and the fluctuationsare E_ij
! are promoted to Taylor maps

one_turn_map=ray                          ! (H)
 write(mfmap,*); write(mfmap,*) " Map produced by code " ; write(mfmap,*);  ! (I)
 call print(one_turn_map,mfmap)

call  c_normal(one_turn_map,normal_form)    ! (J)

write(mf,*);write(mf,*);
write(mf,*) " Result from the normal form algorithm for the code ";
write(mf,*)


write(mf,*) " Normal form result for tune in radians =        ", &    ! (K)
-aimag(normal_form%ker%f(1)%v(1).sub.'1000')
write(mf,*) " Normal form result for tune shift in radians = ", &    ! (L)
-aimag(normal_form%ker%f(3)%v(1).sub.'1011')
write(mf,*)
```

```
call kill(one_turn_map, drift_map, quad_map,id)
call kill(normal_form)

call kill(ray)



close(mfmap)
close(mf)

call ptc_end

end program modulated_map
```

Appendix J
Program modulated_map_Jordan

```
program modulated_map
use madx_ptc_module
use pointer_lattice
use c_TPSA
implicit none

    interface
       subroutine build_lattice_als(ALS,MIS,error,exact,sl,thin,onecell)
         use madx_ptc_module
         use pointer_lattice
         implicit none
         type(layout), target :: ALS
         logical(lp) mis
         real(dp),optional :: error(6)
         logical, optional :: exact,sl,thin,onecell
       end subroutine build_lattice_als
    end interface

type(layout), pointer:: ALS
real(dp) prec,closed_orbit(6)
real(dp) L,Kq,k0,mu_mod,beta,dmu,mu_x,circ

type(internal_state),target :: state
logical(lp) :: mis=.false.
type(c_damap)  one_turn_map, quasi_diagonal, diagonal,a_ac, id
type(c_normal_form) normal_form
integer :: pos =1
integer i,map_order,mf1,mfmap
type(probe) ray_closed
type(probe_8) ray
type(real_8) y(6)
complex(dp) g1,g2,al1,al2
!!!!!!!!!!!!!!!!!!!!!


c_verbose=.false.
prec=1.d-10 ! for printing
longprint=.false.

call ptc_ini_no_append
call append_empty_layout(m_u)
ALS=>m_u%start


call kanalnummer(mfmap,"maps.txt")
```


E. Forest, *From Tracking Code to Analysis*,
DOI 10.1007/978-4-431-55803-3

```
state= nocavity0 + modulation0 !

map_order=1
call init_all(state,map_order,0)


call alloc(y)
call alloc(one_turn_map,quasi_diagonal, diagonal,id,a_ac)
call alloc(normal_form)
call alloc(ray)


call build_lattice_als(ALS,mis,exact=.false.)

!!!! circ is the circumference of the ring !!!!
call get_length(als,circ)
!!!! AC_modulate.txt sets the magnet QF1 as a modulated magnet !!!!
 call kanalnummer(mf1,file="AC_modulation.txt")
  write(mf1,*) "select layout"
  write(mf1,*) 1
  write(mf1,*) " MODULATE"
  write(mf1,*) " BEND1"
  write(mf1,*) "1.d0 0 0        !DC_ac,A_ac,theta_ac"
  write(mf1,*) "1.d0   1        ! D_ac,n_ac  "
  write(mf1,*) "1 0.001d0 0       ! n d_bn(n) d_an(n)   "  ! (A)
  write(mf1,*) "0  0 0 "
  write(mf1,*) " return "
 close(mf1)
 call read_ptc_command77("AC_modulation.txt")

!!!! set a modulation clock !!!!!!
mu_mod=twopi*0.12345d0;
ray_closed%ac%om=mu_mod/circ         ! (B1)
ray_closed%ac%x=0.d0 ;               ! (B2)
write(6,*) " Modulation tune in radians =",circ*ray_closed%ac%om

closed_orbit=0.d0;                                                       ! (C)

call find_orbit_x(als,closed_orbit(1:6),STATE,1.e-8_dp,fibre1=pos)    ! (D)

ray_closed=closed_orbit      ! (E)

id=1;
! ray= closed orbit + identity map

ray=id+ray_closed;           ! (F)

call propagate(als,RAY,state,fibre1=pos)  ! (G)

 one_turn_map=ray                         ! (H)
 write(mfmap,*); write(mfmap,*) " Map produced by code " ; write(mfmap,*);
 call print(one_turn_map,mfmap)

do_linear_ac_longitudinal=.false.          ! (I)
call  c_normal(one_turn_map,normal_form)    ! (J1)
id=normal_form%a_t*from_phasor()
do_linear_ac_longitudinal=.true.
call  c_normal(one_turn_map,normal_form)    ! (J2)
normal_form%a_t=normal_form%a_t*from_phasor()

write(mfmap,*);write(mfmap,*);
write(mfmap,*) " Correct A (M=ARA^-1) from the algorithm for the code ";
write(mfmap,*);write(mfmap,*);

call  print(normal_form%a_t,mfmap,prec)
```

```
diagonal=normal_form%a_t**(-1)*one_turn_map*normal_form%a_t    ! (K1)

quasi_diagonal=id**(-1)*one_turn_map*id                        ! (K2)

g1=quasi_diagonal%v(6).sub.'00000010'                          ! (L1)
g2=quasi_diagonal%v(6).sub.'00000001'                          ! (L2)

al1=-g1/(1.d0-(quasi_diagonal%v(7).sub.'00000010'))            ! (L3)
al2=-g2/(1.d0-(quasi_diagonal%v(8).sub.'00000001'))            ! (L4)

a_ac=1

a_ac%v(6)=a_ac%v(6)+(al1.cmono.'00000010')+(al2.cmono.'00000001') ! (L5)

!!!!!!        Print all the resulting maps         !!!!!!

write(mfmap,*);write(mfmap,*);
write(mfmap,*) " Correct R (M=ARA^-1) from the algorithm for the code ";
write(mfmap,*);write(mfmap,*);


call  print(diagonal,mfmap,prec)

write(mfmap,*);write(mfmap,*);
write(mfmap,*) " Quasi-diagonal R (M=ARA^-1) from the algorithm for the code ";
write(mfmap,*);write(mfmap,*);

call  print(quasi_diagonal,mfmap,prec)

quasi_diagonal=a_ac**(-1)*quasi_diagonal*a_ac

write(mfmap,*);write(mfmap,*);
write(mfmap,*) " Quasi-diagonal should now be diagonal ";
write(mfmap,*);write(mfmap,*);


call  print(quasi_diagonal,mfmap,prec)

call kill(y)
call kill(one_turn_map,quasi_diagonal, diagonal,id,a_ac)
call kill(normal_form)
call kill(ray)

 close(mfmap)
call ptc_end(graphics_maybe=1,flat_file=.false.)

end program modulated_map
```

Appendix K
Program one_resonance_map

```
program one_resonance_map
use madx_ptc_module
use pointer_lattice
use c_TPSA
implicit none

    interface
       subroutine build_lattice_als(ALS,MIS,error,exact,sl,thin,onecell)
         use madx_ptc_module
         use pointer_lattice
         implicit none
         type(layout), target :: ALS
         logical(lp) mis
         real(dp),optional :: error(6)
         logical, optional :: exact,sl,thin,onecell
       end subroutine build_lattice_als
    end interface

type(layout), pointer:: ALS
real(dp) prec,closed_orbit(6)
real(dp) dmu,mdotmu,adotmu

type(internal_state),target :: state
logical(lp) :: mis=.false.
type(c_damap)  one_turn_map, three_turn_map,id,co_moving_map
type(c_normal_form) normal_form
type(c_vector_field) Fh,F2
type(c_taylor) h3,h3t,h2,h3c,jm,ja,jx,jy
integer :: pos =1, nind(11)
integer i,map_order,mf
type(probe) ray_closed
type(probe_8) ray
type(fibre), pointer :: p
character*48 :: command_gino
!!!!!!!!!!!!!!!!!!!!!


c_verbose=.false.
prec=1.d-8 ! for printing
use_info = .true.
longprint=.false.

state=only_2d0

call ptc_ini_no_append
call append_empty_layout(m_u)
```


E. Forest, *From Tracking Code to Analysis*,
DOI 10.1007/978-4-431-55803-3

```
ALS=>m_u%start

call build_lattice_als(ALS,mis,exact=.false.)

!!!!Fitting the tune to nu_x=0.334 !!!!
 call kanalnummer(mf,file="fit_tune.txt")
  write(mf,*) "select layout"
  write(mf,*) "  1  "
  write(mf,*) "set families"
  write(mf,*)  "2 "
  write(mf,*) "1 NO "
  write(mf,*) "QF"
  write(mf,*) " 2, 1 "
  write(mf,*) "1 NO "
  write(mf,*) "QD"
  write(mf,*) " 2, 2 "
  write(mf,*) "FITTUNE"
  write(mf,*) " 0.0000000001 "
  write(mf,*) " 0.334 , 0.2712345 "
  write(mf,*) "deallocate families"
  write(mf,*) " return "
 close(mf)
 p=>als%start; call move_to(ALS,p,"SF",pos);write(6,*) pos
 call add(p,3,0,1.d0);
 call read_ptc_command77("fit_tune.txt")

call kanalnummer(mf,"result_of_3nu_x.txt")

map_order=6
call init_all(state,map_order,0)

call alloc(one_turn_map, three_turn_map,id,co_moving_map)
call alloc(normal_form); call alloc(ray)
call alloc(Fh);call alloc(F2);call alloc(h3,h3t,h3c,h2);

closed_orbit=0.d0;
call find_orbit_x(als,closed_orbit(1:6),STATE,1.e-8_dp,fibre1=pos)

ray_closed=closed_orbit
id=1;
! ray= closed orbit + identity map
ray=id+ray_closed;

call propagate(als,RAY,state,fibre1=pos)

one_turn_map=ray                                          ! (1)
write(mf,*);write(mf,*)" Map produced by code ";write(mf,*);
call print(one_turn_map,mf,prec)

three_turn_map= one_turn_map**3         !3-turn map     ! (2)

call  c_normal(three_turn_map,normal_form)              ! (3)
write(mf,*);write(mf,*)"tune ",normal_form%tune(1);write(mf,*);

Fh=0
do i=1,normal_form%ker%n     ! Rotation so all exponents commute
 Fh=normal_form%ker%f(i)+Fh                               ! (4)
enddo

h3=(cgetpb(Fh)*to_phasor())*normal_form%a_t**(-1)      ! (5)

write(mf,*);write(mf,*)" Invariant of the 3-turn map";write(mf,*);
call print(h3,mf,prec)

Write(mf,*); Write(mf,*) " Checking that it is indeed invariant"
Write(mf,*) " h3*one_turn_map - h3  ";write(mf,*);
h3t=h3*one_turn_map
```

```
h3t=h3t-h3

call print(h3t,mf,prec)

write(mf,*);write(mf,*)" Invariant using the Logarithm ";write(mf,*);

Fh=log(three_turn_map)                                    ! (6)
h3=getpb(Fh)

call print(h3,mf,prec)

h3t=h3*one_turn_map
h3t=h3t-h3

Write(mf,*)
Write(mf,*) " Checking that it is indeed invariant"
Write(mf,*) " h3*one_turn_map - h3  ";write(mf,*);
call print(h3t,mf,prec)

normal_form%nres=0;normal_form%m=0;
do i=1,map_order+1                                        ! (7a)
  if(mod(i,3)==0) then
   normal_form%nres=normal_form%nres+1
   normal_form%m(1,normal_form%nres)=i
  endif
enddo

call  c_normal(one_turn_map,normal_form)                  ! (7b)

h2=(pi/3.d0)*((1.d0.cmono.1)**2+(1.d0.cmono.2)**2)       ! (8a)
F2=getvectorfield(h2)                                     ! (8b)

id=normal_form%a_t**(-1)*one_turn_map*normal_form%a_t ! (8c)
co_moving_map=exp(F2)*id                                  ! (8d)
Fh=log(co_moving_map)                                     ! (8e)
h3c=getpb(Fh)*normal_form%a_t**(-1)                       ! (8f)

write(mf,*); write(mf,*) " Invariant of the co-moving map "; write(mf,*);

call print(h3c,mf,prec)

h3t=h3c*one_turn_map
h3t=h3t-h3c

Write(mf,*)
Write(mf,*) " Checking that it is indeed invariant   "
Write(mf,*) " h3*one_turn_map - h3  ";write(mf,*);
call print(h3t,mf,prec)
call kill(one_turn_map, three_turn_map,id,co_moving_map)
call kill(normal_form);
call kill(ray)
call kill(Fh);
call kill(F2);
call kill(h3,h3t,h3c,h2);

!!!!Fitting the tune to nu_x=0.37123 nu_y=0.3135  !!!!
!!!!   nu_x +2 nu_y = 0.99823
 call kanalnummer(mf,file="fit_tune.txt")
  write(mf,*) "select layout"
  write(mf,*) "  1  "
  write(mf,*) "set families"
  write(mf,*)  "2 "
  write(mf,*) "1 NO "
  write(mf,*) "QF"
  write(mf,*) " 2, 1 "
  write(mf,*) "1 NO "
  write(mf,*) "QD"
```

```
  write(mf,*) " 2, 2 "
  write(mf,*) "FITTUNE"
  write(mf,*) " 0.0000000001 "
  write(mf,*) " 0.37123 , 0.3135 "
  write(mf,*) "deallocate families"
  write(mf,*) " return "
 close(mf)
 call read_ptc_command77("fit_tune.txt")
close(mf)

call kanalnummer(mf,"result_of_nu_x+2nu_y.txt")
write(mf,*)" Results of the nu_x +2 nu_y = 1 resonance";write(mf,*);

map_order=4
state=only_4d0

call init_all(state,map_order,0)


call alloc(one_turn_map, three_turn_map,id,co_moving_map)
call alloc(normal_form); call alloc(ray)
call alloc(Fh);call alloc(F2);
call alloc(h3,h3t,h3c,h2,jm,ja,jx,jy);

closed_orbit=0.d0;
call find_orbit_x(als,closed_orbit(1:6),STATE,1.e-8_dp,fibre1=pos)

ray_closed=closed_orbit
id=1;
! ray= closed orbit + identity map
ray=id+ray_closed;

call propagate(als,RAY,state,fibre1=pos)

one_turn_map=ray                                        ! (A)

 normal_form%m=0;
 normal_form%nres=2
 normal_form%m(1,1)=1 ; normal_form%m(2,1)=2;           ! (B1)
 normal_form%m(1,2)=2 ; normal_form%m(2,2)=4;           ! (B2)
 normal_form%positive=.false.

call  c_normal(one_turn_map,normal_form)                ! (C)

mdotmu=2*pi/5.d0                                                ! (D1)
adotmu=2*pi*(2*normal_form%tune(1)-normal_form%tune(2))/5.d0 ! (D2)

jx=(0.5d0.cmono.'2')+(0.5d0.cmono.'02')               ! (E1)
jy=(0.5d0.cmono.'002')+(0.5d0.cmono.'0002')           ! (E2)
jm=(jx+2*jy)                                          ! (E3)
ja=(2*jx-jy)                                          ! (E4)

h2=mdotmu*jm+adotmu*ja                                  ! (F1)
F2=getvectorfield(h2)                                   ! (F2)

id=normal_form%a_t**(-1)*one_turn_map*normal_form%a_t ! (F3)
co_moving_map=exp(F2,id)                                ! (F4)

Fh=log(co_moving_map)                                   ! (F5)
h3c=getpb(Fh)*normal_form%a_t**(-1)                     ! (F6)

h3t=h3c*one_turn_map
h3t=h3t-h3c                                             ! (G)

call print(h3c,mf,prec)
```

```
Write(mf,*)
Write(mf,*) " Checking that it is indeed invariant   "
Write(mf,*) " h3*one_turn_map - h3  ";write(mf,*);
call print(h3t,mf,prec)

close(mf)


call ptc_end(graphics_maybe=1,flat_file=.false.)

end program one_resonance_map
```

Appendix L
Program very_damped_map

```
program very_damped_map
use madx_ptc_module
use pointer_lattice
use c_TPSA
implicit none
type(internal_state),target :: state
type(c_damap)  m,a1,m_n,R_c
type(c_vector_field) T_prime,vf
type(c_taylor) K,x,p
type(c_normal_form) N,N_N
real(dp) prec,alpha,nu,radius,z(2),rad0,r1,r1p
complex(dp) zz(2)
integer map_order,io,km,ji(2),ns,mf,a,i,mf1
real(dp), allocatable :: co(:)
logical :: normalise=.false.

prec=1.d-13 ! for printing
longprint=.false.
c_verbose=.false.

state=only_2d0

write(6,*) " map order and alpha "
read(5,*) map_order, alpha
allocate(co(0:map_order/2))
call init_all(state,map_order,0)
call alloc(m,a1,m_n,R_c)
call alloc(T_prime);call alloc(vf);call alloc(K,x,p);
call alloc(N);call alloc(N_N)

nu=0.4433d0
x=1.d0.cmono.1                        !(1a)
p=1.d0.cmono.2                        !(1b)

 k = (1.0_dp+alpha-x**2)*(cos(twopi*nu)*x + sin(twopi*nu)*p)  ! (2a)
 p = cos(twopi*nu)*p - sin(twopi*nu)*x                         ! (2b)
 x = k                                             ! (2c)

M%v(1)=x                   ! (3a)
M%v(2)=p                   ! (3b)

call  c_normal(m,n)        ! (4)
```


E. Forest, *From Tracking Code to Analysis*,
DOI 10.1007/978-4-431-55803-3

```
if(normalise) then          ! (5a)
  remove_tune_shift=.true.
  m_n=n%a_t**(-1)*m*n%a_t
  call  c_normal(m_n,n_n)
  m_n=to_phasor()*m_n*from_phasor()
  a1= to_phasor()*n_n%a_t**(-1)*from_phasor()
  call  flatten_c_factored_lie(n_n%ker,vf)
  T_prime=a1*vf
else                        ! (5b)
  m_n=to_phasor()*n%a_t**(-1)*m*n%a_t*from_phasor()
  vf%v(1)=+i_*(1.d0.cmono.1)*(twopi*n%tune(1))
  vf%v(2)=-i_*(1.d0.cmono.2)*(twopi*n%tune(1))
  R_c=exp(vf,m_n)
  T_prime=log(R_c)
endif

radius=sqrt(4*alpha/(2*cos(twopi*nu)**2+1))  ! (6)
rad0=radius; write(6,*) " naive average radius = ",radius;

a=int(map_order/2)
do i=0,a; ji(1)=i+1;ji(2)=i;co(i)=T_prime%v(1).sub.ji;enddo; ! (7)

 radius=rad0**2    ! Newton search for equilibrium radius
do i=1,10
  r1=0;r1p=0;
 do io=0,a; r1 =co(io)*radius**io+r1; enddo;
 do io=1,a; r1p =io*co(io)*radius**(io-1)+r1p; enddo;              ! (8)
 radius=radius-r1/r1p; write(6,*) sqrt(radius);
enddo
radius=sqrt(radius)
write(6,*) map_order,"th order radius for alpha = ",alpha,"tune =",nu
write(6,*) "radius =" ,radius

call kanalnummer(mf,"plot.dat")
call kanalnummer(mf1,"naive.dat")
ns=1000
do i=1,ns
m%v(1)=radius*cos(twopi*i/ns)   ! results of perturbation theory              ! (9a)
m%v(2)=radius*sin(twopi*i/ns)
m=n%a_t.o.m !A_t(z) where z=(radius*cos(twopi*i/ns),radius*sin(twopi*i/ns)) ! (9b)
z(1)=m%v(1).sub.'0'
z(2)=m%v(2).sub.'0'
write(mf,*) z
 z(1)=rad0*cos(twopi*i/ns)   ! naive result
 z(2)=rad0*sin(twopi*i/ns)
write(mf1,*) z
enddo
close(mf)
close(mf1)

end program very_damped_map
```

Appendix M
Program spin_phase_advance_isf

```
program spin_phase_advance_isf
use madx_ptc_module
use pointer_lattice
implicit none

type(probe) xs0,xs1,XST
type(probe_8) xs
type(layout), pointer :: als
INTEGER MF,mft,mfisf,I,N,k,pos,no,kp,nturn,mfa
TYPE(FIBRE),POINTER:: P
type(internal_state) state
real(dp)  prec,cut,n_isf(3), closed(6), x(6), theta0
logical first
logical :: mis=.false.,thin=.false.
type(c_damap) c_map,c_spin0,U,U_c,D,f,A,b,R ,id_s,D_tilde
type(c_taylor) phase(3),nu_spin, fonction,fonction_FLOQUET,a12,a34
type(c_ray) cray
type(c_normal_form) c_n
TYPE(c_spinor) ISF,S_ISF,ISFoM,dISF,O
type(spinor) ISF_strobo   ! real spinor
type(c_vector_field) h_vector_field
type(c_taylor)  h_poisson_bracket
integer expo(4)
real(dp) DX_AVERAGE_DCS,betax_1,betax_2
    interface
       subroutine build_lattice_als(ALS,MIS,error,exact,sl,thin,onecell)
         use madx_ptc_module
         use pointer_lattice
         implicit none
         type(layout), target :: ALS
         logical(lp) mis
         real(dp),optional :: error(6)
         logical, optional :: exact,sl,thin,onecell
       end subroutine build_lattice_als
    end interface

!-----------------------------------

first=.true.;Lmax = 10.d0;use_info = .true.;prec=1.d-16;thin=.false.

call ptc_ini_no_append
call append_empty_layout(m_u)
ALS=>m_u%start
```


E. Forest, *From Tracking Code to Analysis*,
DOI 10.1007/978-4-431-55803-3

```
call build_lattice_als(ALS,mis,exact=.false.,thin=thin)

p=>als%start

!!!!!  Make the lattice linear by zeroing all sextupoles !!!!!!
!!!
courant_snyder_teng_edwards=.true.
time_lie_choice=.true.
p=>als%start
do i=1,als%n
 IF(P%MAG%P%NMUL>=3) THEN
  if(first) then
   CALL ADD(P,2,0,0.1D0)
   call make_it_knob(p%magp%bn(2),1) ! (0)
   first=.false.
  endif
 ENDIF

 p=>p%next
enddo

x(1:3)=1.d-5; x(6)=1.d-6; x(5)=1.d-3; x(4)=0.d0!
cut=4.d0;
call MESS_UP_ALIGNMENT(als,x,cut)  ! (1)
x=0.d0;DX_AVERAGE_DCS=0.d0;
closed=0.d0

state=nocavity0    ! (2)

CALL FIND_ORBIT(ALS,CLOSED,1,STATE,c_1d_5)  ! (3)

state=state+SPIN0

write(6,*) " If order>=4, the phase advance will jump by 100 positions"
write(6,*) " so the fractional tunes will be off"
write(6,*) " but canonisation results will be examined at position 100 "
write(6,*)
write(6,*) " with no=1,2,3 a full Twiss will be done "
write(6,*) " with the computation <x^2> thrown in if no>1 "
write(6,*);write(6,*);
write(6,*) " Give order no : 1,2,3,4,..."
read(5,*) no


call init_all(STATE,no,1)

call alloc(c_map)
call alloc(c_n)
call alloc(c_spin0)
CALL ALLOC(ISF); call alloc(S_ISF); call alloc(ISFoM);call alloc(dISF);call alloc(O);
call alloc(U,U_c,D,f,A,b,R,id_s,D_tilde)
call alloc(phase);call alloc(xs);
call alloc(nu_spin,fonction,fonction_FLOQUET,a12,a34,h_poisson_bracket)
call alloc(h_vector_field)

call kanalnummer(mft,"spin_twiss.txt")
if(thin) call kanalnummer(mfa,"analytical_x_average.txt")

!!!! Polymorphic probe is created in the usual manner
   XS0=CLOSED
   ID_S=1
   XS=XS0+ID_S
```

```
!!!! get spin polymorphic probe after one turn
CALL propagate(ALS,XS,+STATE,FIBRE1=1)  ! (4)
!!!!!!!!!!!!!!!!!!!!!!!!!!!!!!!!!!!!!!!!!!!!!!!!

! Copy probe_8 into a complex damap
c_map=XS ! (5)
!!!!!!!!!!!!!!!!!!!!!!!!!!!!!!!!!!!!!!!!!!!!!!!!

call c_normal(c_map,c_n,dospin=my_true)  ! (6)

 write(6,'(4(1x,g21.14))') c_n%tune(1:3), c_n%spin_tune

U=c_n%As*c_n%A_t    ! att=c_n%A_t*c_n%As

! id_s is a rotation
id_s=U**(-1)*c_map*U ! (7a)

! a trick to compute the fractional tunes and time slip
call c_full_canonise(id_s,U_c,D,f,A,b,R,phase,nu_spin) ! (7b)



write(mft,*);write(mft,*) " Fractional tune x"; write(mft,*);

call c_clean_taylor(phase(1),phase(1),prec)
call print(phase(1),mft)

write(mft,*);write(mft,*) " Fractional tune y"; write(mft,*);
call c_clean_taylor(phase(2),phase(2),prec)
call print(phase(2),mft)

if(state%nocavity) then
 write(mft,*);write(mft,*) "  Time "; write(mft,*);
 call c_clean_taylor(phase(3),phase(2),prec)
 call print(phase(3),mft)
endif

write(mft,*);write(mft,*) " Fractional tune spin"; write(mft,*);
call c_clean_taylor(nu_spin,nu_spin,prec)
call print(nu_spin,mft)

U=c_n%As*c_n%A_t ! Non-descript U exiting normal form ! (8a)

call c_full_canonise(U,U_c,D,F,A,b,R,phase,nu_spin) ! (8b)

phase(1)=0.d0 ;phase(2)=0.d0 ; ;phase(3)=0.d0; nu_spin=0.d0;

   XS=XS0+U_c   ! (8c)

p => als%start

do i=1,als%n

 CALL propagate(ALS,XS,+STATE,FIBRE1=i,fibre2=i+1)  ! (9a)

if((mod(i,100) == 0.or.i==als%n.or.i==1).or.no<=3) then
  if(mod(i,100)==0) write(6,*) " Position ",i

  xs0=xs ! Saving orbit  ! (9b)
  U=XS ! copying in map  ! (9c)

  ! U = U_c o  R = D o f o A o b o R
```

```
 call c_full_canonise(U,U_c,D,F,A,b,R,phase,nu_spin) ! (10)

 if(no>=4) then   ! (A0)  special interlude
   call kanalnummer(mf,"check_canonisation.txt")

   write(mf,*);Write(mf,*) " Time slip factor "; write(mf,*);
   h_vector_field=log(F)
   h_poisson_bracket=getpb(h_vector_field) ! (A1)
   write(mf,*);Write(mf,*) "Fixed point map"; write(mf,*);
   call print(F,mf,prec)
   write(mf,*);Write(mf,*) " Lie exponent of the fixed point map"; write(mf,*);
   call print(h_poisson_bracket,mf,prec)


   do k=1,4
     id_s%v(k)=F%v(k)-(1.d0.cmono.k) ! (A2)
   enddo
   h_poisson_bracket=h_poisson_bracket &  ! (A3)
  +(id_s%v(1)*(1.d0.cmono.2)-id_s%v(2)*(1.d0.cmono.1)) &
  +(id_s%v(3)*(1.d0.cmono.4)-id_s%v(4)*(1.d0.cmono.3))
   write(mf,*);Write(mf,*) "  Comparing with the canonical form "; write(mf,*);
   call print(h_poisson_bracket,mf,prec)

  expo=0;expo(2)=1;
  a12=A%v(1).par.expo   ! (B1)
  expo=0;expo(4)=1;
  a34=A%v(3).par.expo   ! (B2)
   write(mf,*); Write(mf,*) " Checking Courant-Snyder-Teng-Edwards "
   write(mf,*);Write(mf,*) " A_12 should be zero "; write(mf,*);
   call print(a12,mf,prec)
   write(mf,*);Write(mf,*) " A_34 should be zero "; write(mf,*);
   call print(a34,mf,prec)

   h_vector_field=log(b)    ! (C1)
   h_poisson_bracket=getpb(h_vector_field) ! (C2)
   h_poisson_bracket=h_poisson_bracket*from_phasor()  ! (C3)

   write(mf,*);Write(mf,*) " Lie exponent of the nonlinear part "; write(mf,*);
   call print(h_poisson_bracket,mf,prec)

  U_c=f*A*b
  D_tilde=to_phasor()*U_c**(-1)*D*U_c*from_phasor()  ! (D1)
  O=log(D_tilde%s) ! (D2)

   write(mf,*);Write(mf,*) " Vertical spinor O_y of the canonised D~ "; write(mf,*);
   call print(O%v(2),mf,prec)
 close(mf)
  endif

  !!!!!!!!!!!!!!!   doing something  !!!!!!!!!!!!!!!!!

if(no>1) then
  fonction =2*(1.d0.cmono.1)**2   ! 2*x**2  (Ea)
  call C_AVERAGE(fonction,U_c,fonction_FLOQUET) ! (Eb)
endif
 write(mft,*) "position, Element ", i, p%mag%name

 betax_1=(U_c%v(1).sub.'1000')**2+(U_c%v(1).sub.'0100')**2 ! (Ec)
 betax_2=(U_c%v(1).sub.'0010')**2+(U_c%v(1).sub.'0001')**2 ! (Ed)
 write(mft,*) " Ripken Beta_x_1 Beta_x_2 ",betax_1,betax_2
 write(mft,*) " 2< x^2 > "
if(no>1) then
  call print(fonction_FLOQUET,mft)
```

```
 else
  write(mft,*) "Unfortunately no information if no = 1"
 endif


ISF=2   ! (Fa)
ISF=D%s*ISF ! (Fb)

Write(mft,*) " ISF vector n "
call print(ISF,mft)

write(mft,*) " phase x"
call c_clean_taylor(phase(1),phase(1),prec)
call print(phase(1),mft)

write(mft,*) " phase y"
call c_clean_taylor(phase(2),phase(2),prec)
call print(phase(2),mft)


if(state%nocavity) then
 write(mft,*) "  Time "
 call c_clean_taylor(phase(3),phase(3),prec)
 call print(phase(3),mft)
endif

write(mft,*) " phase spin"
call c_clean_taylor(nu_spin,nu_spin,prec)
call print(nu_spin,mft)
!!!!!!!!!!!!!!!!!!!!!!!!!!!!!!!!!!!!!!!!!!!!!!!!!!!!!

If(thin) then
 if(p%mag%name(1:2)=="SF".or.p%mag%name(1:2)=="SD") then
  DX_AVERAGE_DCS=(betax_1)**1.5_DP*p%mag%BN(3)/4.0_DP &    ! (Fa)
    *(-SIN(PHASE(1)*TWOPI)+SIN((PHASE(1)-c_n%TUNE(1))*TWOPI)) &
    /(1.0_DP-COS(c_n%TUNE(1)*TWOPI)) + DX_AVERAGE_DCS
 endif
endif
write(mft,*) "!!!!!!!!!!!!!!!!!!!!!!!!!!!!!!!!!!!!!!!!!!!!!!!!!!!!!"

XS=XS0+U_c ! (11)
endif

p=>p%next
enddo

if(thin) then ;

 DX_AVERAGE_DCS=DX_AVERAGE_DCS*SQRT(betax_1) ! (Fb)
 write(mfa,'(a11,F20.13,a20)')' d<x>/dCS ', dx_average_dCS, " < ---- analytical  "

 if(no==2.or.no==3) then

 fonction =(1.d0.cmono.1)  ! x
  write(mfa,*);write(mfa,*) "Full <x> "
  call C_AVERAGE(fonction,U_c,fonction_FLOQUET) ! (Fc)
  call print(fonction_FLOQUET,mfa)
  write(mfa,*);write(mfa,*) "Full x-dispersion  "
  call print(F%v(1),mfa)
 endif
endif

call kanalnummer(mfisf,"checking_isf.txt")
```

```
write(mfisf,*) "!!!!  Exploring the ISF  at the end of the lattice !!!!"

Write(mfisf,*) " Testing Barber's Equation S ISF = ISF o m "

S_ISF = c_map%s*ISF    !  (12a)
ISFoM = ISF*c_map      !  (12b)

Write(mfisf,*) "  |S ISF-  ISF o m|/ |S ISF| "

do i=1,3
dISF%v(i)=S_ISF%v(i)-ISFoM%v(i)
write(mfisf,*) i,full_abs(dISF%v(i)),full_abs(dISF%v(i))/full_abs(S_ISF%v(i)) ! (12c)
enddo

x=0.d0
x(1)=0.001d0 ;x(3)=0.001d0 ;     ! (13a)
xs1=closed+x

xst=0
cray%x=0.d0
cray%x(1:6)=x
do i=1,3
 n_isf(i) = ISF%v(i).o.cray  ! (13b)
enddo

Write(6,*) "  Stroboscopic Average 5000 turns : patience "
nturn=5000
kp=1000
call stroboscopic_average(als,xs1,xst,1,STATE,nturn,kp,ISF_strobo,mfisf) ! (14c)

Write(mfisf,*) "  Stroboscopic Average "

write(mfisf,*);
write(mfisf,'(a19,4(1x,g20.13),a19,i4)') " ISF  for x(1:4) = " &
,x(1:4), " number of turns = ", nturn
write(mfisf,'(a24,3(1x,g20.13))') " Stroboscopic average ",ISF_strobo
write(mfisf,'(a24,3(1x,g20.13))') " From the normal form ",n_isf
write(mfisf,'(a4,20x,3(1x,g20.13))')" n0 ",  real(ISF%v(1).sub.'0'), &
real(ISF%v(2).sub.'0'),real(ISF%v(3).sub.'0')

close(mfisf)
close(mft)
if(thin) close(mfa)

 call ptc_end(graphics_maybe=1,flat_file=.false.)

end program spin_phase_advance_isf
```

Appendix N
Program hamitonian_guignard_cs.f90

```
program Guignard_Hamiltonian_cs
use madx_ptc_module
use pointer_lattice
use c_TPSA
implicit none

    interface
       subroutine build_lattice_als(ALS,MIS,error,exact,sl,thin,onecell)
         use madx_ptc_module
         use pointer_lattice
         implicit none
         type(layout), target :: ALS
         logical(lp) mis
         real(dp),optional :: error(6)
         logical, optional :: exact,sl,thin,onecell
       end subroutine build_lattice_als
    end interface

type(layout), pointer:: ALS
real(dp) prec,closed_orbit(6)
real(dp) circ,ds,s,ds_ave,intp(2)

type(internal_state),target :: state
logical(lp) :: mis=.false.
type(c_damap)  one_turn_map,id_s,U_c,D,f,A,b,R,U
type(c_normal_form) normal_form
type(c_vector_field) Gh
type(c_taylor) h
type(c_taylor), allocatable :: hn(:)
integer :: pos =1
integer i,map_order,mf,k,ns,n_mode
type(probe) ray_closed
type(probe_8) ray
type(fibre), pointer :: p
type(integration_node), pointer :: t
character*48 :: command_gino
logical int_step,used_ds_ave,asym
integer icase
real(dp), allocatable:: theta(:),bet(:),ht(:)
real(dp) hv
!!!!!!!!!!!!!!!!!!!!!


c_verbose=.false.
prec=1.d-6 ! for printing
use_info = .true.
```


E. Forest, *From Tracking Code to Analysis*,
DOI 10.1007/978-4-431-55803-3

```
longprint=.false.

Write(6,*) " Random errors -> t, no errors -> f"
read(5,*) mis

1 write(6,*) " Choose the state "
if(.not.mis) write(6,*) "only_2d0  -> 1-d-f map if no errors -> type 1"
write(6,*) "only_4d0  -> 2-d-f map               -> type 2"
write(6,*) "delta0    -> 2-d-f map + delta       -> type 3"
write(6,*) "nocavity0 -> 3-d-f map if no cavity -> type 4"
write(6,*) "default0  -> 3-d-f map if    cavity -> type 5   <- SLOW "
read(5,*) i

    select case(i)
    case(1)
      state=only_2d0
    case(2)
      state=only_4d0
    case(3)
      state=delta0
    case(4)
      state=nocavity0
    case(5)
      state=default0
    case default
    write(6,*) "Choose between 1 to 5"
    goto 1
    end select



call ptc_ini_no_append
call append_empty_layout(m_u)
ALS=>m_u%start
call build_lattice_als(ALS,mis,exact=.false.)

!!!!Fitting the tune and controlling the step size  !!!!
 call kanalnummer(mf,file="fit_tune.txt")
  write(mf,*) "select layout"
  write(mf,*) "  1  "
  write(mf,*) "L MAX  "
  write(mf,*) "0.3 "      ! (1) all ds of the order  0.3 metre
  write(mf,*) "CUTTING ALGORITHM "
  write(mf,*) "2 "                 ! Drifts are cut as well
  write(mf,*) "LIMIT FOR CUTTING "
  write(mf,*) "10000 10000 "
  write(mf,*) "THIN LENS "
  write(mf,*) "1000.d0 "
  write(mf,*) "set families"
  write(mf,*)  "2 "
  write(mf,*) "1 NO "
  write(mf,*) "QF"
  write(mf,*) " 2, 1 "
  write(mf,*) "1 NO "
  write(mf,*) "QD"
  write(mf,*) " 2, 2 "
  write(mf,*) "FITTUNE"
  write(mf,*) " 0.0000000001 "
  write(mf,*) " 0.3678 , 0.2712345 "
  write(mf,*) "deallocate families"
  write(mf,*) " return"
 close(mf)

write(6,*) " Do you want to break the symmetry of the lattice "
write(6,*) " by mispowering a single sextupole ?  "
write(6,*) " Yes -> t    No -> f  "
read(5,*) asym
```

```
if(asym) then
 p=>als%start; call move_to(ALS,p,"SF",pos);  ! (2a)
 call add(p,3,1,10d0); !call add(p,2,1,.01d0);    ! (2b)
endif

 call read_ptc_command77("fit_tune.txt")

 call MAKE_NODE_LAYOUT(als)

call kanalnummer(mf,"guignard_hamiltonian.txt")

write(6,*) " int_step t or f "
read(5,*) int_step      ! (3)
write(6,*) " Uniform ds -> t "
read(5,*) used_ds_ave   ! (4)

write(6,*) " case 1,2  "
write(6,*) " case = 1 -> linear transformation "
write(6,*) " case = 2 -> full nonlinear transformation: not interesting "
read(5,*) icase         ! (5)

write(6,*) " Number of Fourier modes: make it 12 or more "
read(5,*) n_mode
allocate(hn(0:n_mode))     ! (6)

map_order=4                ! (7)

call init_all(state,map_order,0)

call alloc(one_turn_map, id_s,U_c,D,f,A,b,R,U)
call alloc(normal_form); call alloc(ray);
call alloc(Gh);
call alloc(hn); call alloc(h);


closed_orbit=0.d0;
call find_orbit_x(als,closed_orbit(1:6),STATE,1.e-8_dp,fibre1=1)

ray_closed=closed_orbit
id_s=1;
ray=id_s+ray_closed;

call propagate(als,RAY,state,fibre1=1)   ! (8a)

one_turn_map=ray                         ! (8b)

call c_normal(one_turn_map,normal_form)  ! (8c)

call  c_canonise(normal_form%A_t ,U_c,f,A,b) ! (9a)

id_s=U_c                                     ! (9b)


if(used_ds_ave) then
 circ=twopi
else
 call GET_LENGTH(als,circ)
endif

if(int_step) then          ! (10)
 ns=als%t%n
else
 ns=als%n
endif

allocate(bet(0:ns),theta(0:ns),ht(0:ns))
```

```
bet=0.d0
theta=0.d0
ds_ave=twopi/ns              ! (11)
ds=0.d0

if(icase==1) U_c=U_c.cut.2

s=0.d0
ray=U_c+ray_closed;         ! (12)

p=>als%start                ! (13a)
t=>p%t1                     ! (13b)

do i=1,ns

if(mod(i,100)==0) then
  write(6,*) ns-i, " steps remaining "
endif

if(used_ds_ave) then
   ds=ds_ave                 ! (14a)
else
  if(int_step) then
   ds=twopi*t%s(5)/circ      ! (14b)
  else
   ds=twopi*p%mag%p%ld/circ ! (14c)
  endif
endif


if(int_step) then
 call propagate(als,RAY,state,node1=i,node2=i+1)    ! (15a)
else
 call propagate(als,RAY,state,fibre1=i,fibre2=i+1) ! (15b)
endif

  ray_closed=ray ! Saving orbit
  U=ray ! copying in map

  U_c=U

bet(i)=1.d0/((u_c%v(1).sub.'10')**2+(u_c%v(1).sub.'01')**2) ! (15c)


if(icase==1) U_c=U_c.cut.2       ! (16a)
call  c_canonise(U_c,U_c,f,A,b) ! (16b)
U=U_c**(-1)*U  ! (16c)

  Gh=log(U)       ! (17a)
  h=getpb(Gh)     ! (17b)


! Checking convergence of the logarithm
a=exp(-(Gh.cut.2),(U.sub.1))   ! (18)
do k=1,c_%nd2
 if(abs(full_abs(a%v(k))-1)>1.d-5) then
    call print(a,6)
    write(6,*);write(6,*) "Log failed at element ",i, p%mag%name
    stop
 endif
enddo

s=s+ds;  theta(i)=s
do k=0,n_mode
 hn(k)=hn(k)-exp(-i_*k*theta(i))*h/circ ! (19)
enddo
```

```
ray=ray_closed+U_c

if(int_step) then   ! (20)
 t=>t%next
 p=>t%parent_fibre
else
 p=>p%next
endif

enddo

if(asym) then
 write(mf,*) " Results for an asymmetric ring "
else
 write(mf,*) " Results for 12-fold symmetric ring "
endif
if(.not.used_ds_ave) then
write(mf,*) " (Circumference/pi) x Hamitonian of the ring Guignard style"
else
 write(mf,*) " 1/pi x Hamitonian of the ring Guignard style"
endif
write(mf,*) "  "

do k=0,n_mode
write(mf,*);write(mf,*) k;write(mf,*);
hn(k)=hn(k)*from_phasor()                    ! (21a)
call print((circ/pi)*hn(k),mf,prec)          ! (21b)

enddo

do i=0,ns

hv=(hn(0).sub.'11')
do k=1,n_mode
 hv=hv + 2*(hn(k).sub.'11')*exp(i_*k*theta(i))  ! (22)
enddo


 ht(i) = hv
enddo




write(mf,*) " Tracked Canonical Transformation"
write(mf,*) "  "
call print(U_c,mf,prec)
write(mf,*) " Original Canonical Transformation "
write(mf,*) "  "
call print(id_s,mf,prec)

write(mf,*) " Products with inverse    "
write(mf,*) "  "

u_c=u_c*id_s**(-1)            ! (23)

call print(u_c,mf,prec)

close(mf)

if(.not.used_ds_ave) then
 call kanalnummer(mf,"plot_linear_H.dat")

 do i=0,ns
  write(mf,*) theta(i),bet(i),ht(i)*2  ! 2J= (x + i p)*(x-i p) (24)
 enddo
```

```
close(mf)
endif

call ptc_end(graphics_maybe=1,flat_file=.false.)

end program Guignard_Hamiltonian_cs
```

Appendix O
Program hamitonian_guignard.f90

```
program Guignard_Hamiltonian
use madx_ptc_module
use pointer_lattice
use c_TPSA
implicit none

    interface
       subroutine build_lattice_als(ALS,MIS,error,exact,sl,thin,onecell)
         use madx_ptc_module
         use pointer_lattice
         implicit none
         type(layout), target :: ALS
         logical(lp) mis
         real(dp),optional :: error(6)
         logical, optional :: exact,sl,thin,onecell
       end subroutine build_lattice_als
    end interface

type(layout), pointer:: ALS
real(dp) prec,closed_orbit(6)
real(dp) circ,ds,s,ds_ave,intp(3)

type(internal_state),target :: state
logical(lp) :: mis=.false.
type(c_damap)  one_turn_map,id_s,U_c,D,f,A,b,R,U
type(c_normal_form) normal_form
type(c_vector_field) Gh,f_lin,f_non,h_left,f_co_moving
type(c_taylor) h
type(c_taylor), allocatable :: hn(:)
integer :: pos =1
integer i,map_order,mf,k,ns,n_mode
type(probe) ray_closed
type(probe_8) ray
type(fibre), pointer :: p
type(integration_node), pointer :: t
character*48 :: command_gino
logical int_step,used_ds_ave,asym
integer icase
real(dp) hv
!!!!!!!!!!!!!!!!!!!!!


c_verbose=.false.
prec=1.d-6 ! for printing
use_info = .true.
longprint=.false.
```


E. Forest, *From Tracking Code to Analysis*,
DOI 10.1007/978-4-431-55803-3

```
Write(6,*) " Random errors -> t, no errors -> f"
read(5,*) mis

1 write(6,*) " Choose the state "
if(.not.mis) write(6,*) "only_2d0  -> 1-d-f map if no errors -> type 1"
write(6,*) "only_4d0  -> 2-d-f map                -> type 2"
write(6,*) "delta0    -> 2-d-f map + delta        -> type 3"
write(6,*) "nocavity0 -> 3-d-f map if no cavity -> type 4"
write(6,*) "default0  -> 3-d-f map if    cavity -> type 5   <- SLOW "
read(5,*) i

    select case(i)
    case(1)
      state=only_2d0
    case(2)
      state=only_4d0
    case(3)
      state=delta0
    case(4)
      state=nocavity0
    case(5)
      state=default0
    case default
    write(6,*) "Choose between 1 to 5"
    goto 1
    end select

call ptc_ini_no_append
call append_empty_layout(m_u)
ALS=>m_u%start

call build_lattice_als(ALS,mis,exact=.false.)

!!!!Fitting the tune and controlling the step size  !!!!
 call kanalnummer(mf,file="fit_tune.txt")
  write(mf,*) "select layout"
  write(mf,*) "  1  "
  write(mf,*) "L MAX  "
  write(mf,*) "0.3 "      ! (1)     ! all ds < 0.3
  write(mf,*) "CUTTING ALGORITHM "
  write(mf,*) "2 "                  ! Drifts are cut as well
  write(mf,*) "LIMIT FOR CUTTING "
  write(mf,*) "10000 10000 "
  write(mf,*) "THIN LENS "
  write(mf,*) "1000.d0 "
  write(mf,*) "set families"
  write(mf,*)  "2 "
  write(mf,*) "1 NO "
  write(mf,*) "QF"
  write(mf,*) " 2, 1 "
  write(mf,*) "1 NO "
  write(mf,*) "QD"
  write(mf,*) " 2, 2 "
  write(mf,*) "FITTUNE"
  write(mf,*) " 0.0000000001 "
  write(mf,*) " 0.3678 , 0.2712345 "
  write(mf,*) "deallocate families"
  write(mf,*) " return"
 close(mf)

write(6,*) " Do you want to break the symmetry of the lattice "
write(6,*) " by mispowering a single sextupole ?  "
write(6,*) " Yes -> t    No -> f  "
read(5,*) asym
```

```
if(asym) then
 p=>als%start; call move_to(ALS,p,"SF",pos);  ! (2a)
 call add(p,3,1,10d0); !call add(p,2,1,.01d0);     ! (2b)
endif

 call read_ptc_command77("fit_tune.txt")

 call MAKE_NODE_LAYOUT(als)

call kanalnummer(mf,"guignard_hamiltonian.txt")

write(6,*) " int_step t or f "
read(5,*) int_step      ! (3)
write(6,*) " Uniform ds -> t "
read(5,*) used_ds_ave   ! (4)

write(6,*) " case 1,2  "
write(6,*) " case = 1 -> linear transformation "
write(6,*) " case = 2 -> full nonlinear transformation: not interesting "
read(5,*) icase         ! (5)


write(6,*) " Number of Fourier modes: make it 12 or more "
read(5,*) n_mode
allocate(hn(0:n_mode))     ! (6)



map_order=4                ! (7)
call init_all(state,map_order,0)

call alloc(one_turn_map, id_s,U_c,D,f,A,b,R,U)
call alloc(normal_form); call alloc(ray);
call alloc(f_non);call alloc(f_lin)
call alloc(Gh); call alloc(h_left)
call alloc(hn); call alloc(h);call alloc(f_co_moving);


closed_orbit=0.d0;
call find_orbit_x(als,closed_orbit(1:6),STATE,1.e-8_dp,fibre1=1)

ray_closed=closed_orbit
id_s=1;
ray=id_s+ray_closed;

call propagate(als,RAY,state,fibre1=1)    ! (8a)

one_turn_map=ray                          ! (8b)

call c_normal(one_turn_map,normal_form)  ! (8c)
U=normal_form%A_t    ! att=c_n%A_t*c_n%As

intp=0
if(normal_form%nres==0) then
 intp(1)=14
 intp(2)=8
 intp(3)=0
endif

! id_s is a rotation
id_s=U**(-1)*one_turn_map*U ! (8d)

call extract_linear_from_normalised(id_s,b,a,f_lin,f_non,intp) ! (8e)

call  c_canonise(normal_form%A_t ,U_c,f,A,b) ! (9a)
id_s=U_c                                     ! (9b)
```

```
 if(icase==1) then
   h_left=f_lin
 else
   h_left=f_non+f_lin
 endif
h_left=(1.d0/twopi)*h_left

if(used_ds_ave) then
 circ=twopi
else
 call GET_LENGTH(als,circ)
endif

if(int_step) then            ! (10)
 ns=als%t%n
else
 ns=als%n
endif

ds_ave=twopi/ns              ! (11)
ds=0.d0

if(icase==1) U_c=U_c.cut.2

s=0.d0
ray=U_c+ray_closed;         ! (12)

p=>als%start                ! (13a)
t=>p%t1                     ! (13b)

do i=1,ns

if(mod(i,100)==0) then
  write(6,*) ns-i, " steps remaining "
endif

if(used_ds_ave) then
   ds=ds_ave                 ! (14a)
else
  if(int_step) then
   ds=twopi*t%s(5)/circ      ! (14b)
  else
   ds=twopi*p%mag%p%ld/circ ! (14c)
  endif
endif

if(int_step) then
 call propagate(als,RAY,state,node1=i,node2=i+1)   ! (15a)
else
 call propagate(als,RAY,state,fibre1=i,fibre2=i+1) ! (15b)
endif

  ray_closed=ray ! Saving orbit  ! (16a)
  U=ray ! copying in map  ! (16b)

  U_c=exp(-ds*h_left,U)         !  (16c)
   if(icase==1) U_c=U_c.cut.2  !  (16d)
  U=U_c**(-1)*U                 !  (16e)

  Gh=log(U)       ! (17a)
  h=getpb(Gh)     ! (17b)


! Checking convergence of the logarithm
a=exp(-(Gh.cut.2),(U.sub.1))  ! (18)
do k=1,c_%nd2
 if(abs(full_abs(a%v(k))-1)>1.d-5) then
```

```
    call print(a,6)
    write(6,*);write(6,*) "Log failed at element ",i, p%mag%name
    stop
 endif
enddo

s=s+ds;
do k=0,n_mode
 hn(k)=hn(k)-exp(-i_*k*s)*h/circ ! (19)
enddo

ray=ray_closed+U_c

if(int_step) then   ! (20)
 t=>t%next
 p=>t%parent_fibre
else
 p=>p%next
endif

enddo

if(asym) then
 write(mf,*) " Results for an asymmetric ring "
else
 write(mf,*) " Results for 12-fold symmetric ring "
endif
if(.not.used_ds_ave) then
write(mf,*) " (Circumference/pi) x Hamitonian of the ring Guignard style"
else
write(mf,*) " 1/pi x Hamitonian of the ring Guignard style"
endif
write(mf,*) "  "

do k=0,n_mode
write(mf,*);write(mf,*) k;write(mf,*);
hn(k)=hn(k)*from_phasor()                  ! (21a)
call print((circ/pi)*hn(k),mf,prec)        ! (21b)
enddo


write(mf,*) " Tracked Canonical Transformation"
write(mf,*) "  "
call print(U_c,mf,prec)
write(mf,*) " Original Canonical Transformation "
write(mf,*) "  "
call print(id_s,mf,prec)

write(mf,*) " Products with inverse   "
write(mf,*) "  "

u_c=u_c*id_s**(-1)           ! (23)

call print(u_c,mf,prec)

close(mf)

call ptc_end(graphics_maybe=1,flat_file=.false.)

end program Guignard_Hamiltonian
```

Appendix P
Program hamiltonian_guignard_1df.f90

```
program Guignard_normal_form
use madx_ptc_module
use pointer_lattice
use c_TPSA
implicit none

    interface
       subroutine build_lattice_als(ALS,MIS,error,exact,sl,thin,onecell)
         use madx_ptc_module
         use pointer_lattice
         implicit none
         type(layout), target :: ALS
         logical(lp) mis
         real(dp),optional :: error(6)
         logical, optional :: exact,sl,thin,onecell
       end subroutine build_lattice_als
    end interface

type(layout), pointer:: ALS
real(dp) prec,closed_orbit(6)
real(dp) circ,ds,s,ds_ave,intp(3),ts

type(internal_state),target :: state
logical(lp) :: mis=.false.
type(c_damap)  one_turn_map,id_s,U_c,U,A,fi,b
type(c_normal_form) normal_form
type(c_vector_field) logN,f_lin,f_non,h_left
integer :: pos =1
integer i,map_order,mf,ns,n_mode,km
type(probe) ray_closed
type(probe_8) ray
type(fibre), pointer :: p
type(integration_node), pointer :: t
character*48 :: command_gino
logical int_step,used_ds_ave,asym
integer icase
type(c_vector_field_fourier) G,F,K
!!!!!!!!!!!!!!!!!!!!!


c_verbose=.false.
prec=1.d-6 ! for printing
```


E. Forest, *From Tracking Code to Analysis*,
DOI 10.1007/978-4-431-55803-3

```
use_info = .true.
longprint=.false.

mis=.false.
state=only_2d0


call ptc_ini_no_append
call append_empty_layout(m_u)
ALS=>m_u%start

call build_lattice_als(ALS,mis,exact=.false.,onecell=.true.) ! (0)

!!!!Fitting the tune and controlling the step size  !!!!
 call kanalnummer(mf,file="fit_tune.txt")
  write(mf,*) "select layout"
  write(mf,*) "  1  "
  write(mf,*) "L MAX  "
  write(mf,*) "0.3 "      ! (1)     ! all ds < 0.3
  write(mf,*) "CUTTING ALGORITHM "
  write(mf,*) "2 "                  ! Drifts are cut as well
  write(mf,*) "LIMIT FOR CUTTING "
  write(mf,*) "10000 10000 "
  write(mf,*) "THIN LENS "
  write(mf,*) "1000.d0 "
  write(mf,*) " return"
 close(mf)

 p=>als%start; call move_to(ALS,p,"SF",pos);  ! (2)
 call add(p,3,1,10d0);


 call read_ptc_command77("fit_tune.txt")

 call MAKE_NODE_LAYOUT(als)

call kanalnummer(mf,"guignard_hamiltonian.txt")

int_step=.true.
used_ds_ave=.true.
icase=1

write(6,*) " Enter number of Fourier modes"
read(5,*) n_mode
n_fourier=n_mode

map_order=4                ! (3)
call init_all(state,map_order,0)

call alloc(one_turn_map, id_s,U_c,A,U,fi,b)
call alloc(normal_form); call alloc(ray);
call alloc(f_non);call alloc(f_lin)
call alloc(logN); call alloc(h_left)

call alloc(G);call alloc(F);call alloc(K);


closed_orbit=0.d0;
call find_orbit_x(als,closed_orbit(1:6),STATE,1.e-8_dp,fibre1=1)
```

```
ray_closed=closed_orbit
id_s=1;
ray=id_s+ray_closed;

call propagate(als,RAY,state,fibre1=1)    ! (4)

one_turn_map=ray                          ! (5)

call c_normal(one_turn_map,normal_form)   ! (6)
U=normal_form%A_t     ! att=c_n%A_t*c_n%As

intp=0
intp(1)=1

! id_s is a rotation
id_s=U**(-1)*one_turn_map*U ! (7)

call extract_linear_from_normalised(id_s,b,a,f_lin,f_non,intp) ! (8)

call  c_canonise(normal_form%A_t ,U_c,fi,A,b) ! (9)


h_left=f_lin
h_left=(1.d0/twopi)*h_left
f_lin= (1.d0/twopi)*(from_phasor()*f_lin)
f_non= (1.d0/twopi)*(from_phasor()*f_non)    ! (10)
prec=1.d-10
write(mf,*); write(mf,*) " Results of one-turn map Normalisation"
write(mf,*) " ____________________________lin__________________________________"
call print(f_lin,mf,prec)
write(mf,*) " ____________________________non__________________________________"
call print(f_non,mf,prec)
write(mf,*) " ______________________________________________________________";
write(mf,*);

circ=twopi
ns=als%t%n
ds_ave=twopi/ns          ! (11)
ds=ds_ave                !

U_c=U_c.cut.2

s=0.d0
ray=U_c+ray_closed;        ! (12)

p=>als%start               ! (13a)
t=>p%t1                    ! (13b)

do i=1,ns

 if(mod(i,1000)==0) then
   write(6,*) ns-i, " steps remaining "
 endif

 call propagate(als,RAY,state,node1=i,node2=i+1)    ! (14)


  ray_closed=ray ! Saving orbit  ! (15a)
  U=ray ! copying in map  ! (15b)

  U_c=exp(-ds*h_left,U)        !  (15c)
```

```
  U_c=U_c.cut.2                    !  (15d)
  U=U_c**(-1)*U                    !  (15e)

  logN=log(U)       ! (16)

! Checking convergence of the logarithm
a=exp(-(logN.cut.2),(U.sub.1))  ! (17)
do km=1,c_%nd2
 if(abs(full_abs(a%v(km))-1)>1.d-5) then
    call print(a,6)
    write(6,*);write(6,*) "Log failed at element ",i, p%mag%name
    stop
 endif
enddo

s=s+ds;
do km=-n_mode,n_mode
 G%f(km)=G%f(km)+(exp(-i_*km*s)/circ)*logN ! (18)
enddo

 ray=ray_closed+U_c                      ! (19)

 t=>t%next
 p=>t%parent_fibre
enddo

U=from_phasor()
call transform_vector_field_fourier_by_map(G,G,u) ! (20)

prec=1.d-5
write(mf,*); write(mf,*) " Results of Guignard Normalisation"
write(mf,*) " One exponent k=0 with",n_fourier, "modes"
call normalise_vector_field_fourier(G,F,K)     ! (21)
call c_clean_vector_field_fourier(K,K,prec)
call print(K%f(0),mf)

write(mf,*) " Factored k=0 with",n_fourier, "modes"
K=G
call normalise_vector_field_fourier_factored(k)     ! (22)
call c_clean_vector_field_fourier(K,K,prec)
call print(K%f(0),mf)

close(mf)

! call kanalnummer(mf,file="plot.dat")
! do i=0,n_mode
!  write(6,*) " doing mode",i
! k=g
! n_fourier=i
! call normalise_vector_field_fourier_factored(K)
! ts=imag(K%f(0)%v(1).sub.'21')
! write(mf,*) i, ts
! enddo
!close(mf)

call ptc_end(graphics_maybe=1,flat_file=.false.)

end program Guignard_normal_form
```

Appendix Q
Program hamiltonian_guignard_1df_x.f90

```
program Guignard_normal_form_average_x
use madx_ptc_module
use pointer_lattice
use c_TPSA
implicit none

    interface
       subroutine build_lattice_als(ALS,MIS,error,exact,sl,thin,onecell)
         use madx_ptc_module
         use pointer_lattice
         implicit none
         type(layout), target :: ALS
         logical(lp) mis
         real(dp),optional :: error(6)
         logical, optional :: exact,sl,thin,onecell
       end subroutine build_lattice_als
    end interface

type(layout), pointer:: ALS
real(dp) prec,closed_orbit(6)
real(dp) circ,ds,s,ds_ave,intp(3),ts
type(internal_state),target :: state
logical(lp) :: mis=.false.
type(c_taylor) fonction,fonction_FLOQUET,phase(3)
type(c_damap)  one_turn_map,id_s,U_c,U,A,fi,b,a_cs
type(c_normal_form) normal_form
type(c_vector_field) logN,f_lin,f_non,h_left
integer :: pos =1
integer i,map_order,mf,mf1,ns,n_mode,km
type(probe) ray_closed
type(probe_8) ray,ray_cs_twiss
type(fibre), pointer :: p
type(integration_node), pointer :: t
character*120 :: dsc
logical :: doit=.true., used_ds_ave
type(c_vector_field_fourier) G,F,K,f1
real(dp) DX_AVERAGE_DCS,betax_1,theta
complex(dp) coe
!!!!!!!!!!!!!!!!!!!!!


c_verbose=.false.
prec=1.d-6 ! for printing
use_info = .true.
```


E. Forest, *From Tracking Code to Analysis*,
DOI 10.1007/978-4-431-55803-3

```
longprint=.false.

mis=.false.
state=only_2d0


call ptc_ini_no_append
call append_empty_layout(m_u)
ALS=>m_u%start

call build_lattice_als(ALS,mis,exact=.false.,thin=.true.,onecell=.true.)

write(6,*) " Give integration step ds "
write(6,*) " > 3 and nothing is cut; each step is a full magnet"
write(6,*) "  real fun starts around ds=0.5 "
read(5,*) ts
write(dsc,*) ts

!!!!Fitting the tune and controlling the step size  !!!!
 call kanalnummer(mf,file="fit_tune.txt")
  write(mf,*) "select layout"
  write(mf,*) "  1  "
  write(mf,*) "L MAX  "
  write(mf,'(a120)')dsc  !  all ds < 0.9
  write(mf,*) "CUTTING ALGORITHM "
  write(mf,*) "2 "                 ! Drifts are cut as well
  write(mf,*) "LIMIT FOR CUTTING "
  write(mf,*) "10000 10000 "
  write(mf,*) "THIN LENS "
  write(mf,*) "1000.d0 "
!  write(mf,*) "MISALIGN EVERYTHING"
!  write(mf,*) "0 0 0 0 0.1   0  5 "
  write(mf,*) "return"
 close(mf)


 call read_ptc_command77("fit_tune.txt")


courant_snyder_teng_edwards=.true.
time_lie_choice=.true.

p=>als%start

 call MAKE_NODE_LAYOUT(als)

call kanalnummer(mf,"guignard_hamiltonian.txt")
call kanalnummer(mf1,"guignard_canonical_transformation.txt")

Write(6,*) " Constant phase advance per step        ---> t "
write(6,*) " Courant Snyder phase advance per step ----> f "
read(5,*) used_ds_ave


write(6,*) " Enter number of Fourier modes"
read(5,*) n_mode

n_fourier=n_mode

map_order=4                 ! (3)
call init_all(state,map_order,0)

call alloc(one_turn_map, id_s,U_c,A,U,fi,b,a_cs)
call alloc(normal_form); call alloc(ray);
```

```
call alloc(f_non);call alloc(f_lin)
call alloc(logN); call alloc(h_left)
call alloc(fonction,fonction_FLOQUET)
call alloc(G);call alloc(F);call alloc(K);
call alloc(phase);call alloc(ray_cs_twiss)

DX_AVERAGE_DCS=0.d0
closed_orbit=0.d0;
call find_orbit_x(als,closed_orbit(1:6),STATE,1.e-8_dp,fibre1=1)

ray_closed=closed_orbit
id_s=1;
ray=id_s+ray_closed;

call propagate(als,RAY,state,fibre1=1)    ! (4)

one_turn_map=ray                          ! (5)

call c_normal(one_turn_map,normal_form)   ! (6)
U=normal_form%A_t


fonction=1.0_dp.cmono.1
  call C_AVERAGE(fonction,U,fonction_FLOQUET) ! (7)
a_cs=U.sub.1
call c_canonise(a_cs,a_cs)


intp=0
intp(1)=1

! id_s is a rotation
id_s=U**(-1)*one_turn_map*U
call extract_linear_from_normalised(id_s,b,a,f_lin,f_non,intp) ! (8)

call  c_canonise(normal_form%A_t ,U_c)        ! (9a)
id_s=U_c.sub.1                                      ! (9b)
h_left=(1.d0/twopi)*f_lin


if(used_ds_ave) then
  circ=twopi
else
  call GET_LENGTH(als,circ)
endif

ns=als%t%n
ds_ave=twopi/ns          ! (11)
ds=ds_ave                !
if(used_ds_ave) then
 Write(mf1,*) " Constant phase advance per step "
 write(mf1,*) "dtheta =",ds_ave, ' radians '
 write(mf1,*) "ds <= ",ts, " metres "
else
 write(mf1,*) " Approximate Courant Snyder phase advance per step "
 write(mf1,*) "ds <= ",ts
endif

U_c=U_c.cut.2

s=0.d0
ray=U_c+ray_closed;          ! (12)
ray_cs_twiss=ray_closed+a_cs

p=>als%start                 ! (13a)
```

```
t=>p%t1                    ! (13b)

do i=1,ns

 if(mod(i,ns/10)==0) then
   write(6,*) ns-i, " steps remaining "
 endif

 call propagate(als,ray,state,node1=i,node2=i+1)    ! (14a)

 call propagate(als,ray_cs_twiss,state,node1=i,node2=i+1)    ! (14b)


  ray_closed=ray ! Saving orbit  ! (15a)
  a_cs=ray_cs_twiss
  a_cs=a_cs.sub.1
  call c_full_canonise(a_cs,a_cs,phase=phase)

  U=ray ! copying in map  ! (15b)

if(used_ds_ave) then
 U_c=exp(-ds*h_left,U)          !  (16a)
 U_c=U_c.cut.2                  !  (16b)
 U=U_c**(-1)*U                  !  (16c)
else
 ds=twopi*t%s(5)/circ
 U_c=U.cut.2                    !  (16d)
 call  c_canonise(U_c,U_c)      !  (16e)
 U=U_c**(-1)*U                  !  (16f)
endif
  logN=log(U)                   !  (17)


! Checking convergence of the logarithm
a=exp(-(logN.cut.2),(U.sub.1))  ! (18)
do km=1,c_%nd2
 if(abs(full_abs(a%v(km))-1)>1.d-5) then
    call print(a,6)
    write(6,*);write(6,*) "Log failed at element ",i, p%mag%name
    stop
 endif
enddo

s=s+ds;
do km=-n_mode,n_mode
 G%f(km)=G%f(km)+(exp(-i_*km*s)/twopi)*logN ! (19)
enddo

  betax_1=(a_cs%v(1).sub.'10')**2+(a_cs%v(1).sub.'01')**2  ! (20a)
 if((p%mag%name(1:2)=="SF".or.p%mag%name(1:2)=="SD").and.t%cas==case0) then
  DX_AVERAGE_DCS=(betax_1)**1.5_DP*p%mag%BN(3)/4.0_DP &     ! (20b)
    *(-SIN(PHASE(1)*TWOPI)+SIN((PHASE(1)-normal_form%TUNE(1))*TWOPI)) &
    /(1.0_DP-COS(normal_form%TUNE(1)*TWOPI)) + DX_AVERAGE_DCS
 endif

 ray=ray_closed+U_c                          ! (21)
 ray_cs_twiss=ray_closed+a_cs
 t=>t%next
 p=>t%parent_fibre
enddo

U=from_phasor()
call transform_vector_field_fourier_by_map(G,G,u) ! (22)
```

```
prec=1.d-5



call alloc(f1)
if(used_ds_ave.and.doit) then
 call normalise_vector_field_fourier(G,F,K)      ! (23a)
else
 n_extra=50
 call normalise_vector_field_fourier(G,F,K,F1)  ! (23b)
endif

prec=1.d-11

call c_clean_vector_field_fourier(K,K,prec)
call c_clean_vector_field_fourier(g,g,prec)

write(mf1,*) " Original force "
 call print_vector_field_fourier(g,mf1)
write(mf1,*) " Normalised force "
 call print_vector_field_fourier(k,mf1)

  write(mf,*);write(mf,*)"Results for <x> ";write(mf,*)n_fourier,"modes"; write(mf,*);

  betax_1=(a_cs%v(1).sub.'10')**2+(a_cs%v(1).sub.'01')**2
 DX_AVERAGE_DCS=DX_AVERAGE_DCS*SQRT(betax_1)            ! (26) or (20c)
 write(mf,'(a11,F20.13,a20)')' d<x>/dCS ', dx_average_dCS, " < ---- analytical"
  write(mf,*); write(mf,*) "Result of map normal form "
  call print(fonction_FLOQUET,mf)           ! (27)

theta=0.d0
call c_evaluate_vector_field_fourier(f1,theta,f_non)


f_non=to_phasor()*f_non   !  Turns f_non in Cartesian basis
id_s=exp(f_non,a_cs)
call c_evaluate_vector_field_fourier(f,theta,f_non)

f_non=to_phasor()*f_non !  Turns f_non in Cartesian basis
A=exp(f_non,id_s)


  if(used_ds_ave) then
   write(mf,*);write(mf,*) "Constant linear phase advance"
   write(mf,*) "dtheta =",ds_ave, ' radians '
   write(mf,*) "ds <= ",ts, " metres "
  else
   write(mf,*);write(mf,*) "Courant-Snyder linear phase advance "
   write(mf,*) "ds <= ",ts, " metres "
  endif
 write(mf,*);write(mf,*) "Result of Guignard normal form "
 fonction=1.0_dp.cmono.1
  call C_AVERAGE(fonction,A,fonction_FLOQUET)   ! (28)
  call print(fonction_FLOQUET,mf)                  ! (29)
close(mf)

 write(mf1,*) " Linear Courant-Snyder transformation at point of calculation for <x> "
call print(a_cs,mf1)
if(.not.used_ds_ave.or.(.not.doit)) then
 write(mf1,*) " F1 : removing the s-dependence of the phase advance"
 call print_vector_field_fourier(f1,mf1)
endif
 write(mf1,*) " F : nonlinear transformation"
 call print_vector_field_fourier(f,mf1)
```

```
close(mf1)
call ptc_end(graphics_maybe=1,flat_file=.false.)

end program Guignard_normal_form_average_x
```

Index of Links to Useful Concepts and Formulae

E. Forest, *From Tracking Code to Analysis*,
DOI 10.1007/978-4-431-55803-3

MIX
Papier aus verantwortungsvollen Quellen
Paper from responsible sources
FSC® C105338

If you have any concerns about our products,
you can contact us on
ProductSafety@springernature.com

In case Publisher is established outside the EU,
the EU authorized representative is:
Springer Nature Customer Service Center GmbH
Europaplatz 3, 69115 Heidelberg, Germany

Printed by Libri Plureos GmbH
in Hamburg, Germany